Genomic Diversity

Applications in Human Population Genetics

Genomic Diversity

Applications in Human Population Genetics

Edited by

Surinder S. Papiha

University of Newcastle upon Tyne
Newcastle upon Tyne, United Kingdom

Ranjan Deka

University of Cincinnati
Cincinnati, Ohio

and

Ranajit Chakraborty

University of Texas Health Science Center
Houston, Texas

Kluwer Academic / Plenum Publishers
New York, Boston, Dordrecht, London, Moscow

Library of Congress Cataloging-in-Publication Data

Genomic diversity : applications in human population genetics / edited by Surinder S. Papiha, Ranjan Deka, and Ranajit Chakraborty.
p. cm.
Includes bibliographical references and index.
ISBN 0-306-46295-8
1. Human genetics--Variation. 2. Human population genetics. I. Papiha, Surinder Singh. II. Deka, Ranjan. III. Chakraborty, Ranajit, 1946-

QH455 .G469 1999
599.93'5--dc21

99-043453

Proceedings of a symposium on Molecular Anthropology in the Twenty-First Century, held during the 14th International Congress of the Association of Anthropological and Ethnological Sciences, held July 26–August 1, 1998, in Williamsburg, Virginia

ISBN: 0-306-46295-8

233 Spring Street, New York, N.Y. 10013

http.//www.wkap.nl

10 9 8 7 6 5 4 3 2 1

A C.I.P. record for this book is available from the Library of Congress.

Printed in the United States of America

PREFACE

The disciplines of human population genetics and anthropological genetics had their origin in the early part of the 20th century. One of the central themes of these subjects has been the reconstruction and understanding of the origin and past history of modern humans, *Homo sapiens sapiens.* This has been greatly aided by, in addition to paleontological, archaeological, and linguistic evidence, data on the distribution of single gene characters (Blood groups and protein markers) in contemporary populations. Thus, analysis of genetic variation in extant populations has received considerable attention. By the 1980s, when the use of serological and biochemical polymorphisms reached their plateaus, newly discovered molecular techniques provided a new direction and a whole battery of powerful polymorphic systems were introduced for assaying genetic diversity. Restriction fragment length polymorphisms (RFLPs), single nucleotide polymorphisms (SNPs), variable number of tandem repeats (VNTRs), short tandem repeats (STRs), Alu insertion polymorphisms and a vast range of haplotype variants based on nuclear, mitochondrial and Y-chromosome DNA markers, all proved to be extremely useful in phylogenetic and evolutionary studies. Although thousands of molecular polymorphisms have already been described, genomic markers and methodologies are being continuously developed and improved for better understanding of the problems of human population genetics. So much has already been achieved that this outgoing century will be remembered for its advances in molecular technology, which revolutionized the biomedical, genetical, and evolutionary research.

This volume with original data in many chapters will try to highlight the extent of genetic variation in contemporary populations. Importance of the analysis of this variation is substantial in not only gaining insight into our past but it has tremendous impact on other fields in human genetics. Recently, unraveling the genetic etiologies of complex diseases has become a central theme in genetic and biomedical research. Such undertakings for identification of susceptibility loci will obviously be based on details of genetic variation in human populations and population genetic principles.

The first chapter outlines the development of classical and molecular markers and then compares the findings of these two types of markers in understanding the origin and relationship among certain selected populations of the Indian subcontinent. Chapters 2 and 3 examine the potential use of the molecular markers in the genetic differentiation of the native South Amerindians. Deka (Chapter 4) provides an account of the unstable DNA polymorphic loci, trinucleotide repeats and their evolution. Tyler-Smith (Chapter 5) introduces the development of the Y-chromosome markers and their nomenclature, while in Chapters 6–8 the use of Y-chromosome markers to study the origin of some of the world populations is presented. In Chapters 9–12, mitochondrial DNA (mtDNA) studies among the populations of Europe, India and Southwest Pacific provide an excellent example of the use of these markers in evolutionary genetics. Kolman (Chapter 13) discusses the possible use of ancient DNA technology in molecular anthropology. Two chapters (Chapters 14 and 15) are devoted to population genetics of Alu elements. Deininger and colleagues have discussed the evolution, origin and structure of Alu elements and their use as genetic markers in microevolutionary studies in Chapter 14, while in Chapter 15, a worldwide distribution of an Alu insertion polymorphism in the progesterone receptor gene is presented by Herrera and colleagues. Chakraborty in the final chapter discusses some of the statistical methods being used while dealing with molecular markers.

The current trends and the future of human population and evolutionary genetics were presented at the Symposium 'Molecular Anthropology in Twenty first Century' held during the 14th International Congress of the Association of Anthropological and Ethnological Sciences at Williumsburgh on July-August 1998 and the idea of compiling this volume was agreed during the symposium. However, it took some time for several busy laboratories to send in their contributions. We would like to express our appreciation to Kluwer Academic/Plenum Publishers for their patience while the manuscripts were finalized.

S. S. Papiha
R. Deka
R. Chakraborty

CONTENTS

CLASSICAL TO MOLECULAR POLYMORPHISMS

Population Genetic Studies from the Indian Sub-Continent

S. S. Papiha[1] and S. S. Mastana[2]

[1]Department of Biochemistry and Genetics
University of Newcastle upon Tyne, UK
[2]Department of Human Sciences
University of Loughborough, UK

ABSTRACT

Earlier part of the 20th century has seen the development of several serological and biochemical polymorphisms and their use in the studies of human populations. More recently, several types of molecular markers (single nucleotide polymorphisms, SNPs; restriction fragment length polymorphisms, RFLPs; minisatellites or variable number of tandem repeats, VNTRs; mircosatellites or short tandem repeats, STRs; mitochondrial and Y chromosomal markers) have been discovered in the human genome. This paper traces the development of various types of single gene markers and illustrates the use of some of these polymorphisms in the study of evolutionary histories of the populations of the Indian Subcontinent. Four VNTR loci (D2S44, D7S22, D12S11 and D1S80) have been studied to trace the origin of the Sinhalese of Sri Lanka, while six STR loci (CSF1PO, TPOX, THO1, F13A01, FESFPS and vWA) were analyzed to understand the evolutionary relationship and genetic differentiation among six diverse and isolated populations of India. For all genetic variation analyses, correspondence analysis, genetic distance and gene diversity studies, VNTR and STR data demonstrate that these loci are powerful markers for genetic differentiation and evolutionary studies. The molecular data also conclusively support the findings of the classical markers.

1. INTRODUCTION

"Whatever it is you're studying you are better off if you have a mutation".

This anonymous phrase is a fundamental reality. Mutations in the genetic material are the ultimate source of genetic variation, without which there would have been no

Genomic Diversity: Applications in Human Population Genetics
edited by Papiha, Deka, and Chakraborty, Kluwer Academic / Plenum Publishers, New York, 1999.

evolution. Each population is a unit consisting of a large number of individuals and the way in which the genes they carry are distributed and embodied within a population provides its genetic structure. The pool of genes at any one time held by the individuals making up a population, is its genetic constitution (Roberts 1987). A gene pool can be described in terms of gene frequencies and the variation in the gene frequencies in man over a given time scale provides fundamental knowledge of human micro-evolution.

Studies of the variation in gene frequencies of single gene characters started in the beginning of the 20th century when the ABO blood groups were discovered (Landsteiner 1901). Hirszfeld and Hirszfeld (1919) described varying frequencies of the A, B and O blood groups among soldiers of different nations and this observation proved of immense anthropological importance, for it was the first stepping stone towards the use of single gene systems to explore the history of human populations. This exciting development followed the discovery of many more polymorphic blood group systems, MN and P (Landsteiner and Levine 1927), Se, H, A3 and M2 in the 1930s, Rh C, c, E, Kell, Lewis and Kidd in 1940's and Duffy in 1950 (Cutbush and Mollison 1950; Race and Sanger 1975). In the early half of this century the study of the red cell antigens remained an active field of investigation and during this time more than two hundred antigens were detected on the surface of red cell membranes. These antigens have been classified into some 15 blood group systems.

This rapid progress in the field of serological genetics slowed by 1950, but by this time new biochemical techniques were being developed to study other new polymorphisms. An extensive genetic variation, which provided human biochemical individuality, was largely demonstrated by analyses of enzymes and proteins by zone electrophoresis methods. In the 1940's using this technique, Pauling and his colleagues (1949) identified the first mutant gene product in man by distinguishing haemoglobin S from the normal adult haemoglobin A. This was the start of the discovery of a whole array of biochemical polymorphisms and during the period between 1950 and 1970, very many were discovered (Harris and Hopkinson 1976). Studies of human polymorphisms by Harris and his colleagues at the Galton Laboratory, London, and similar work in Drosophila by Lewontin and Hubby (1966) showed that genetic polymorphisms are much more common than previously thought. Despite the limitations of electrophoretic techniques, which detect only one third of the total biochemical variation produced by mutations, the variation observed was enormous.

For both these serological and biochemical polymorphisms, now known as conventional or classical polymorphisms, the pattern of discovery followed a similar trend—slow start, then rapid progress and subsequently a plateau. The serological techniques developed over these years were also responsible for the discovery of two more highly polymorphic classical genetic systems in man. Immunoglobulin allotypes (GM, KM and AM) and the protein antigen located on the surface of human white cells (B and T lymphocytes) known as HLA (human leukocyte antigens) were both discovered during the middle part of the century (Dausset 1954; Grubb and Laurell 1956; Payne *et al.* 1964). All these four types of classical markers reveal variations at the protein level generated by so-called "neutral" mutations. The time span between 1940 and 1990 was employed mainly in collecting information on geographical and racial gene frequency variation among the peoples of the old and new worlds and the data are now embodied in several text books (Mourant *et al.* 1976; Steinberg and Cook 1981; Tills *et al.* 1983; Roychoudhury and Nei 1988; Bhasin *et al.* 1996; Cavalli-Sforza *et al.* 1994) and a series of books on histocompatibility testing workshops.

The blood group and protein polymorphism data thus collected have indicated that genetic diversity among local human populations within continents is much larger than that between continents.

2. TYPES OF MOLECULAR MARKERS

The human genome consists of 3,300 Mb of nuclear DNA and 16.6 Kb of extranuclear mitochondrial DNA. The transmission of classical markers requires stability of the coding regions of DNA involved in translation into proteins. These coding regions constitute less than 5% of the total nuclear genome. The remaining nuclear genome consists of unique or low copy number sequences and moderate to highly repetitive sequences. The great bulk of this non-coding genome, whose biological function is still not clearly defined, is likewise vulnerable to mutations. The technical advances in molecular genetics, such as the discovery of large numbers of restriction endonucleases, transfer of restriction enzyme digested DNA to nylon membranes, cloning of random DNA fragments and amplification of DNA by the techniques of polymerase chain reaction (PCR) all have contributed in revealing an array of new molecular markers from this non-coding region. These new DNA markers not only gave a new look to the investigation of human genetic diversity, but brought to the forefront a new and important era in the application of human population genetics to the field of forensic genetics and molecular medicine.

Many types of polymorphism have been observed in the non-coding parts of the human genome. This paper illustrates the use of these molecular markers in studies of human population genetics.

2.1. Restrictions Fragment Length Polymorphisms (RFLPs)

Meselson and Yucan (1968) discovered enzyme restriction endonuclease and it was soon realised that such enzymes recognise, and cut at, specific DNA sequences. Depending on the nature of the sequence recognised, the restriction enzyme may cut the target DNA at several places. The smallest and simplest DNA mutation, the substitution of a single nucleotide, can destroy or create the recognition site of a restriction enzyme thus giving a larger or smaller cleaved fragment of target DNA. Such a change in fragment size was first observed in the human β globin region with the enzyme HpaI (Kan and Dozy 1978) and the polymorphism observed was termed a restriction fragment length polymorphism (RFLP). Numerous RFLP variants occurred throughout the human genome, which could be detected by arbitrary cloned DNA fragments (Wyman and White 1980). This led Botstein *et al.* (1980) to suggest that these molecular markers could be used to map the entire human genome, which was not possible with the classical markers.

The introduction of these RFLP markers had an important effect on studies of both human population genetics and biological anthropology. Between the years 1980 and 1990 more than 3,000 polymorphic sequences were reported, most of them being bi-allelic in nature and their polymorphic information content (PIC) value less than 0.38. These systems, therefore, were not highly informative either for gene mapping or for population genetic studies unless a large number were analysed simultaneously. Some population studies with multiple RFLP systems have been reported providing better understanding of the origins of different racial groups and the relationship between them (Wainscoat *et al.* 1986; Summer 1987; Bowcock *et al.* 1987, 1991a, 1991b; Kidd *et al.* 1991; Papiha *et al.* 1991; Kidd *et al.* 1993 and Lin *et al.* 1994). Among these, the largest studies included analyses of five human populations (Biaka and Mbuti Pygmies, Melanesians, Chinese and Europeans) for a set of 100 RFLPs. A second study extended the data to eight populations (including Japanese, New Guineans and Australians), but their analyses were limited to seventy-nine DNA polymorphisms. These population studies concluded that most two allele systems are present in virtually all human populations and that population specific (private or semi-private) alleles are very few. The genetic differentiation index (F_{ST}) and heterozygosity between known genes and anonymous DNA markers differ though not significantly, suggesting very little influence of natural selection.

Interestingly, unlike other studies, which showed the highest genetic diversity among African populations, these RFLP studies exhibited greater heterozygosity in the European sample. Since the RFLPs analysed in these studies were initially detected with high frequency in populations of European origin, this ambiguity could be attributed to ascertainment-bias of polymorphisms investigated (Lin *et al.* 1994). The phylogenetic relationships revealed by population tree construction, agreed with the conclusions of previous studies that the first evolutionary split happened between the African and non-African populations and that the common ancestor of modern humans originated in or near Africa.

Several million-biallelic or single nucleotide polymorphisms (BA/SNPs) are available in the human genome. It is now possible to analyse, simultaneously, several hundred of them by the so-called 'chip' hybridization technology (Wang *et al.* 1998). Uniform analysis of vast numbers of randomly selected BA/SNPs on various populations will certainly provide a better understanding of genetic affinities in the near future.

2.2. Insertion/Deletion Polymorphisms

The insertion or deletion, like the bi-allelic or single nucleotide polymorphism may involve simple mutations of a single nucleotide, but larger deletion/insertion changes (>30 bp) are also not uncommon. One such polymorphism, which is more frequently observed and has been extensively studied in population genetics, is an *Alu* insertion polymorphism. This will be discussed in detail in a subsequent paper.

2.3. Minisatellite or Variable Number Tandem Repeat (VNTR) Polymorphisms

Jeffreys and colleagues (1985), working with the human myoglobin gene came across a 33 bp sequence, which when hybridised with some restriction enzyme digested total genomic DNA, produced a complex pattern of multiple DNA bands. This ladder of DNA fragments was individual-specific and was called a DNA fingerprint. Each band of the DNA fingerprint represents a sequence repeated in tandem fashion and can be considered as one allele at a number of independent loci in the human genome. These VNTR alleles, also known as minisatellite alleles, show co-dominant Mendelian inheritance and each locus-specific VNTR has numerous alleles, showing a very high level of heterozygosity (Wong *et al.* 1987).

Many have questioned whether this new class of loci can be used in studying genetic differentiation between populations while employing classic population genetic models (Lander 1989; Cohen 1990; Kidd *et al.* 1991). Even before these, there were some studies, which demonstrated that traditional population genetic models could be applied to

VNTR data. For example, Clark (1987), using data on two VNTR loci, HRAS1 and D14S1, from three major groups (Baird *et al.* 1986) suggested that the observed allelic variation shows an acceptable fit to the mutation drift model (neutral expectations). VNTR alleles show an exceptionally high mutation rate compared to classical genetic markers, and Jeffreys *et al.* (1988) tried to explain the heterozygosity and mutation rates at these loci using infinite allele mutation models or a stepwise mutation model. More recently Jin and Chakraborty (1995) argued from existing data that, although the exact mechanism of mutation at the VNTR loci is unknown, the infinite allele model and the stepwise mutation model do provide some indications for assessing population substructuring. They predicted that the loci with higher mutation rates and high heterozygosity in subpopulations with no migration should exhibit a smaller proportional heterozygosity deficiency. However, their data on six VNTR loci in major cosmopolitan populations of the USA, showed a large and significant heterozygosity deficiency for VNTR loci. This discordant finding was attributed to the empirical limitations of RFLP analysis, which occasionally fails to detect extremely small or large size VNTR alleles. Chakraborty (1990), in his earlier work, also suggested that even a single VNTR locus can provide information concerning substructuring within a population with a statistical power greater than several classical genetic markers studied simultaneously.

Various concerns regarding the use of VNTR data in population genetics may partly be due to the limited amount of population data available for VNTR loci on genetically well characterised ethnic groups. In fact, for certain regions like the Indian subcontinent, population data on VNTR are very limited indeed. Therefore, we provide here, a case study of the origin of the Sinhalese of Sri Lanka (Formerly known as Ceylon) through VNTR analyses.

Anthropologically speaking Sri Lanka is one of the most interesting islands in the world. Its culture is based on that of India, but its population comprises representatives of several races. The total population of Sri-Lanka in the 1981 census was 14,846,750 of which 74% were Sinhalese, 12.8% Sri-Lankan Tamils, 5.4% Indian Tamils, 7.1% Moor and 0.7% minority groups including the Burghers and Malays. The role of migration in the genesis of the Sinhalese people is uncertain, but it is believed that they entered Sri Lanka from India at an early date. Ethnohistorians trace the first Sinhalese settlement back to the 5th century BC when Prince Vijaya, their legendary 'founding father' arrived from North India near the present day region of West Bengal. Vijaya and his followers colonised the northern dry zone region of Sri Lanka where their descendants produced a flourishing civilisation. The possible routes of migration from Bengal are shown in the map (Fig. 1).

Several studies have tried to trace the origin and affinities of the Sinhalese using classical genetic markers. Simple univariate comparisons of blood groups, red cell enzyme (Roberts *et al.* 1972a, b) and glucose 6 phosphate dehydrogenase deficiency (Abeyaratne *et al.* 1976) showed appreciable differences in frequency between Sinhalese and Tamils, as well as local frequency variation within Sri Lanka. Kirk (1976) applied multivariate statistics on the gene frequency data of the Sinhalese, Tamils, Keralites, Bengali, Gujarati and Punjabis and concluded that the Sinhalese are closest to the East Indian populations of Bengal. However, Saha (1988) using data on the Sinhalese and Tamil sampled in Singapore, disagreed with Kirk's findings and showed Sinhalese affinity with South Indian populations. Recently, Kshatriya (1995) studied the genetic affinity of the Sinhalese applying genetic distance and mixture analyses to published data on fifteen loci, and his analyses not only showed a close affinity with the Tamils but also calculated that 75% of Sinhalese genes originated from Tamil admixture. In view of these contradictory

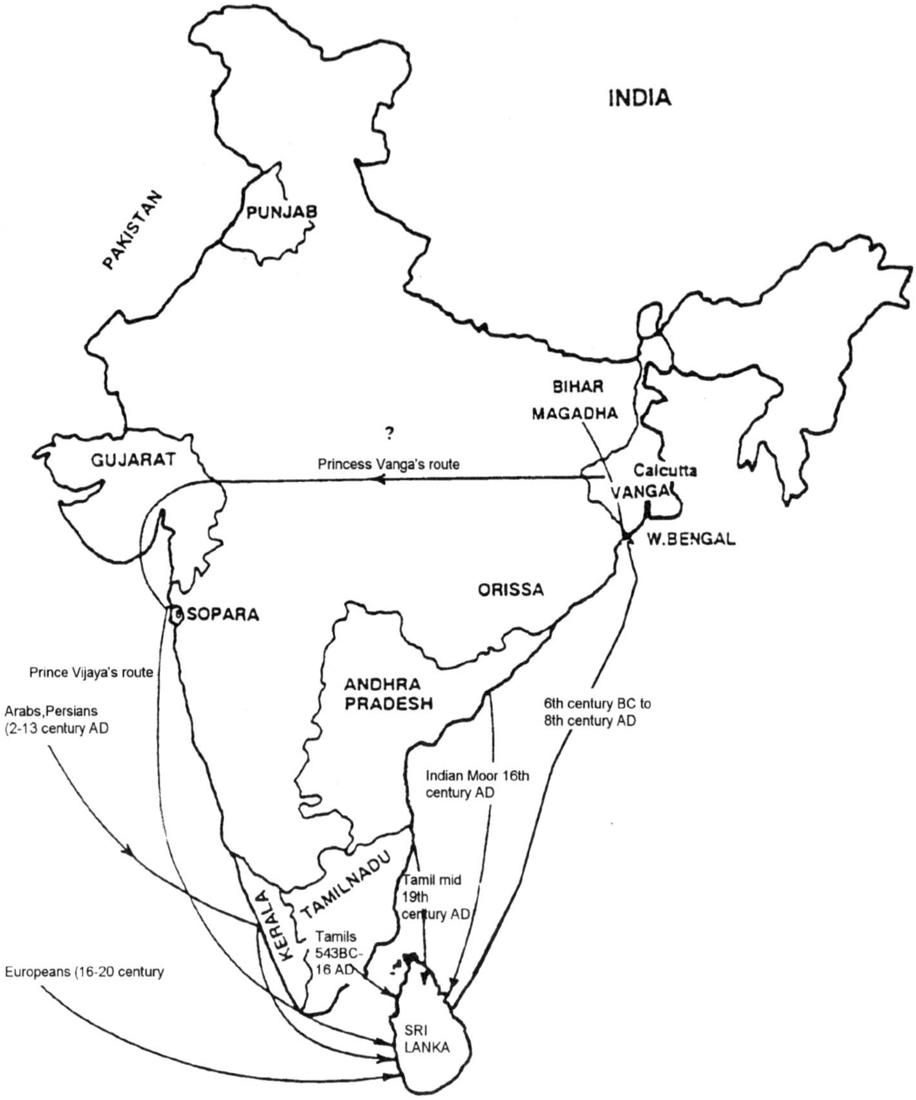

Figure 1. Map of India and Sri Lanka showing possible routes of migrations of Sinhalese.

observations, we analyse here Sinhalese affinity using VNTR markers and compare these findings with our classical marker data obtained on the same populations.

The populations included in the present analyses are Punjabi (North-West India), Brahmin and Gujarati (Western India), Muslim (South India), Bengali (North-East India) and Sinhalese (Sri Lanka). Detailed ethnographic descriptions and the data on classical markers on these populations have already been published elsewhere (Mastana and Papiha 1994; Kshatriya 1995 and Papiha *et al.* 1996a).

Genomic DNA was extracted from frozen peripheral blood leukocytes using the standard phenol-chloroform method. Ten microgram DNA samples were digested to completion with restriction enzyme HinfI, and the DNA fragments transferred to nylon membrane by the Southern blotting technique. The membranes were hybridised with radioactive locus-specific probes YNH24 (D2S44), g3 (D7S22) and MS43a (D12S11). After hybridisation and autoradiography, the DNA fragment sizes were measured by computer using a programme based on the method by Elder and Southern (1983). The details of the methods used in our laboratory have been published (Papiha *et al.* 1996b). For genotyping the samples for locus D1S80, the DNA amplification and typing were performed following the standard procedure described in detail by Deka *et al.* (1994).

For probes g3 and MS43a, the fragment sizes of alleles ranged from 1–15 Kb whilst for YNH24 they ranged from 1.5 to 10 Kb. The data for these three loci were therefore grouped for most of the analyses using the fixed bin method, following the fixed bin boundaries as listed in Budowle *et al.* (1991). For the fixed-bin and D1S80 locus, allele frequencies were calculated by the gene counting method. Comparisons with Hardy-Weinberg expectations were tested by three different procedures: χ^2 based on total heterozygosity at the locus, the likelihood ratio (Weir 1992) and the exact test suggested by Guo and Thompson (1992). The details of these tests have already been described (Papiha *et al.* 1996b). To compare VNTR data with classical markers, allele frequencies of 11 polymorphic systems (ABO, Rh, MNS, Fy, K, HP, ACP, ESD, PGD, PGM and ADA) were used. Measures of genetic variations analysed were heterozygosity (H) and genetic diversity indices G_{ST} and R_{ST}. The genetic distance was computed using the Nei's method (Nei 1978). Genetic affinity was analysed by correspondence analysis using gene frequency data and by constructing dendrograms from distance matrices using the neighbourhood joining method.

Figures 2–4 compare the distribution of fixed bin allele frequencies of g3, MS43a, YNH24 loci, and Fig. 5 compares the allele frequencies at the D1S80 locus in the six populations studied. Visual comparison shows considerable variation between the populations. To test for Hardy-Weinberg equilibrium (HWE) analysis was carried out both with and without non-detectable allele frequency (r) as described by Chakraborty *et al.* (1994). There was some population departure from HWE, for g3 (Brahmin, Gujarati and Sinhalese) and MS43a and YNH24 (Sinhalese), but when r was included in the computation no deviation was observed. For most populations except at the g3 locus, (r) was less than 4% and these values were similar to that described in other studies (Steinberger *et al.* 1993 and Chakraborty *et al.* 1994).

Heterozygosity (H) was lowest (<40%) for the protein loci but considerably higher for all four VNTR loci, in the range of 76–94%. The lowest H at 76% was for the D1S80 locus. The indices of genetic differentiation G_{ST} and R_{ST} for VNTRs and classical markers are given in Table 1. The G_{ST} and R_{ST} values for the four VNTR systems were 0.0135 and 0.0121 respectively, these values being slightly higher than those for the 11 classical markers, 0.0119 and 0.0109, respectively. G_{ST} and R_{ST} values for the 15 systems were similar, 0.0125 and 0.0115, respectively. These values are close to the F_{ST} values reported for different populations and regions of the Indian sub-continent (Papiha 1996). Estimates of genetic diversity for both molecular and classical markers showed moderate values suggesting that the differentiation between these groups may well be maintained by genetic drift.

The fixed bin and D1S80 allele frequencies data were subjected to correspondence analyses (Fig. 6). The first dimension clearly separates the two closely related populations of Bengali and Sinhalese from the other four groups. For further analyses of genetic affinity between these population, Nei's standard (D_A) and Shriver's genetic distance (D_{SW}) matrices, based on data from four VNTR loci, and Nei's standard genetic distance (D_A) matrices based on 11 classical and 15 combined systems, were reduced to two dimensional dendrograms, using the neighbourhood joining method (Fig. 7).

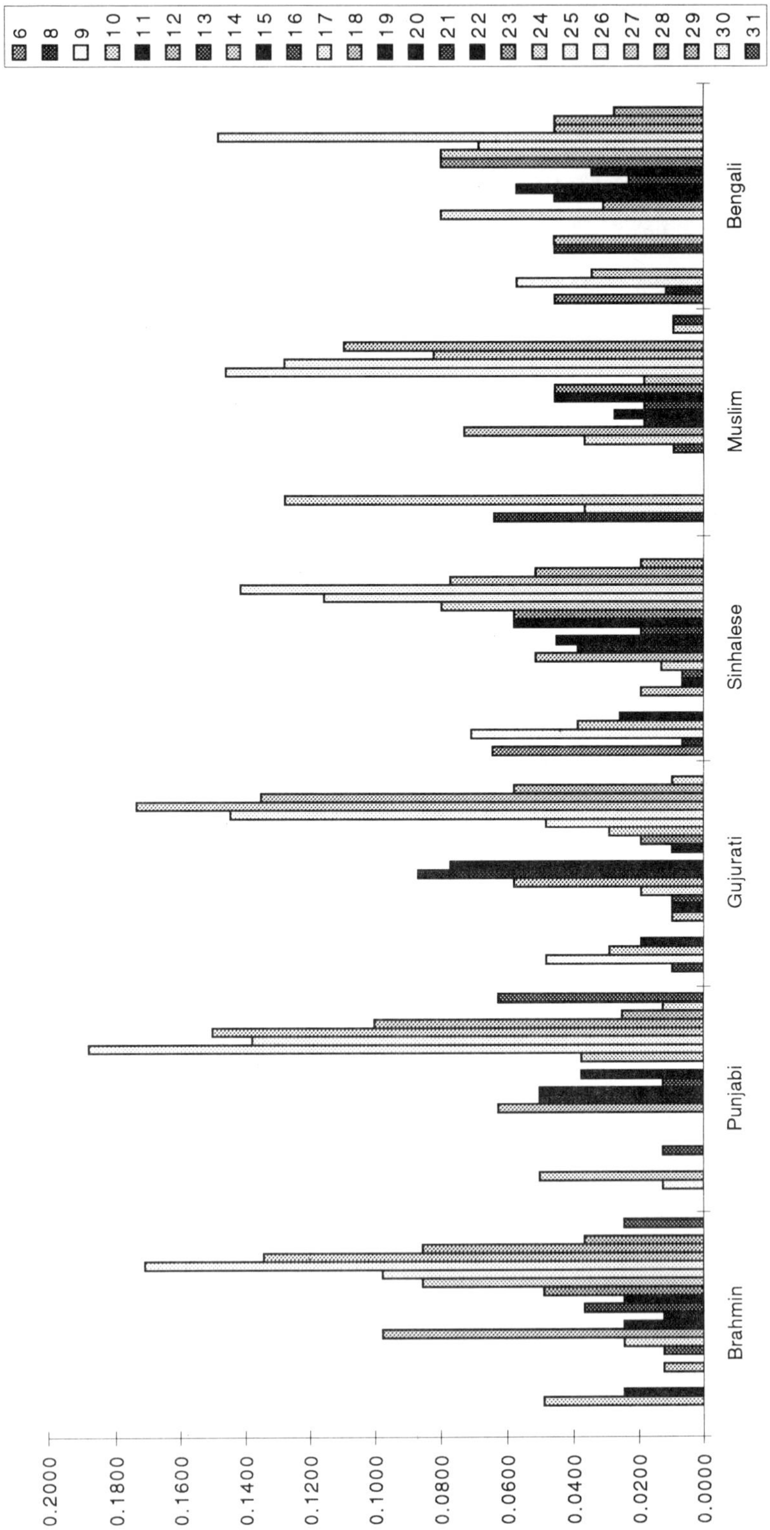

Figure 2. Distribution of g3 bin frequencies in Six populations of Indian sub-continent.

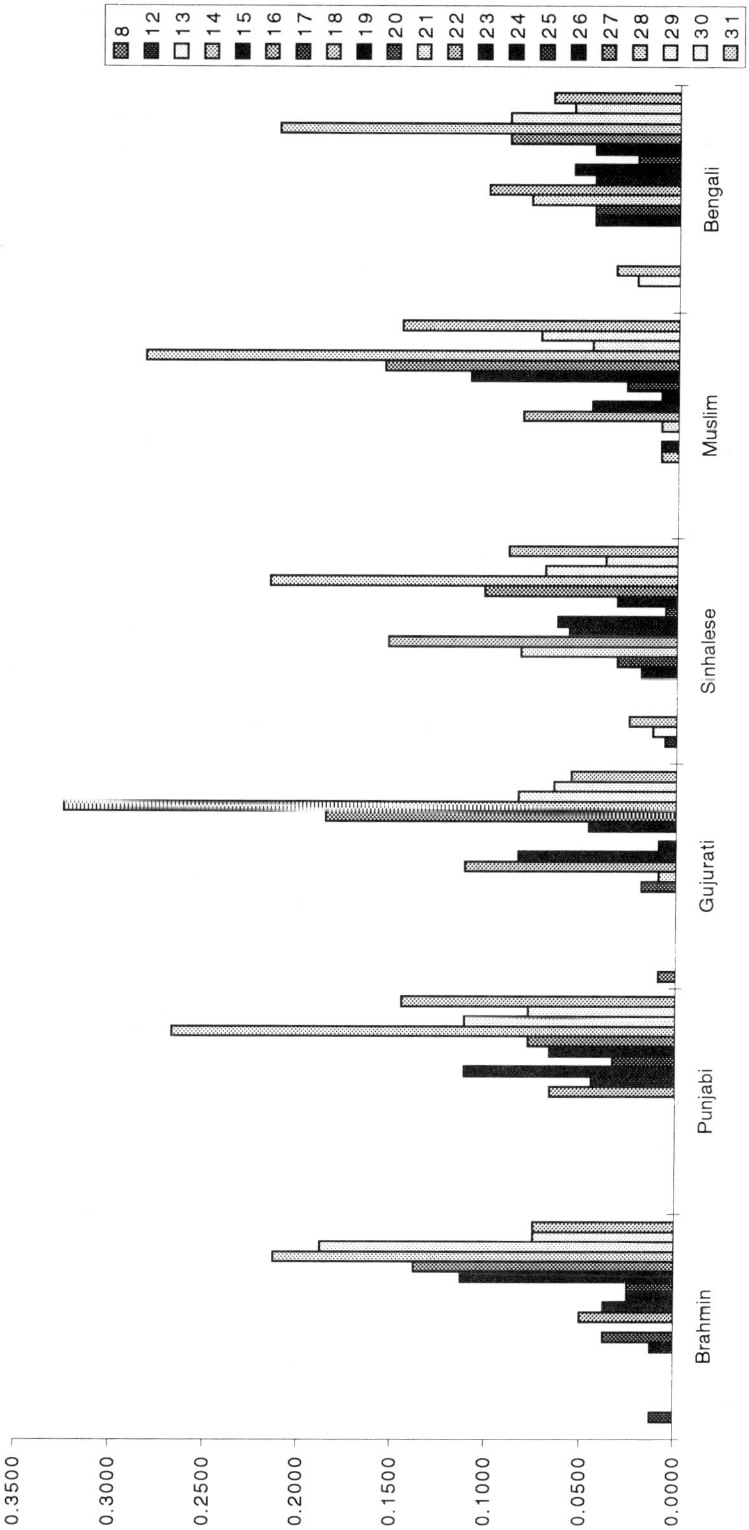

Figure 3. Distribution of MS43A bin frequencies in Six populations of Indian sub-continent.

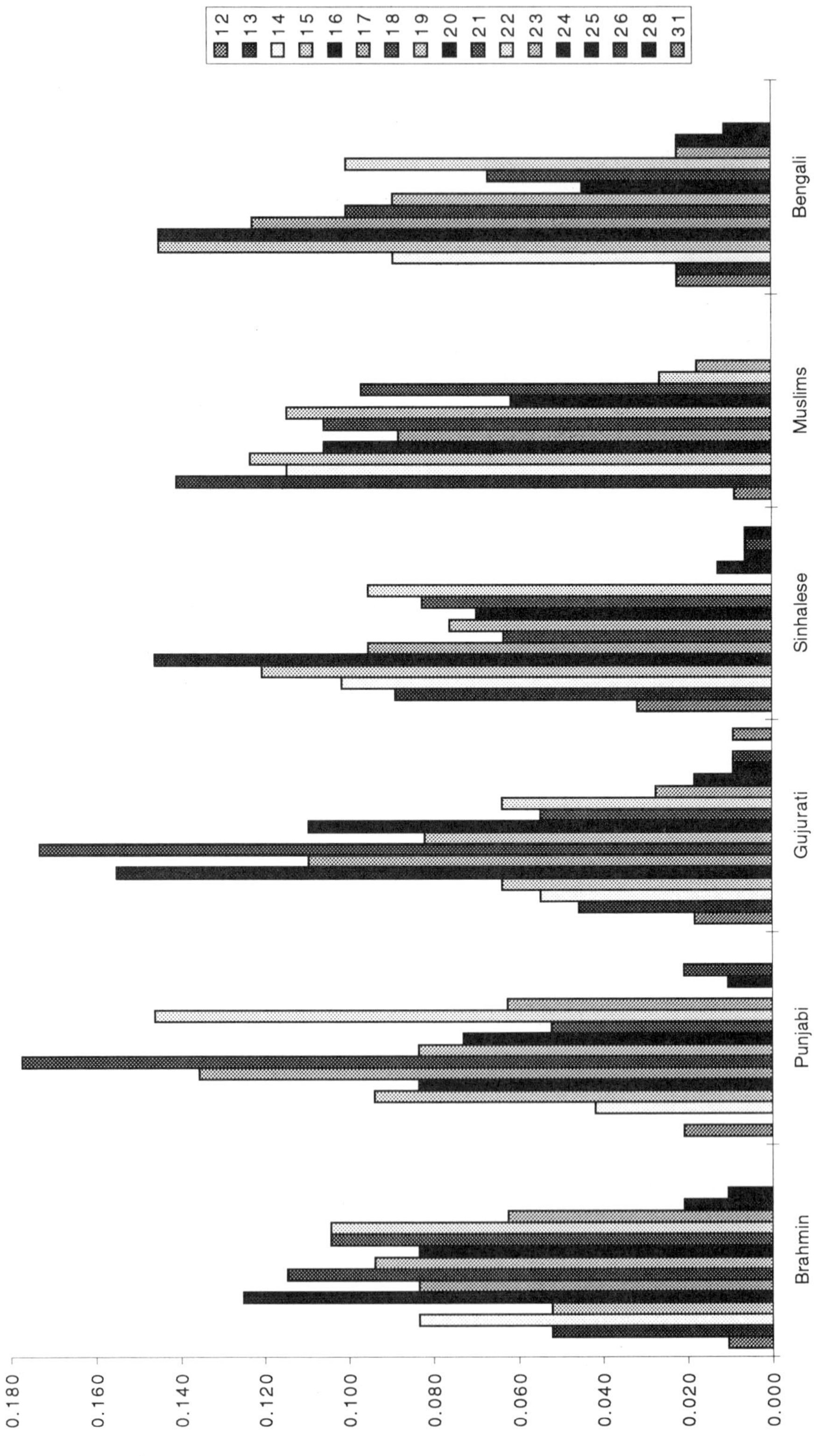

Figure 4. Distribution of YNH24 Bin Frequencies in Six populations of Indian Subcontinent.

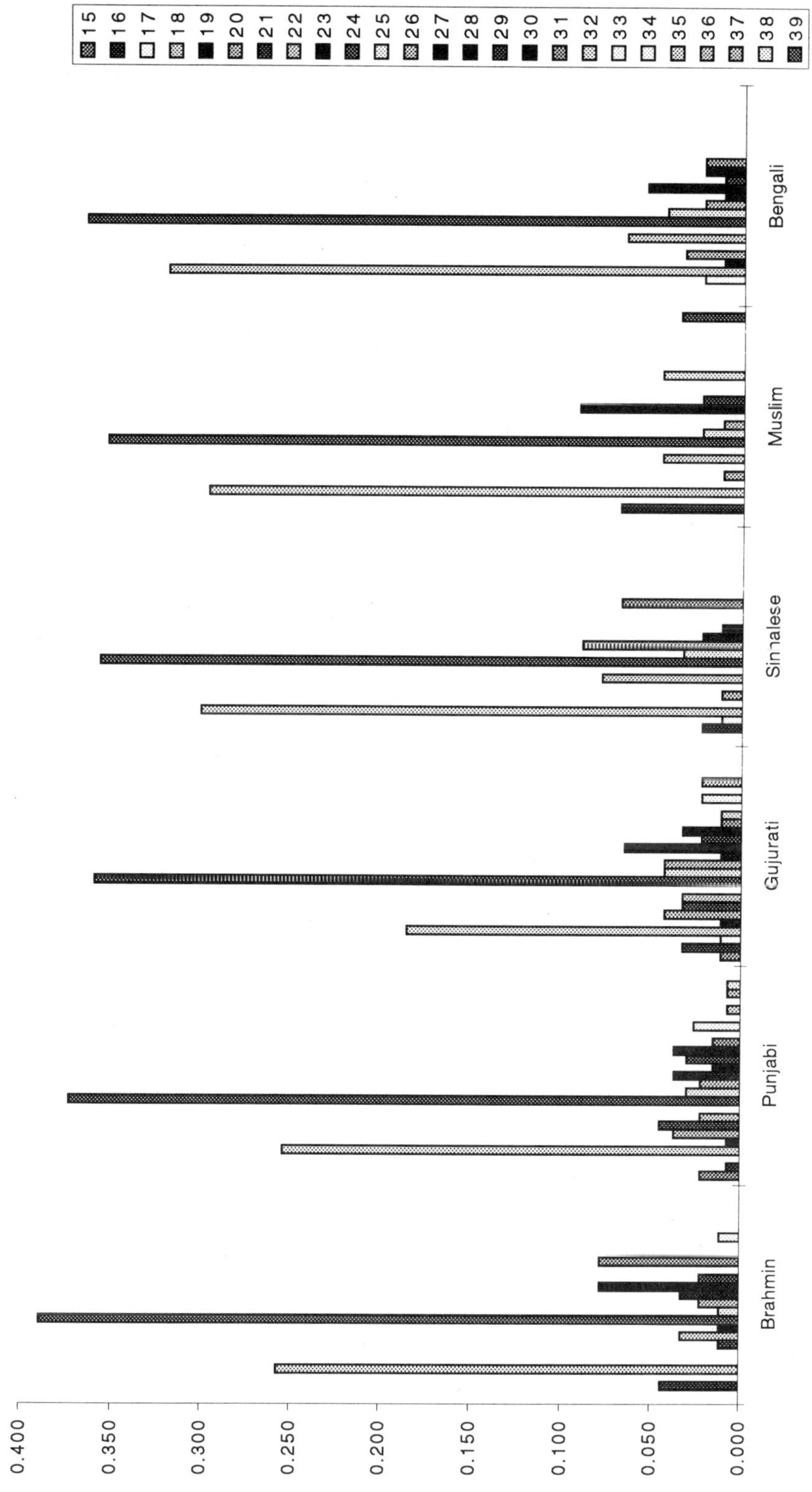

Figure 5. Distribution of D1S80 Allele frequencies in Six populations of Indian sub-continent.

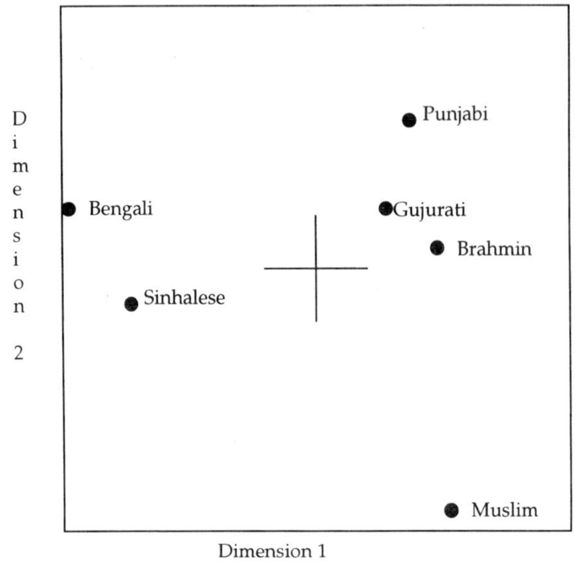

Figure 6. Correspondence Analysis of 4 VNTR loci.

For the four VNTR data both Sinhalese and Bengali showed a single cluster which separates quite early in the evolutionary time scale. For both these matrices the affinity between Bengali and Sinhalese is supported by high boot strap values. The data of 11 classical markers and 15 combined markers both further support the affinity between Bengali and Sinhalese populations, though the relationship of the other four groups slightly changes.

Affinity of Sinhalese with the population of north-eastern India was further confirmed by admixture analysis. This was carried out by the method of Long (1991), first using the allele frequencies of thirteen classical systems. The first analyses applied to two parental populations (Tamil and Bengali) and the Sinhalese showed that 70% of the gene pool originated from the Bengali (Table 2). However, when the analyses included Gujarati (Western India) as the third parental population, the proportion of the Bengali genes in Sinhalese increased to 72%. VNTR data were available on Bengali, Punjabi and Gujarati and when these three populations were considered as parental populations, the VNTR results indicated that 82% of the Sinhalese genes originated from Bengali admixture.

In conclusion, in our present investigation into the origin of the Sinhalese, the hypervariable molecular markers, VNTRs have proved extremely useful. For all genetic variation analyses, correspondence analysis, genetic distance, genetic differentiation and

Table 1. Genetic diversity values G_{ST} and R_{ST} for different types of markers in six populations of the Indian Sub-continent

Type of Marker	G_{ST}[1]	R_{ST}[2]
4 VNTRs	0.0135	0.0121
11 CLASSICAL MARKERS	0.0119	0.0109
ALL 15 SYSTEMS	0.0125	0.0115

1. Nei (1973).
2. Harpending and Jenkins (1973).

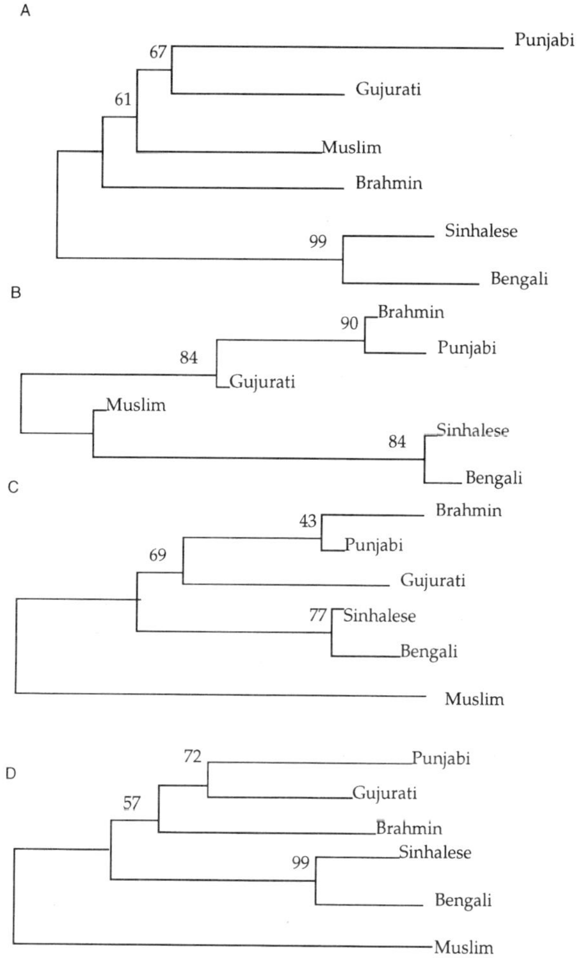

Figure 7. Dendrograms of Six populations derived from (a) Nei's DA distance for 4 VNTR loci, (b) Shriver's DSW distance of 4 VNTR loci, (c) DA distance for 11 conventional systems and (d) DA distance for 15 systems.

admixture analyses, VNTR data conclusively support findings of the classical markers. Furthermore, they confirm that the present day gene pool of the Sinhalese of Sri-Lanka seems to have originated largely via migration from the northeastern region of India.

2.4. Microsatellites or Short Tandem Repeat Polymorphisms (STRPs)

Technique of polymerase chain reaction (PCR) have been particularly useful in typing markers with smaller core sizes (di-, tri- or tetranucleotides) popularly known as Short Tandem Repeats (STRs) or microsatellite polymorphisms. VNTRs with their high heterozygosity, vast number of alleles and especially their high mutation rate make them more useful to explore recent population history. Complementing these, STRs are used to analyse intra- and inter-population affinities and evolutionary histories (Bowcock *et al.* 1994; Deka *et al.* 1995; Jorde *et al.* 1995, 1997 and Perez-Lezaun *et al.* 1997).

Table 2. Admixture analyses in the Sinhalese gene pool based on different types of markers

Sinhalese: (2 parental populations) 13 classical systems	
Tamil	0.2997 ± 0.015
Begali	0.7003 ± 0.015
Sinhalese: (3 parental populations) 13 classical systems	
Tamil	0.1638 ± 0.039
Gujarati	0.1182 ± 0.052
Bengali	0.7182 ± 0.055
Sinhalese: (3 parental populations) 4 VNTRs	
Bengali	0.8209 ± 0.029
Punjabi	0.0252 ± 0.086
Gujarati	0.1539 ± 0.067

The most widely distributed microsatellites in the human genome are the $(CA)_n$ repeats which have been used extensively for mapping human disease genes. Trinucleotide repeats are polymorphic but some of these repeat loci are very unstable and their expansion is associated with several diseases. STRs based on a 4-bp motif have proved more useful in population genetic studies. Not only are these tetranucleotide markers wide spread throughout the human genome, but many loci show different allele sizes and can thus be simultaneously genotyped on the same gel using multiplex PCR method.

The generation of multiple alleles at the STR loci fits with the Stepwise Mutation Model. With their fairly moderate mutation rate and high heterozygosity STR markers are the preferred choice for population genetic studies. Perez-Lezaun *et al.* (1997) analysed 20 tetranucleotide STRs to study the genetic differentiation and affinity among 16 world populations. Their results indicate that STR allele distribution and their variation among human populations are maintained through genetic drift. To examine genetic diversity in Africa, Jorde *et al.* (1997) examined 60 STR loci among human population samples from three major continents. Their analyses suggest African populations have about 20% greater diversity at STR loci and these microsatellites are indeed very useful DNA markers for evolutionary studies. At the same time the work of Barbujani *et al.* (1997) with 30 STR and 79 polymorphic loci in 16 populations of the world indicates that 84.4% of genetic variation is found between the members of the same population, 10% being due to the difference between continents. These estimates and the genetic affinity studies based on the frequencies of STRs have also shown general concordance with the findings of studies of classical and other nuclear markers (Roychoudhury and Nei 1988 and Jorde *et al.* 1995).

Religious, linguistic, cultural and geographical barriers influence the dynamics of genetic diversity in the populations of India. It is of interest to enquire what the STR loci reveal of the evolutionary relationships among such strongly isolated populations?

Six tetranucleotide STR loci (CSF1PO, TPOX, THO1, F13A01, FESFPS and vWA) were genotyped on samples from three tribal groups of Andhra Pradesh (AP), South India (Koya, Lambadi and Chenchu), two caste groups of Western India (Brahmin and Patel-Gujarati) and a tribal group of the highland state of Himachal Pradesh (HP) North India (Kanet). The six loci were amplified by a multiplex polymerase chain reaction. The CTT triplex amplified the loci CSFIPO, TPOX and THOI while the FFv triplex included the loci F13AO1, FESFPS and vWA. Individuals were genotyped by com-

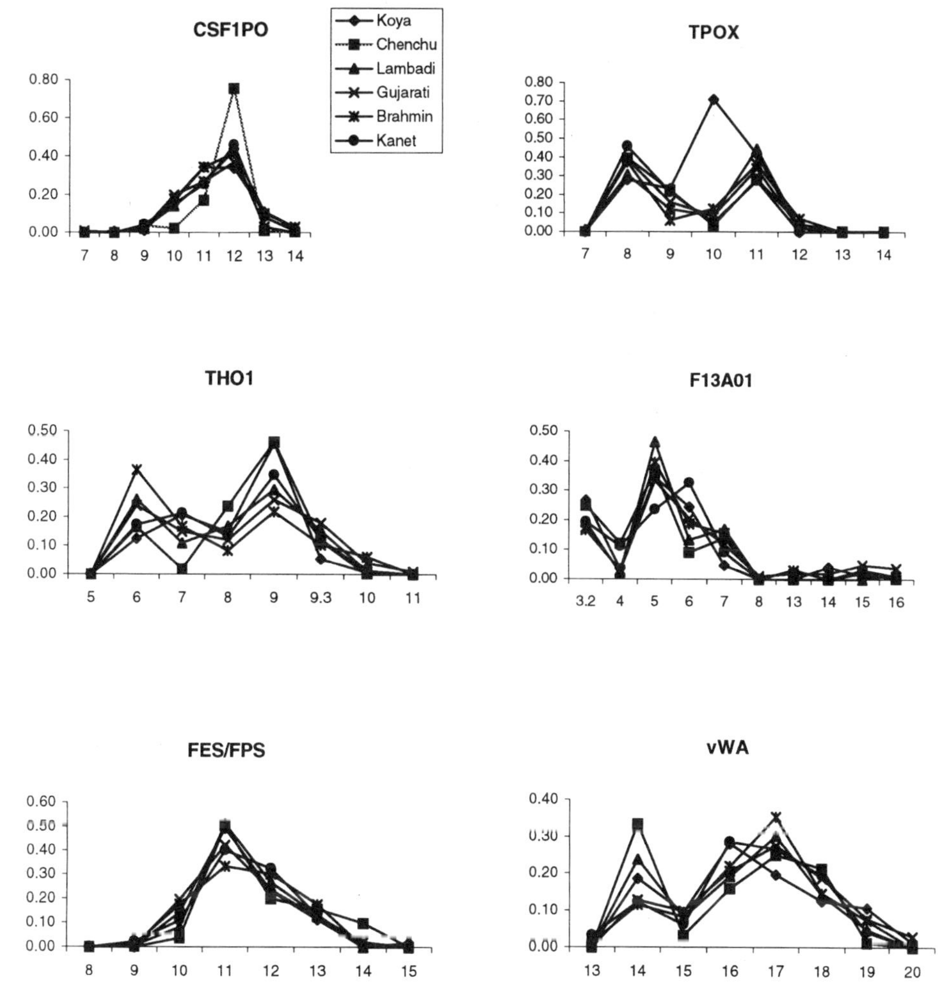

Figure 8. Distribution of STR frequencies in Six Indian Populations. Legend is given in CSF1PO.

paring their amplified alleles with a ladder provided in the GeneprintTH STR system obtained from Promega.

The allele frequency distribution of the six STR in the six populations is given in Fig. 8. The CSFIPO, F13AO1 and FESFPS show a nearly unimodal allele frequency distribution while the other three loci, TPOX, THO1 and vWA, are clearly bimodal. Each locus is observed to have a wide variation in allele frequencies amongst the populations studied. For example at the TPOX locus there is a significant level of diversity between Koya *vs* Lambadi ($\chi^2 = 8.6$; df 3; $p < 0.05$), Koya *vs* Kanet ($\chi^2 = 9.64$; df 3; $p < 0.025$), Chenchu *vs* Lambadi ($\chi^2 = 13.7$; df 3; $p < 0.01$) and Lambadi *vs* Kanet ($\chi^2 = 8.83$; df 3; $p < 0.025$). All tribal populations show significant variation at all other loci, but differences between the caste populations are less prominent.

Average heterozygosity (H) is the simplest measure of genetic differentiation. The average heterozygosity values for the six STR loci and compared in Table 3 with those of 14 classical protein markers analysed previously (Mastana and Papiha 1994; Papiha

Table 3. Average heterozygosity with standard errors for six populations of India

Populations	Conventional system (14) H ± SE	STRs (6) H ± SE
Koya (AP)	0.332 ± 0.054	0.733 ± 0.022
Chenchu (AP)	0.314 ± 0.052	0.668 ± 0.058
Lambardi (AP)	0.363 ± 0.049	0.725 ± 0.023
Gujarati	0.378 ± 0.056	0.776 ± 0.021
Brahmin	0.388 ± 0.054	0.751 ± 0.017
Kenet (HP)	0.378 ± 0.058	0.749 ± 0.022

et al. 1997). Average heterozygosity ranges from 0.667 ± 0.058 in the Chenchu population to 0.775 ± 0.021 in Gujarati. The lowest heterozygosity was found in the three tribal groups of Andhra Pradesh, which reflects the fact that the genetic structure of these populations is highly influenced by inbreeding. Average heterozygosity for the 14 classical protein systems ranges from 0.314 ± 0.052 in the Chenchu to 0.388 ± 0.054 in Brahmins. Once again the average heterozygosity of 14 protein systems was lowest in the tribal populations. Gene diversity estimates were calculated by the method of Nei (1973) where G_{ST} is the rate of gene differentiation among populations, the extent of total genetic variation is H_T and H_S is the sub population heterozygosity. If population differentiation occurs primarily by genetic drift in isolation or partial isolation, the population G_{ST} value would be expected to increase as the average geographical distance between populations increases.

Gene diversity indices G_{ST}, H_T and H_S for the six populations based on the data on six STR loci are given in Table 4 alongside the values obtained for 14 classical polymorphic systems. For classical systems the average values for 14 systems are $H_T = 0.369$, $H_S = 0.356$ and $G_{ST} = 0.035$, while for 6 STR systems $H_T = 0.739$, $H_S = 0.722$ and $G_{ST} = 0.024$. Some variation may be due to population size, type and number of systems studied, but the genetic diversity observed from classical and STR markers shows a similar trend. There is a greater variation in sub-populations indicating that the infrastructure of these populations is maintained by genetic drift.

Another approach to studying genetic differentiation was provided by genetic distance analyses. Nei's genetic distance (D_A) matrix was reduced to a two dimensional dendrogram by the UPGMA method. For the six STRs, 14 classical markers and the 20 systems combined, the dendrograms are given in Fig. 9. In all the three dendrograms, geographically close caste populations show the greatest affinity. However, the tribal populations from the same region each show a distinct gene pool with very little affinity for each other. The clustering of two geographically isolated groups Koya and Kanet in the dendrogram of six STRs may be due to the small number of STRs studied. These

Table 4. Genetic diversity for STRs classical system in six populations of India

Type of Markers	H_T	H_S	G_{ST}
6 STRs	0.739	0.772	0.024
14 Classical systems	0.369	0.356	0.035

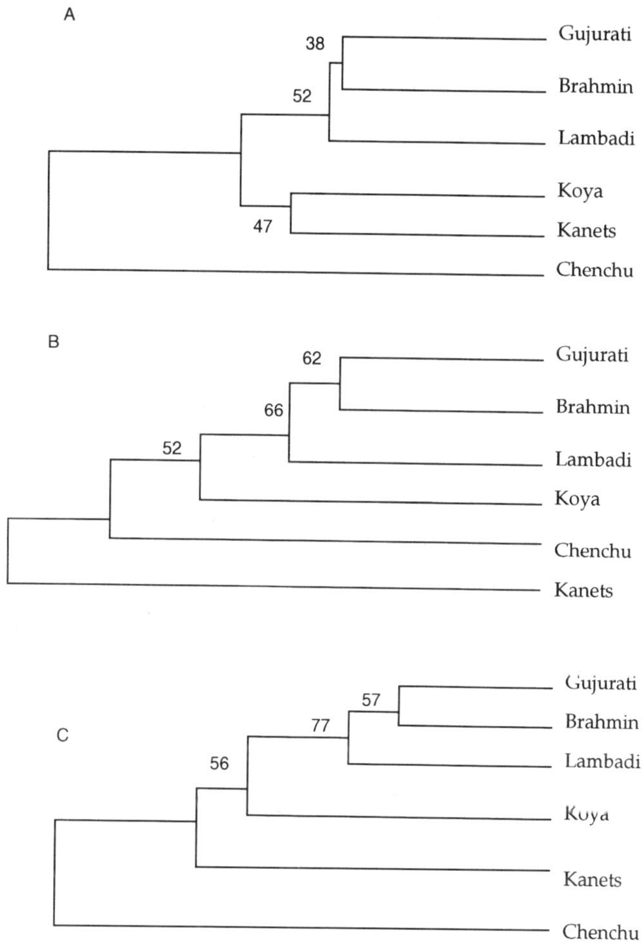

Figure 9. UPGMA dendrograms derived from Nei's DA distance for a) Six STR loci b) 14 conventional systems and c) All systems (20).

results demonstrate that molecular STR systems are additional and powerful markers for genetic differentiation and evolutionary studies.

2.5. Mitochondrial and Y Chromosome DNA Markers

Despite some limitations both mitochondrial DNA (mtDNA) and Y chromosome markers have shown considerable potential in evolutionary studies. Apart from the pseudoautosomal region of the Y chromosome both the Y and mtDNA, being haploid, show no recombination. For both Y and mtDNA there is uniparental transmission; mitochondrial being transmitted maternally and Y exclusively paternally. Both the Y and the mtDNA provide essential data, important in understanding the difference between male and female migration and their contribution to present day evolution. Since there is no recombination, both mitochondrial and Y-chromosomes behave as single units and various markers along the whole Y and mtDNA stretch are inherited as single blocks.

This synteny of markers generates haplotypes and the frequency of these shows great diversity in human populations. All four types of polymorphism insertion/deletion, single nucleotide polymorphisms, mini and microsatellites have been found on the DNA of the Y chromosome. From haplotypes deduced from these markers, and using reasonable assumptions about the rate at which different types of mutations occur, one can estimate the date for the most recent common ancestor of the Y chromosomes. Hammer (1995) sequenced 2,400-bp from the same Y chromosome region for 16 ethnically diverse humans and four chimpanzees. They were able to calculate the date of the common ancestral human Y chromosome as 188,000 years before present, with 95% confidence limits from 51,000 to 411,000 years. The YAP+ insertion was dated to 141,000 years before the present time. Whenever for the Y-chromosome a significant number of individuals in a population are analysed, the frequency of occurrence of the different haplotypes can be used both to distinguish populations and to shed light on the sub-structures within populations.

Studies of mtDNA have also shown considerable potential in the study of human evolution. Earlier work involved the variation in frequency of restriction fragment sites found along the whole mtDNA (Horai *et al.* 1984). However, more recent studies have combined the sequence of the control region (D loop) which shows extensive variability in its two hypervariable segments (Vigilant *et al.* 1991; Ward *et al.* 1991 and Torroni *et al.* 1994, 1996 and Torroni 1997). Analyses of restriction site polymorphisms and sequencing of the D loop have identified common genetic variants, grouped in several haplogroups called A, B, C, D, E, F, G, H, I, J, K, L (L1, L2, L3 and L3a), M, T, U, V, W and X (Torroni *et al.* 1996 and Watson *et al.* 1997). Some of these haplogroups are continental and population specific. For example, L1, L2 and L3 are specifically from the African continent. It is suggested that a sub-cluster L3—L3a, was the starting point of the expansion dating from 60–70,000 years ago, from which all non-African mtDNA lineages originated. Asian-specific haplogoups are A, B, F and M. The M supergroup is further split between C, D, E and G. Native Americans have mainly A, B, C and D lineages. The European-specific mtDNA haplogroups are H, I, J, K, T, U, V, W and X. The H haplogroup is the most frequent with a frequency around 40–50%. The V is its sister group with an expansion time of 14,000 years starting somewhere near, or in, Iberia. U is the most diverse and, like M in Asia, has several distinguishable subgroups. J is considered to be of neolithic origin in Europe, while T is similar in age to H dating from around 20–30,000 years ago. Haplogroup X is an interesting marker. This marker has been found in low frequency in modern native Americans and has also been found in skeletal remains of Amerindians, so indicating an ancient link between Eurasians and Native Americans (Morell 1998). The distribution of the haplotype data combined with the complete sequence analysis of mtDNA of humans and most closely related primates, suggests that the origin of modern humans or *Homo sapiens sapiens,* may be about 150,000–200,000 years before the present. This estimate is supported by data on the Y chromosome and other types of genetic markers.

In conclusion, this century has seen remarkable progress in the study of classical markers and recently in the development of new molecular markers. We are now in a position to make a careful selection of the markers and type of analysis that will best answer the questions of the origin of modern humans and the expansion of their ancestral stocks. Studies of the Y chromosome, mtDNA and of the haplotypes of nuclear markers, as investigated by Tishkoff *et al.* (1996) may be the most profitable way to approach population genetics in the twenty-first century.

3. ACKNOWLEDGMENTS

We wish to thank Drs. D F Roberts, D J Pritchard and Pawanjit Singh for their suggestions. Thanks to Miss C J Coulter for help in the laboratory analysis. The work was supported by small grant from University of Newcastle upon Tyne

4. REFERENCES

Abeyaratne KP, Premawansa S, Rajapakse L, Roberts DF, Papiha SS (1976). A survey Of Glucose-6-phosphate dehydrogenase deficiency in the North Central Province of Sri Lanka (formerly Ceylon) Am. J Phys. Anthropol. 44:135–138.

Baird M, Balazs I, Giusti A *et al.* (1986). Allele frequency distribution of two highly polymorphic DNA sequences in three ethnic groups and its application to the determination of paternity. Am. J. Hum. Genet. 39:489–501.

Barbujani G, Magagni A, Minch E, Cavalli-Sforza LL (1997). An apportionment of human DNA diversity. Proc. Natl. Acad. Sci. USA 94:4516–4519.

Bhasin MK, Walter H, Danker-Hopfe H (1994). People of India. Publishers Kamla-Raj Enterprises Delhi.

Botstein D, White RL, Skolnick M, Davis RW (1980). Construction of genetic linkage map in man using restriction fragment length polymorphisms. Am. J. Hum. Genet. 32:314–331.

Bowcock AM, Bucci C, Herbert JM, Kidd JR *et al.* (1987). Study of 47 DNA markers in five populations from four continents. Gene Geography 1:47–64.

Bowcock AM, Herbert JM, Mountain JL, Kidd JR *et al.* (1991a). Study of an additional 58 DNA markers in five human populations from four continents. Gene Geography 5:155–173.

Bowcock AM, Kidd JR, Mountain JL, Herbert JM *et al.* (1991b). Drift admixture and selection in human evolution: A study with DNA polymorphisms. Proc. Natl. Acad. Sci. USA. 88:839–843.

Bowcock AM, Ruiz-Linares A, Tomfohrde J, Minch E, Kidd JR, Cavalli-Sforza LL (1994). High resolution of human evolutionary trees with polymorphic microsatellites. Nature 368:455–457.

Budowle B, Guisti AM, Waye JS *et al.* (1991). Fixed bin analysis for statistical evaluation of continuous distributions of allelic data from VNTR loci for use in forensic comparisons. Am. J. Hum. Genet. 48:841–855.

Cavalli-Sforza LL, Menozzi P, Piazza A (1994). The history and geography of human genes. Princeton University Press, New Jersey.

Chakraborty R (1990). Genetic profile of cosmopolitan populations. Effects of hidden subdivision. Anthrop. Anz. 48:313–331.

Chakraborty R, Zhong Y, Jin L, Budowle B (1994). Nondetectability of restriction fragments and independence of DNA-fragment sizes within and between loci in RFLP typing of DNA. Am. J. Hum. Genet. 41:948–956.

Clark AG (1987). Neutrality tests of highly polymorphic restriction fragment length polymorphisms Am. J. Hum. Genet. 41:948–956.

Cohen JE (1990). DNA fingerprinting for forensic identification: potential effects of data interpretation of subpopulation of heterogeneity and band number variability. Am. J. Hum. Genet. 46:358–368.

Cutbush M, Mollison PL (1950). The Duffy blood group system. Heredity 4:383–389.

Dausset J (1954). Leuko-agglutinins IV: Leukoagglutinins and blood transfusion. Vox Sanguinis 4:190.

Deka R, Decroo S, Jin L, McGarvey ST *et al.* (1994). Population genetics characteristic of the DIS80 locus in seven human populations. Hum. Genet. 94:252–258.

Deka R, Shriver MD, Yu LM, Ferrell RE, Chakraborty R (1995). Intra- and inter-population diversity at short tandem repeat loci in diverse populations of the world. Electrophoresis 16:1659–1664.

Elder JK, Southern EM (1983). Measurements of DNA length by gel electrophoresis II. Comparison of methods for relating mobility to fragment length Anal. Biochem. 128:227–231.

Grubb R, Laurell AB (1956). Hereditary serological human serum groups. Acta Pathological Microbiologica Scandinavica 39:390.

Guo SW, Thompson EA (1992). Performing the Exact Test of Hardy-Weinberg proportion for Multiple Alleles. Biometrics 48:361–372.

Hammer MF (1995). A recent common ancestry for human Y chromosomes. Nature 378:376–378.

Harpending HC, Jenkins T (1973). Genetic distance among Southern African populations. In: Methods and Theories of Anthropological genetics. Crawford MH, Workman PL (Eds). University of New Mexico Press pp177–199.

Harris H, Hopkinson DA (1976). Handbook of enzyme electrophoresis in human genetics North-Holland, Amsterdam.

Hirszfeld L, Hirszfeld H (1919). Serological differences between the blood of different races: The results of researches on the Macedonian front. Lancet ii:675–679.

Horai S, Gojobori T, Matsunaga E (1984). Mitochondrial DNA polymorphisms in Japanese I? Analyses with restriction enzymes of six base pair recognition. Hum. Genet. 68:324–332.

Jeffreys AJ, Royle V, Wilson V *et al.* (1988). Spontaneous mutation rates to new length alleles at tandem-repetitive hypervariable loci in human DNA. Nature 332:278–281.

Jeffreys AJ, Wilson V, Thein SL (1985). Hypervariable microsatellite region in human DNA. Nature 314:67–73.

Jin L, Chakraborty R (1995). Population structure, stepwise mutations, heterozygote deficiency and their implications in DNA forensics. Heredity 74:274–285.

Jorde LB, Bamshad MJ, Watkins WS, Zenger R, Fraley AE, Krakowiak PA, Carpenter KD, Soodyall H, Jenkins T, Rogers AR (1995). Origins and affinities of modern humans: a comparison of mitochondrial and nuclear genetic data. Am. J. Hum. Genet. 57:538.

Jorde LB, Roger AR, Bamshad M, Watkins WS *et al.* (1997). Microsatellite diversity and the demographic history of modern human. Proc. Natl. acad. Sci. USA 94:3100–3103.

Kan YW, Dozy AM (1978). Polymorphism of DNA sequence adjacent to human β-globulin structural gene: relationship to sickle mutation. Proc. Natl. Acad. Sci. USA. 75:5631–5635.

Kidd JR, Black FL, Weiss KM, Balazs I, Kidd KK (1991). Studies of three Amerindian populations using nuclear DNA polymorphisms. Human Biology 63:775–794.

Kidd JR, Pakstis AJ, Kidd KK (1993). Global level of DNA variation. Proceedings of the 4th International Symposium in Human Identification. Madison Promega 21–30.

Kirk RL (1976). The legend of Prince Vijaya: A study of Sinhalese origin. Amer. J. Phys. Anthropol. 45:91–100.

Kshatriya GK (1995). Genetic affinities of Sri Lankan populations. Hum. Biol. 87:843–864.

Lander ES (1989). DNA fingerprinting on trial. Nature 339:P501–505.

Landsteiner K (1901). Ueber agglutinationserscheimun normalen menschlichengen Blutes Wien. Klin. Wschr. 14:1132–1134.

Landsteiner K, Levine P (1927). A new agglutinable factor differentiating individual human bloods. Proceedings of the Society for Experimental Biology, New York 24:600–602.

Lewontin RC, Hubby JL (1966). A molecular approach to the study of genetic heterozygosity in natural population Amount of variation and degree of heterozygosity in natural population of Drosophila pseudo obscura. Genetics 54:595–609.

Lin AA, Herbert JM, Mountain JL, Cavalli-Sforza LL (1994). Comparison of 79 DNA polymorphisms tested in Australians, Japanese and Papua New Guineans with those of the other five populations. Gene Geography 8:191–214.

Long JC (1991). The genetic structure of Admixed populations. Genetics 121:417–428.

Mastana SS, Papiha SS (1994). Genetic structure and microdifferentiation among four endogamous groups of Maharastra, Western India. Ann. Hum. Biol. 21:261–262.

Meselson M, Yucan R (1968). DNA restriction enzyme from *E. coli*. Nature 217:1110–1114.

Morell V (1998). Gene may link ancient Eurasians, Native Americans. Science 280:520.

Mourant AE, Kopec AC, Domaniewska-Sobczak K (1976). The distribution of the human blood groups and other polymorphisms. Oxford University Press, London.

Nei M (1973). Analysis of gene diversity in subdivided populations. Proc. Natl. Acad. Sci. USA 70:3321–3323.

Nei M (1978). Estimation of average heterozygosity and genetic distance from a small number of individuals. Genetics 89:583–590.

Papiha SS (1996). Genetic variation in India. Hum. Biol. 68:607–628.

Papiha SS, Mastana SS, Jayasekara R (1996a). Genetic variation in Sri Lanka. Hum. Biol. 68:707–737.

Papiha SS, Mastana SS, Purandare CA, Jayasekara R, Chakraborty R (1996b). Population genetic study of three VNTR loci (D2544, D7S22 and D12S11) in five ethnically defined populations of the Indian subcontinent. Hum. Biol. 68:819–835.

Papiha SS, Mastana SS, Roberts DF, Onyemelukwke GC, Bhattacharya SS (1991). Population variations in molecular polymorphisms of the short arm of the human X chromosome. Amer. J. Phy. Anthrop. 85:329–334.

Papiha SS, Singh BN, Lanchbury JS, Mastana SS, Sunharao Y (1997). Genetic study of the tribal populations of Andhra Pradesh, South India. Hum. Biol. 69:171–199.

Pauling L, Itano HA, Singer SJ, Wells IC (1949). Sickle cell anemia, a molecular disease. Science 110:543–548.

Payne R, Tripp M, Weigle J, Bodmer W, Bodmer J (1964). A new leukocyte iso-antigen system in man. Cold Spring Harbor Quantitative Biology 29:28p5.

Perez-Lezaun A, Calafell T, Mateu E, Comas D, Ruiz-Pachaco R, Bertranpetit J (1997). Microsatellite variation and differentiation of modern human. Hum. Genet. 99:1–7.

Race RR, Sanger R (1975). Blood group in man. Oxford Blackwell Scientific Publication.

Roberts DF (1987). Genetic structure and differentiation of human populations. Anthrop. Anz. 45:277–288.

Roberts DF, Creen CK, Abeyaratne KP (1972a). Blood groups of the Sinhalese Man 7:122–127.

Roberts DF, Papiha SS, Abeyaratne KP (1972b). Red cell enzyme polymorphisms in Ceylon Sinhalese. Am. J. Hum. Genet. 24:181–188.

Roychoudhury AK, Nei M (1988). Human polymorphic genes. World Distribution. New York, Oxford University Press.

Saha N (1988). The blood genetic markers in Sri Lankan populations: A reappraisal of the legend of Prince Vijaya. Amer. J. Phys. Anthropol. 76:217–225.

Steinberg AG, Cook CE (1981). The distribution of the human immunoglobulin allotypes. Oxford University Press, Oxford.

Steinberger EM, Thompson LD, Hartman JM (1993). On the use of excess homozygosity for subpopulation detection. A letter). Amer. J. Hum. Genet. 52:1275–1277.

Summer K (1987). Restriction fragment length polymorphism on the X chromosome in populations from Papua New Guinea, Thailand, North western Australia and China. Gene Geography 1:207–217.

Tills DA, Kopec AC, Tills RE (1983). The distribution of the human blood groups and other polymorphisms. Oxford University Press, Oxford.

Tishkoff SA, Dietzsch E, Speed W, Pakstis AJ *et al.* (1996). Global pattern of linkage disequilibrium at the CD4 locus and modern hupman origin. Science 271:1380–1387.

Torroni A (1997). Classification of mtDNA variation in Europeans: Implication for the peopling of Europe and disease studies. HUGO EC Euroconference. Helsinki 12–16/9/1997.

Torroni A, Huoponen K, Francalacci P, Petrozzi M, Morelli L, Scozzari R, Obinu D, Savontaus ML, Wallace DD (1996). Classification of European mtDNAs from an analysis of three European populations. Genetics 144:1835–1850.

Torroni A, Lott MT, Cabell MF, Chen YS, Lavergne L, Wallace DC (1994). MtDNA and the origin of Caucasian: Identification of ancient Caucasian-specific haplogroups, one of which is prone to a recurrent somatic duplication in the D-loop region. Am. J. Hum. Genet. 55:760–776.

Vigilant L, Stoneking M, Harpending H, Hawkes K, Wilson AC (1991). African population and the evolution of mitochondrial DNA. Science 253:1503–1507.

Wainscoat JS, Hill AVS, Boyce AL, Flint J *et al.* (1986). Evolutionary relationships of human populations from an analysis of nuclear DNA polymorphisms. Nature 319:491–493.

Watson E, Forster P, Richards M, Bandelt HJ (1997). Mitochondrial footprints of human expansions in Africa. Amer. J. Hum. Genet. 61:691–704.

Wang DG, Fan JB, Sias CJ, Berno A *et al.* (1998). Large scale identification, mapping and genotyping of single nucleotide polymorphisms in the human genome. Science 280:1077–1081.

Ward RH, Frazier BL, Dew-Jager K, Phääbo S (1991). Extensive mitochondrial diversity within a single Amerindian tribe. Proc. Natl. Acad. Sci. USA 88:8720–8724.

Weir BS (1992). Independence of VNTR alleles defined as fixed bins. Genetics 130:873–887.

Wong Z, Wilson V, Patel I , Povey S *et al.* (1987). Characterization of a panel of highly variable minisatellites cloned from human DNA. An. Hum. Genet. 51:269–288.

Wyman AR, White R (1980). A highly polymorphic locus in human DNA. Proc. Natl. Acad. Sci USA. 77:6754–6758.

2

VARIABILITY IN nDNA, mtDNA, AND PROTEINS

A Test Case

Mara H. Hutz[1], Sidia M. Callegari-Jacques[1,2], Maria C. Bortolini[1], and Francisco M. Salzano[1]

Departments of [1]Genetics
Biosciences Institute and [2]Statistics
Mathematics Institute
Federal University of Rio Grande do Sul
Porto Alegre, RS, Brazil

ABSTRACT

The human genome is highly heterogeneous, with coding and non-coding (regulatory) regions, segments of highly or moderately repetitive DNA, and remnants of past insertion events. Presently it is an open question which variants and regions most adequately represent the past history of a given set of populations. To test this, we have investigated the genetic variation of five Brazilian Indian tribes in detail using eight nuclear and one mitochondrial DNA systems (nDNA and mtDNA, respectively), as well as 23 protein loci. Three *a priori* hypotheses were elaborated, and the results examined for agreement with them. The Wai Wai, who according to their history should be the most variable of the five populations, actually showed the highest diversity when nDNA and mtDNA markers were considered; but this was not true for the protein systems. As for the population relationships, the nDNA pattern was exactly that expected, the mtDNA and protein-based dendrograms, however, showed some departures from expectation. Therefore, distinct genomic regions may convey different answers to evolutionary questions. The unravelling of the causes for such discrepancies is an interesting problem in itself, and may lead to the establishment of the correct phylogenies.

1. HUMAN GENETIC DIVERSITY

It is estimated that by the end of 2003 the full sequence of the 3 billion base pairs of the human genome will be available (Collins *et al.* 1998). This fantastic achievement, however, would constitute just the beginning of a process which may lead to the understanding of how the 50–100 thousand human genes function and vary. In this connection, it should be remembered that our genetic material is highly heterogeneous, with

Table 2. Information about the DNA systems considered

System according to type	Identification characteristics	Location (chromosome region)
Nuclear DNA		
Haplotypes		
Beta globin	5 RFLPs	11 p 12.5–11 p 12.8
APO B	3 RFLPs 1 insertion/deletion	2 p 23–2 p ter
DRD 2	3 RFLPs 1 microsatellite	11 q 22–11 q 23
LPL	2 RFLPs	8 p 22
VNTRs		
D1S80	Repeat unit: 16 bp	1 p 26–1 p 35
DRD4	Repeat unit: 48 bp	11 p 15.5
Microsatellites		
CD4	Repeat unit: 5 bp	12 p 12–12 p ter
F13A1	Repeat unit: 4 bp	6 p 24–6 p 25
Mitocondrial DNA		
HVS-I	Nucleotide variability by sequence analysis (SNPs)	First 360 base positions

Key to the abbreviations: RFLP: restriction fragment length polymorphism; VNTR: variable number of tandem repeats; bp: base pairs; HVS-I: first hypervariable segment; SNP: single nucleotide polymorphism.

isolation, the Xavante could depart more markedly from the other four populations; and (c) in terms of variability the Wai Wai would present the highest total gene diversity.

The systems studied by us are listed in Tables 2 and 3. Both nuclear and mitochondrial DNA were investigated, with different methods. In the latter case direct sequencing was performed, while in the nDNA restriction techniques and direct observation of amplified repeat fragments were used. The markers occurred in six different chromosomes (nos. 1, 2, 6, 8, 11, and 12), and three types of systems were considered (haplotypes, VNTRs, and microsatellites). As for the protein markers, immunological and electrophoretic techniques were employed for the investigation of loci present in 14 different chromosomes (nos. 1–4, 6–9, 11–13, 16, 18, and X). Blood group antigens, as well as enzymatic and non-enzymatic proteins were examined. The question asked, in relation to all these DNA and protein markers, was whether the patterns observed, both for gene diversity and population relationships, were or were not congruent.

3. NATURE OF THE VARIABILITY FOUND

The genetic systems investigated have peculiarities that could readily differentiated them, individually or in diverse groups. A summary of our observations is presented on Table 4. The number of alleles or haplotypes vary widely among systems. As much as 14 APOB haplotypes were found among members of the five tribes, while several of the protein systems investigated were monomorphic among them (CHE1, G6PD, HBA, HBA2, Kell, PEPA, PEPB). A total of 36 sites were found to vary in the mtDNA HVS-I region, composed of 360 bp. The number of shared haplotypes was much more common among the protein markers (70%–91%), as opposed to the DNA systems (14%–66%). While both APOB and D1S80 showed a high degree of polymorphism, rare alleles were much more common among the former (9 vs 4).

Other information, not given in Table 4, can be added. The average heterozygosity

Table 3. Information about the protein systems considered

System according to type	Identification characteristics	Location (chromosome region)
Blood groups		
ABO	Detected by immunological methods	9 q 34
Duffy	Detected by immunological methods	1 q 12–q 21
Kell	Detected by immunological methods	7 q 33
MNSs	Detected by immunological methods	4 q 28–q 31
P	Detected by immunological methods	Chrom. 6
Rh	Detected by immunological methods	1 p 36.2–p 34
Non-enzymatic proteins		
Hemoglobin A	Detected by electrophoretic methods	11 p 15.5
Hemoglobin A2	Detected by electrophoretic methods	11 p 15.5
Albumin	Detected by electrophoretic methods	4 q 11–q 13
Ceruloplasmin	Detected by electrophoretic methods	3 q 21–q 24
Haptoglobin	Detected by electrophoretic methods	16 q 22.1
Transferrin	Detected by electrophoretic methods	3 q 21
Enzymatic proteins[a]		
ACP	Detected by electrophoretic methods	2 p 23
AK	Detected by electrophoretic methods	9 q 34.1
CA2	Detected by electrophoretic methods	8 q 22
CHE1	Detected by electrophoretic methods	3 q 25.2
ESA	Detected by electrophoretic methods	11 q 13–q 22
ESD	Detected by electrophoretic methods	13 q 14.11
GLO	Detected by electrophoretic methods	6 p 21.3–p 21.2
G6PD	Detected by electrophoretic methods	X q 28
PEPA	Detected by electrophoretic methods	18 q 23
PEPB	Detected by electrophoretic methods	12 q 21
PGD	Detected by electrophoretic methods	1 p 36.2–p 36.13

[a]See the footnote of Table 4 for an explanation of the symbols.

was, as expected, more marked using VNTRs (average 0.62) than employing haplotypes or microsatellites (average 0.45 for both); and no obvious correlation among the three sets of variables was found when their levels of variability were compared in the five populations. As for mtDNA, among the 36 sites which showed variation, those numbered 16256 and 16316 (Anderson *et al.* 1981) were the least (0.013), while 16189 was the most (0.499) variable. The mutation of thymine (T) to cytosine (C) present in the latter site occurred in the five tribes considered and in five other South American Indian populations listed by Bortolini *et al.* (1998); therefore, it can be regarded as characteristic of this ethnic group. Other changes that occurred in all the five tribes studied here and (with one exception) also in the other five were 16111C→T, 16290C→T and 16319G→A (guanine→adenine). Among the protein systems, besides the seven monomorphic already listed, eight (ABO, AK, ALB, CA2, CP, ESA, PGD, TF) showed limited variation, the most polymorphic being the remaining eight (ACP, FY, ESD, GLO, HP, MNSs, P, RH). Four of the latter are blood groups, three enzymatic and one a non-enzymatic protein.

4. GENE DIVERSITY

The same subjects in the five tribes (the DNA data, however, being a subsample of the protein results; see the primary references indicated above) have been studied in

Table 4. Details about the level and peculiarities of the variability observed in the five populations with the systems used

System	No. of alleles/ haplotypes	Shared alleles/ haplotypes N	Shared alleles/ haplotypes %	No. of rare[a] alleles/ haplotypes	Average no. of alleles/haplotypes per tribe
Nuclear DNA					
Haplotypes					
Beta globin	8	2	25	3	3.6
APO B	14	2	14	9	7.8
DRD 2	9	2	22	2	4.4
LPL	3	2	66	0	2.8
VNTRs					
D1S80	11	3	27	4	6.4
DRD4	6	2	33	2	3.8
Microsatellites					
CD4	3	2	66	0	2.2
F13A1	6	2	33	2	4.0
Mitocondrial DNA					
HVS-I	36	17	47	0	14.6
Proteins[b]					
Blood groups (six systems)	16	14	87	2	14.4
Non-enzymatic (six systems)	10	7	70	3	7.8
Enzymatic (11 systems)	19	14	74	4	15.2
The 8 most polymorphic	22	20	91	2	20.4

[a]Haplotypes/alleles with a frequency equal or lower than 2%.
[b]Blood groups: ABO, Duffy (FY), Kell, MNSs, P, Rh. Non-enzymatic: Albumin (ALB), Ceruloplasmin (CP), Hemoglobin A (HBA), Hemoglobin A2 (HBA2), Haptoglobin (HP), Transferrin (TF). Enzymatic: Acid phosphatase (ACP), Adenylate kinase (AK), Carbonic anhydrase 2 (CA2), Esterase A (ESA), Esterase D (ESD), Glucose-6-phosphate dehydrogenase (G6PD), Glyoxalase 1 (GLO), Peptidase A (PEPA), Peptidase B (PEPB), Phosphogluconate dehydrogenase (PGD), Serum cholinesterase 1 (CHE 1).

relation to nDNA, mtDNA and proteins. In relation to proteins we considered both the whole set and a subsample of the eight most polymorphic loci. Results are given in Table 5. The nDNA values are 3–5x as high as those obtained for proteins (23 loci) and mtDNA, and of the same order of magnitude as those obtained when we selected the eight most polymorphic protein loci. Within each set the differences among tribes are not large and did not show marked tendencies, except for the consistent lowest values presented by the Surui. In the mtDNA markers the Surui-Wai Wai difference is statisti-

Table 5. Gene diversity (H_S) in the five populations considering nDNA, mtDNA, and protein data

Tribes	Nuclear DNA	Mitochondrial DNA	Protein: 23 loci	Protein: 8 most polymorphic loci
Wai Wai	0.57 ± 0.06	0.17 ± 0.03	0.17 ± 0.05	0.47 ± 0.04
Xavante	0.56 ± 0.07	0.08 ± 0.02	0.17 ± 0.05	0.48 ± 0.04
Surui	0.41 ± 0.06	0.05 ± 0.02	0.13 ± 0.04	0.37 ± 0.06
Zoró	0.48 ± 0.06	0.12 ± 0.02	0.17 ± 0.05	0.50 ± 0.05
Gavião	0.54 ± 0.07	0.12 ± 0.03	0.18 ± 0.05	0.48 ± 0.05

H_S was calculated as indicated in Nei (1987), using the DISPAN program (Ota 1993).

Table 6. Analysis considering the five populations studied or just the three tupi-mondé groups, and using three sets of markers

Clustering level	Type of system	Gene diversity (H_T)	Relative interpopulation diversity (%)
I			
(Wai Wai, Xavante,	nDNA	0.546	8
Surui, Zoró, Gavião)	mtDNA	0.139	30
	protein	0.173	6
II			
(Surui, Zoró, Gavião)	nDNA	0.490	6
	mtDNA	0.153	12
	protein	0.170	6

H_T was calculated as indicated in Nei (1987), using the DISPAN program (Ota 1993).

cally significant (0.05 vs 0.17). The Wai Wai showed the highest diversity when nDNA and mtDNA are considered (in accordance with hypothesis (c) above), but not in relation to the protein systems.

An analysis was then conducted considering first the five tribes, and then the Tupi-Mondé subset (Table 6). As expected, in two of the three comparisons the Tupi-Mondé showed lower values, but the differences are not impressive. The higher nDNA values already observed in Table 5 are notable again. A new feature in Table 6 is the high (30% vs 6%–8%) interpopulation variation found in the mtDNA results considering the five tribes.

5. POPULATION RELATIONSHIPS

The relationships among the populations using the four sets of data (nDNA, mtDNA, proteins—23 loci, and proteins—8 most polymorphic loci) can be viewed through dendrograms, that are displayed in Fig. 1. The nDNA pattern is exactly that expected with the previously presented working hypotheses (a) and (b). The Tupi-Mondé tribes are set apart, with the Surui (who suffered most with the contact with non-Indians) somewhat separated. Then we have the Xavante and Wai Wai cluster, with the Xavante branch being the longest. In the mtDNA tree the Tupi-Mondé also differentiated in a separate cluster, but now the most distinct from the three are the Zoró (this could be explained by the fact that of the three they were the less influenced by the outside world, but the original hypothesis was formulated differently). The Xavante mtDNA pattern clearly deviate from the others. The protein trees are inconsistent with the hypotheses, either considering the 23 systems or just the eight most polymorphic. In this case, those who depart most from the others are the Surui.

6. CONCLUSIONS AND PROSPECTS

Mountain (1998) listed the set of characteristics that an ideal genetic data set should have. This set should: (a) have many independently evolving, unlinked genetic regions; (b) the nonrecombining genetic regions should be highly variable; (c) mutation rates at each nucleotide site should be known; (d) the ancestral state, or the oldest allele at each

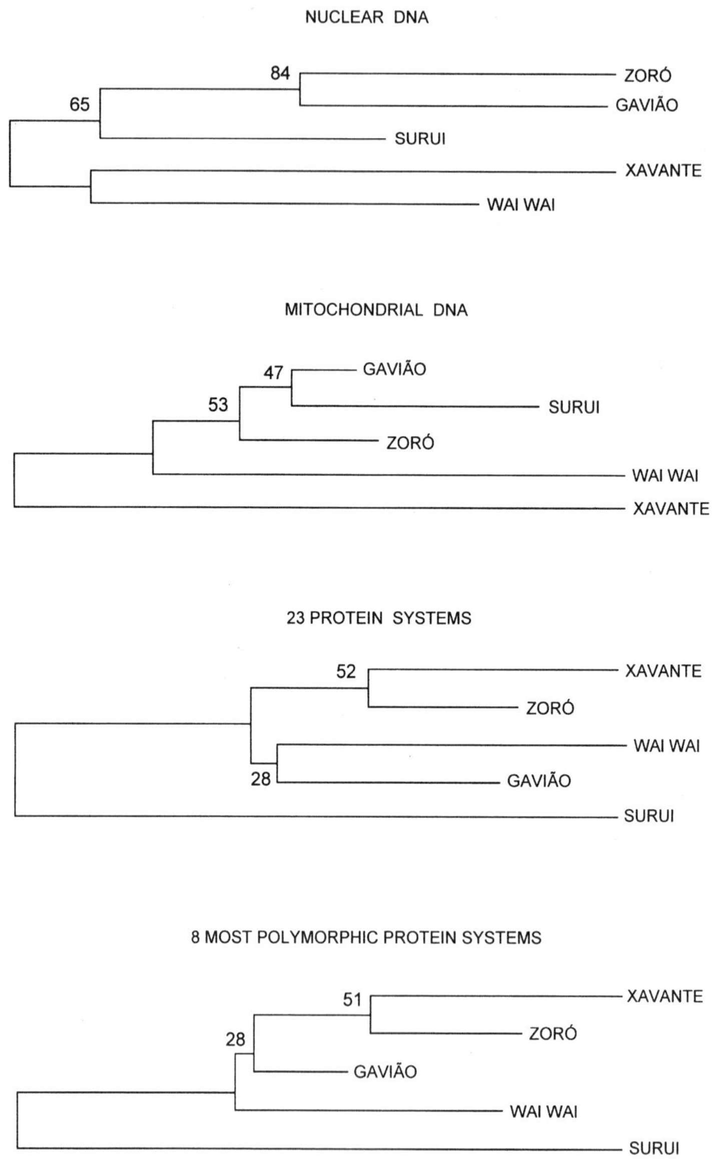

Figure 1. Population relationships obtained with the four sets of data, namely nDNA, mtDNA, proteins (23 loci) and proteins (only the eight most polymorphic loci). D_A distances (Nei *et al.* 1983) were computed, and the relationships among the populations were examined using neighbor-joining trees (Saitou and Nei 1987), which were statistically evaluated through the bootstrap test (Felsenstein 1985) with 3,000 replications (Hedges 1992), using the DISPAN program (Ota 1993). The trees were constructed using the TREEVIEW program (K. Tamura, personal communication).

locus, should be ascertainable; and (e) the DNA regions should provide individuals with equal fitness. If we are interested in differences among male and/or female migration patterns or population sizes, then uniparental markers, such as the nonrecombining regions of the Y, or the mtDNA, should be sought. Mountain herself recognised that this ideal set is essentially unattainable, but we should try to have approximations of it, as our resources permit.

The systems considered in the present review meet some of these demands; for instance, the markers are spread over a wide range of chromosomes. The nDNA regions studied by us are clearly more variable than the mtDNA and protein characteristics investigated, but when we select, from the protein loci, the most variable, agreement between the DNA and protein trees does not improve. Variability *per se*, therefore, is not the answer. There is a long way between DNA and protein, and patterns of expression may be involved in these differences, as well as diverse mutation rates and subtle adaptive inequalities. The trees may also reflect the different evolutionary depths that can be reached using these varied loci.

Should we finish then in a pessimistic mood? Not at all! Progress in the automated investigation of our genome is occurring at a fantastic rate (Collins *et al.* 1998), and what is needed are concerted efforts focused in the understanding of the human variability. A basic program would involve (a) selection of a wide set of markers, which would represent as accurately as possible the human genome; (b) choice of populations for study which would represent the major linguistic stocks and cultural complexes, living in strategic regions; and (c) analysis of the results with the sophisticated computing facilities now available, elaborating, if necessary, mathematical models that could test hypotheses and define the probabilities of specified events.

In elaborating these goals we are aware that they are not easily attainable, not the least due to the increasing difficulty of obtaining samples from specified populations, a consequence of the unfortunate linking of a large portion of the present sequencing efforts with commercial, non-academic objectives. But if appropriate ethics protocols could be established, on the basis of the recent UNESCO Universal Declaration about the Human Genome and Human Rights (UNESCO 1998), it is possible that the suggested program could succeed. Only optimists can make science in Third World countries, and it is on this tone that we close this review.

7. ACKNOWLEDGMENTS

The research reported here would have not been possible without the collaboration of many colleagues and friends, listed in the primary publications. Thanks are due to the Fundação Nacional do Índio for permission to study the Indians and for logistic assistance. The Indian leaders and subjects of the investigation were adequately informed about the aims of the study and gave their approval, which is also gratefully acknowledged. Our current research was also approved by the Brazilian National Comission on Ethics in Research (Decision no. 123/98). Financial assistance was provided by Programa de Apoio a Núcleos de Excelência; Financiadora de Estudos e Projetos; Conselho Nacional de Desenvolvimento Científico e Tecnológico; Fundação de Amparo à Pesquisa do Estado do Rio Grande do Sul; Wenner-Gren Foundation for Anthropological Research; and MacArthur Foundation.

8. REFERENCES

Almeida S, Salzano FM, Hutz MH (1998) A variabilidade dos genes DRD2 e DRD4 em populações indígenas brasileiras. Genet Molec Biol 21(3, Suppl):357.

Anderson S, Bankier AT, Barrel BG, De Brujin MHL, Coulson AR, Drouin J, Eperon IC, *et al.* (1981) Sequence and organization of the human mitochondrial genome. Nature 290:457–474.

Bevilaqua LRM, Matevi VS, Ewald GM, Salzano FM, Coimbra CEA Jr, Santos RV, Hutz MH (1995)

Beta-globin gene cluster haplotype distribution in five Brazilian Indian tribes. Am J Phys Anthropol 98:395–401.

Bogdawa HM (1997) Investigação dos STRs CD4 e F13A1 em três grupos étnicos brasileiros. MSc Dissertation, Federal University of Rio Grande do Sul, Porto Alegre.

Bortolini MC, Baptista C, Callegari-Jacques SM, Weimer TA, Salzano FM (1998) Diversity in protein, nuclear DNA, and mtDNA in South Amerinds—agreement or discrepancy? Ann Hum Genet 62:133–145.

Callegari-Jacques SM, Salzano FM, Weimer TA, Franco MHLP, Mestriner MA, Hutz MH, Schüler L (1996) The Wai Wai Indians of South America: History and genetics. Ann Hum Biol 23:189–201.

Collins FS, Patrinos A, Chakravarti A, Gesteland R, Walters L (1998) New goals for the U.S. human genome project: 1998–2003. Science 282:682–689.

Ewald GM, Callegari-Jacques SM, Coimbra CEA Jr, Santos RV, Gastaldo G, Salzano FM, Hutz MH (1999) Lipoprotein lipase gene polymorphisms in three Brazilian ethnic groups: Association with lipid levels (submitted).

Felsenstein J (1985) Confidence limits on phylogenies: An approach using the bootstrap. Evolution 39:783–791.

Hedges HB (1992) The number of replications needed for accurate estimation of the bootstrap P value in phylogenetic studies. Mol Biol Evol 9:366–369.

Heidrich EM, Hutz MH, Salzano FM, Coimbra CEA Jr, Santos RV (1995) D1S80 locus variability in three Brazilian ethnic groups. Hum Biol 67:311–319.

Hutz MH, Mattevi VS, Callegari-Jacques SM, Salzano FM, Coimbra CEA Jr, Santos RV, Carnese FR, *et al.* (1997) D1S80 locus variability in South American Indians. Ann Hum Biol 24:249–255.

Jorde LB, Bamhad M, Rogers AR (1998) Using mitochondrial and nuclear DNA markers to reconstruct human evolution. BioEssays 20:126–136.

Kaufman L, Vargas AF, Coimbra CEA Jr, Santos RV, Salzano FM, Hutz MH (1999) Apolipoprotein B genetic variability in Brazilian Indians. Hum Biol (in press).

Mountain JL (1998) Molecular evolution and modern human origins. Evol Anthropol 7:21–37.

Nei M, Tajima F, Tateno Y (1983) Accuracy of estimated phylogenetic trees from molecular data. J Mol Evol 19:153–170.

Ota T (1993) DISPAN: Genetic distance and phylogenetic analysis. Institute of Molecular Evolutionary Genetics, The Pennsylvania State University, University Park.

Ricardo B (1996) Povos indígenas no Brasil (1991/1995). Instituto Socioambiental, São Paulo.

Saitou W, Nei M (1987) The neighbour-joining method. A new method for reconstructing phylogenetic trees. Mol Biol Evol 4:406–425.

Salzano FM (1998) Genetic diversity of South American human populations at the DNA and protein levels. J Exp Zool 282:157–163.

Salzano FM, Franco MHLP, Weimer TA, Callegari-Jacques SM, Mestriner MA, Hutz MH, Flowers NM, *et al.* (1997) The Brazilian Xavante Indians revisited: New protein genetic studies. Am J Phys Anthropol 104:23–34.

Salzano FM, Weimer TA, Franco MHLP, Callegari-Jacques SM, Mestriner MA, Hutz MH, Santos RV, *et al.* (1998) Protein genetic studies among the Tupi-Mondé Indians of the Brazilian Amazonia. Am J Hum Biol 10:711–722.

Santos RV (1991) Coping with change in native Amazônia: A bioanthropological study of Gavião, Surui, and Zoró, Tupi-Mondé speaking societies from Brazil. Ph.D. Thesis, Indiana University, Bloomington.

Santos RV, Flowers NM, Coimbra CEA Jr, Gugelmin AS (1996) Human ecology and health in the context of change: The Xavante Indians of Mato Grosso, Brazil. In: Föllér M-L, Hansson LO (eds) Human ecology and health adaptation to a changing world. Göteborg University, Göteborg, pp 94–117.

UNESCO (United Nations Educational, Scientific and Cultural Organization) (1998) Declaración universal sobre el genoma humano y los derechos humanos. Dialogo 23:30–31.

Ward RH, Salzano FM, Bonatto SL, Hutz MH, Coimbra CEA Jr, Santos RV (1996) Mitochondrial DNA polymorphism in three Brazilian Indian tribes. Am J Hum Biol 8:317–323.

3

GENETIC DIVERSITY AMONG FIVE NATIVE AMERICAN TRIBES OF COLOMBIA

Evidence from Nine Autosomal Microsatellites

F. D. Guarino[1,2], L. Federle[1], R. A. H. van Oorschot[2], I. Briceno[3], J. E. Bernal[3], S. S. Papiha[4], M. S. Schanfield[5], and R. J. Mitchell[1]

[1]Department of Genetics and Human Variation
LaTrobe University, Melbourne, Australia
[2]Victoria Forensic Science Centre
Macleod, Melbourne, Australia
[3]Institute of Human Genetics
Pontificia Universidad Javeriana
Bogota, Colombia
[4]Department of Biochemistry and Genetics
University of Newcastle upon Tyne, UK
[5]Analytical Genetic Testing Center
Denver, CO, USA

ABSTRACT

This paper reports upon the validity of using nine highly polymorphic autosomal microsatellites to assess genetic diversity among five tribes of Colombian Indians (Arsario, Kogui, Ijka, Coreguaje, and Wayuu), and also their usefulness in detecting admixture. The tribes investigated exhibit linguistic diversity, inhabit extreme and diverse environments, as well as varying in population size. Previous studies of "classical" genetic markers indicated that the tribes also differ considerably with respect to levels of admixture with both Caucasians and Africans. Our data show that the pattern of genetic variation observed among the native Colombians reflects the following factors: geography, ecology, language, population size and history. Microsatellite data confirm both the genetic similarity of the Arsario and Kogui tribes that live at high altitude, and the absence of admixture with outside groups. Their low levels of heterozygosity are consistent with this. The Ijka are genetically close to the Arsario and Kogui but the presence of extra alleles in the former most likely reflects admixture. The Coreguaje are, by far, the most distinct tribe and all data suggest this a result of drift. The Wayuu and Ijka show clear evidence of admixture with exotic groups, but this is more pronounced in

Genomic Diversity: Applications in Human Population Genetics
edited by Papiha, Deka, and Chakraborty, Kluwer Academic / Plenum Publishers, New York, 1999.

the Wayuu. The nine microsatellites were unable to distinguish between Caucasians and Africans as the source of admixture. Preliminary typing of five Y-chromosome microsatellites is also consistent with this finding.

1. INTRODUCTION

Wyman and White (1980) demonstrated that the human genome contains a large number of polymorphic sequences that later became referred to as Variable Number of Tandem Repeats (VNTRs). These loci are characterised by alleles that differ according to the number of tandem repeats of a particular DNA sequence. Since then a number of these highly polymorphic loci have been used to infer the evolutionary history of Homo sapiens and measure population affinities (Balazs *et al.* 1992; Edwards *et al.* 1992; Bowcock *et al.* 1994; Deka *et al.* 1995). There is still debate concerning the agreement between the diversity patterns revealed by DNA polymorphisms (including these hypervariable loci) and those described by classical blood markers, such as proteins and HLA antigens (Bortolini *et al.* 1998; Chikhi *et al.* 1998).

Microsatellite (Short Tandem Repeat or STR) loci consist of tandemly repeated sequences, containing a core sequence of between 1 to 5 bp and with the total repeat unit usually less than 350 bp, making them very amenable to PCR (Edwards *et al.* 1991; Beckmann and Weber 1992). Minisatellite (VNTR) loci also consist of tandemly repeated sequences, but with a core unit between 6 and 30 bp, and are predominantly found in telomeric regions. In contrast, microsatellites are widely dispersed throughout the human genome, occurring on average every 6 to 10 Kb (Kimpton *et al.* 1993). While the majority of microsatellites do not display the same level of polymorphism as minisatellites, microsatellite loci can be readily co-amplified in the one reaction (multiplexing) and, thereby, compensate for the lower levels of variation by simultaneously allowing the analysis of more than one locus. The multiplex system used in this study, Profiler Plus^{TM}, incorporates nine highly polymorphic microsatellite loci and is used by many forensic laboratories around the world.

Despite possessing numerous advantages, microsatellites are also considered to have some limitations with respect to population and evolutionary studies. Because microsatellites display relatively high mutation rates (between 10^{-3} and 10^{-4}) (Jeffreys *et al.* 1988; Zhivotosky and Feldman 1995) and the majority are thought to evolve via the single-step mutation model (SMM), identical alleles can occur in different populations which have no shared ancestry. Consequently, the more recent the divergence between populations, the more likely it is that shared microsatellite alleles are related by descent and also population differences may be increasingly difficult to detect as the time depth of divergence between them increases. A recent study by Jorde *et al.* (1997) showed that microsatellite loci having smaller variances in allele size were more suitable for detecting differences between major ethnic populations, while loci displaying large size variances (i.e. many alleles) were not. A similar conclusion was reached by Brinkmann *et al.* (1998) who found that microsatellite loci displaying low microvariation can discriminate between major ethnic populations, but are less discriminating within these populations, while loci displaying a large number of alleles are very appropriate for discriminating within these populations. Despite these limitations, microsatellites are still very suitable for studies examining the separation of human lineages or groups over a recent time period (de Knijff *et al.* 1997; Jorde *et al.* 1997; Brinkmann *et al.* 1998).

The focus of this case study is an examination of human microevolution, using nine highly polymorphic microsatellites, among Native American tribes of Colombia who are considered to have inhabited the South American continent for less than 25,000 years. The tribal samples represent not only considerable linguistic diversity, but also very different and extreme ecological zones, including the Amazonian forest, high altitude, and arid desert.

It is most likely that the American continent was populated by migratory waves of populations of Asian descent moving across the Bering Strait, beginning approximately 30,000 years ago (Schanfield 1992; Szathmary 1993; Torroni *et al.* 1993a; Torroni *et al.* 1993b; Crawford 1998) and that Native American populations were well established in South America approximately 10,000 years ago (Salzano and Callegari-Jacques 1988). During the 16th century, present day Colombia was conquered and then colonized by the Spanish. Africans were introduced to Colombia as slaves during the Spanish colonial years, which ended in the early 1800s.

Present day Colombia is populated by approximately 36 million people and remains racially diverse, with more than half the people of mixed Spanish-Native American ancestry (mestizo) and approximately 20% of unmixed European ancestry. A further 14% are of mixed African-Native American descent (mulatto), while 4% of the population is of unmixed African ancestry. Non-admixed Native Americans are estimated at 1% of the total population. There are approximately 80 remaining Native American tribes speaking many different indigenous languages in present day Colombia. In comparison to other South American countries only Brazil possesses a greater number of Native American tribes (Yunis *et al.* 1994).

2. SUBJECTS AND METHODS

2.1. Subjects

In this study the validity of using nine highly polymorphic microsatellites to investigate human microevolution is evaluated, by performing a case study on five Native American tribes from Colombia. All five tribal samples (Fig. 1) comprised blood samples collected from unrelated healthy individuals (Briceno *et al.* 1996a). The Arsario (n = 21) are high altitude dwellers, inhabiting the north western section of the Sierra Nevada de Santa Marta, the highest mountain range in Colombia lying at 19,024 feet. They are members of the Chibcha linguistic stock and have a total population of approximately 2,700 (Yunis *et al.* 1994). The Kogui (n = 21) are also members of the Chibcha linguistic family and inhabit the western ranges of the Sierra Nevada de Santa Marta. They number approximately 2,400 (Yunis *et al.* 1994). The Ijka (n = 14) are also high altitude dwellers, inhabiting the eastern slopes of the Sierra Nevada de Santa Marta, belong to the Chibcha linguistic family, and number approximately 8,700 (Yunis *et al.* 1994). At approximately 1000 A.D. members of the Chibcha linguistic family were forced into higher elevations by the Caribs, who were warlike people and took possession of their land. By the end of the 15th century, the Chibchas were the most advanced indigenous culture, with an economy based on the cultivation of maize and potatoes (Hanratty and Meditz 1990). The Wayuu (n = 14) tribe is comparatively large with respect to the other tribes, numbering between 40,000 and 80,000. They are members of the Arawak linguistic family, most of whom are warlike hunters and fishers (Kline 1983). They are subject to extreme ecological pressures, as they inhabit a desert region in the Guajira peninsula in the north east of Colombia (Yunis *et al.* 1994). The Coreguaje (n = 25)

Figure 1. Location of the 5 Colombian tribes.

inhabit the Amazon forest in southern Colombia and are, therefore, both geographically distant and ecologically distinct from the other tribes studied. The Coreguaje are members of the Tukano linguistic stock and are dispersed into fourteen settlements along three different rivers that are tributary to the Amazon (Briceno *et al.* 1996b). The samples used in this study were collected from the Coreguaje living near the Orteguaza River. The total population of all fourteen Coreguaje settlements is approximately 1,400 (Briceno *et al.* 1996b). A sample of U.S. Native Americans (n = 28) were included in the analysis to investigate the relationship between indigenous groups of both the Americas for these nine microsatellites. They comprise both Amerind and Na-Dene speakers. Each individual self-reported as Native American, but microsatellite and Y chromosome markers have suggested that this sample is admixed (Mitchell *et al.* unpublished).

2.2. Previous Studies on the Five Tribes

Salzano and Callegari-Jacques (1988) investigated the Wayuu and Ijka in addition to other Colombian tribes using a large number of "classical markers" (blood groups, red cell enzymes and red cell proteins) in order to detect admixture and estimate the proportion of Caucasian and African genes respectively (Table 1). From Table 1, three facts are evident. Admixture is present in Colombian tribes, but can be very variable. Secondly, "classical markers" can be used to identify the source of the admixture, and thirdly, the level of admixture with both Caucasian and African populations is much greater in the Wayuu than the Ijka. The Paraujano tribe is located on the Guajira peninsula, in close

Table 1. Estimated admixture with Caucasian and African groups in Colombian Native Americans based on "classical" blood markers (modified from Salzano and Callegari-Jacques, 1998)

Native AmericanTribe	Admixture with Caucasians	Admixture with Africans
Wayuu	0.425	0.390
Ijka	0.058	0.070
Parajuano	0.685	0.189
Secoya	0.000	0.035

proximity to the Wayuu, and are also Arawak speakers. A very high level of Caucasian admixture was detected in the Paraujano, while the amount of African admixture is lower, although still relatively high. The Secoya tribe, like the Coreguaje, inhabit the Amazonian forest and are Tukano speakers. No Caucasian admixture was seen in the Secoya; however, a low level of African admixture was detected.

Yunis *et al.* (1994) typed HLA (DQA1, DQB1, DRB1, DRB3, DRB4 and DRB5) and blood group markers (ABO, MNSs, Rhesus, Kell, Duffy, Kidd, Diego) in the Arsario, Kogui, Ijka and Wayuu tribes. Again, high levels of both Caucasian and African admixture were observed in the Wayuu (Table 2), as evidenced particularly by the cDe (R_0) haplotype, and FY null and Ns alleles. The Ijka showed very low levels of admixture with Caucasians, but high levels of African admixture. The study found no admixture in either the Arsario or Kogui tribes, indicating that these tribes have retained a non-admixed Amerindian gene pool.

The findings of these earlier studies concur with the history of Colombia. This history includes Spanish colonization and the subsequent introduction of African slaves, and later with the effect of trading between the Wayuu and peoples of both African and Dutch ancestry inhabiting islands in the Caribbean (Yunis *et al.* 1994). Further evidence of admixture in the Wayuu comes from the study by Briceno *et al.* (1996a) of variation at the HLA-DPB1 locus in seven Colombian tribes, including all five tribes in this study. No evidence of admixture was found in the Arsario, Kogui, Ijka and Coreguaje tribes, but the exotic allele, *0101, was seen in the Wayuu.

A number of important observations can be made from these earlier studies. Firstly, there is consistent evidence that the Arsario and Kogui tribes are non-admixed, and that their Amerindian component has remained intact. The one study on the Coreguaje also indicates their gene pool as 100% Amerindian. All earlier studies detected relatively high levels of admixture in the Wayuu, and the gene flow from both African and Caucasian

Table 2. Estimated admixture in 4 Native American tribes based on HLA and "classical" blood markers (modified from Yunis *et al.* 1994)

Tribe	Caucasian	African	Amerindian
Arsario	0.0000	0.0000	1.0000
Kogui	0.0000	0.0000	1.0000
Ijka	0.0025	0.2166	0.7809
Wayuu	0.3633	0.2423	0.3934

groups was considerable, but usually the Caucasian component is slightly higher. The findings on the Ijka are more equivocal, with one study finding no evidence of gene flow, yet two studies finding both Caucasian and African components in the gene pool. Those studies reporting admixture in the Ijka differ in the respective quantity; with Salzano and Callegari-Jacques (1988) reporting almost equal, and relatively high, gene flow from Africans and Caucasians, yet Yunis *et al.* (1994) reported very low Caucasian admixture, but a very high African component.

2.3. Methods

DNA was isolated from peripheral blood leukocytes by standard procedure (Sambrook *et al.* 1989). The Profiler Plus™ Multiplex System developed by Applied Biosystems (ABI), and consisting of nine microsatellite loci and the sex identifying locus, amelogenin, was employed in the analysis of all samples (Table 3). DNA amplification followed the Profiler Plus™ protocol, (AmpF*l*STR Profiler Plus™ PCR Amplification Kit: User's Manual, 1997) under which fluorescently labeled primers allow the co-amplification and subsequent detection of the loci. PCR was conducted using a Perkin-Elmer 9600 Thermal Cycler, with the following cycling conditions: 95 °C for 11 mins; followed by 28 cycles of: 94 °C for 1 min, 59 °C for 1 min, 72 °C for 1 min; a final extension of 60 °C for 45 mins and hold at 25 °C. Amplified products were run in a 4.25% acrylamide gel, using 36 cm well-to-read plates on the ABI Prism 377 DNA Sequencer. Genescan™ software was used to analyse the results and data were then imported into Genotyper™, enabling allelic identification.

The A and G statistics program, kindly provided by G. Morgan (22 Eromba Crescent, Ferny Hills, Queensland, 4055) was employed to calculate allele and genotype frequencies based on the observed phenotypes, Hardy-Weinberg Equilibrium (HWE), and heterozygosity. A relationship (R) matrix analysis (Harpending and Jenkins 1973) was used to estimate kinship and genetic distances between tribes as it was specifically designed for use in investigating structure in small populations descended from a common ancestral group. A distance (D) matrix can be readily calculated from the R matrix. Relethford's (1996) modified R matrix methodology, that incorporates the role of drift due to differential population size, was also used. Statistical significance of population clustering and genetic distances was assessed by construction of a UPGMA tree using the Phylogeny Inference Package, PHYLIP (Felsenstein 1993).

Table 3. Details of the autosomal loci scored in the Profiler Plus™ multiplex system (adapted from AmpFlSTR Profiler Plus™ PCR Amplification Kit: User's Manual, 1997)

Locus	Chromosome Location	Number Of Alleles	Approximate Size Range Of Alleles	Dye Label
D3S1358	3p	12	114–142	5-FAM
vWA	12p12-pter	13	157–197	5-FAM
FGA	4q28	32	210–267	5-FAM
D8S1179	8	12	128–168	JOE
D21S11	21	35	189–243	JOE
D18S51	18q21.3	25	273–341	JOE
D5S818	5q21–31	10	135–171	NED
D13S317	13q22–31	9	206–234	NED
D7S820	7q	11	258–294	NED

3. RESULTS

3.1. Allele Frequencies at Autosomal Loci

The D3S1358 locus has a unimodal allele frequency distribution in all six groups with the exception of allele 17 in the Ijka, in which it has a frequency above 50% (Table 4). In all other tribes allele 15 is the mode, and the frequency ranges from 36% in the Kogui, to over 60% in the Coreguaje. The Ijka and Coreguaje display only three alleles (15, 16 and 17), while the Wayuu and U.S. Native Americans are the only populations with all five alleles.

The vWA locus shows a unimodal allele distribution in all groups, with allele 16 the most common in the Arsario, Kogui, Coreguaje and Wayuu tribes (Table 4). However, allele 17 is the most frequent in both the Ijka (46%) and U.S. Native Americans (27%). Allele 15 is absent in the Kogui and Coreguaje, but present at levels under 10% in all other tribes. Allele 19 is absent in the Arsario and Kogui, yet present in the Ijka, Coreguaje, Wayuu and U.S. Native Americans, respectively. Allele 14 is only seen in the Wayuu and U.S. Native Americans.

The FGA locus has a bimodal distribution, and the allele frequencies are very variable across the five Colombian tribes (Table 4). Allele 24 is most common in the Arsario (45%), Kogui (60%) and Wayuu (32%) tribes, while in the Coreguaje and Ijka, the most common alleles are 26 (40%) and 25 (43%) respectively. The U.S. Native American population displays frequencies under 20% for all alleles, including alleles 18 and 20, which are absent in all Colombian tribes. The Wayuu possess eight alleles at the FGA locus, and the Ijka and the Coreguaje show seven alleles. In contrast the Arsario and Kogui only have the same four alleles.

The D8S1179 locus exhibits a bimodal distribution, with allele 13 the most common in the Ijka, Coreguaje, and U.S. Native Americans (Table 4). In the Arsario, Kogui and Wayuu tribes allele 14 is most common. Allele 8 is only found in U.S. Native Americans and at low frequency (5%). Allele 10 occurs at a frequency of ~ 20% in the Coreguaje and Wayuu tribes, is less common in the Arsario (7%), and absent in the Kogui and Ijka. Allele 11 is also found at a similar frequency amongst all tribes; however, it is particularly high in the Arsario (21%) and Ijka (29%). Allele 16 is absent in the Arsario, Ijka and Wayuu, but present in 6% of the Coreguaje and at lower levels in the Kogui and U.S. Native American groups.

The D21S11 locus has a multimodal distribution and the allele frequencies are variable across all tribes (Table 4). Allele 30 is the most common in all tribes except the Coreguaje, where allele 29 is most frequent. Allele 30.2 is only observed in U.S. Native Americans and at a low frequency (4%) and, similarly, allele 34 is observed only in the Arsario. Allele 33 is only in the Arsario and Wayuu tribes and at low frequencies. Allele 33.2 is absent in all tribes except the Coreguaje (8%) and U.S. Native American (4%). Allele 29 is absent in the Ijka, but present in all other tribes, and at a particularly high frequency in the Coreguaje (38%). Allele 28 is absent in Arsario, Kogui and Ijka, yet present in approximately 10% of both Wayuu and U.S. Native Americans, and in 2% of the Coreguaje. The Ijka exhibit a frequency of allele 31 almost three times that seen in the other tribes, and in the Coreguaje, this allele is absent. Conversely, allele 31.2 is observed at high frequencies in the Coreguaje (32%), at much lower levels in the Arsario (12%) and U.S. Native Americans (5%), while it was absent in the other tribes. Allele 32 in the Arsario and Kogui displays frequencies marginally higher than 20%. The Ijka and Wayuu tribes display lower frequencies, while both the Coreguaje and U.S. Native

Table 4. Allele frequencies at the 9 loci in 5 Colombian tribes and in U.S. Native Americans

Locus/ Allele	Arsario n = 21	Kogui n = 21	Ijka n = 14	Wayuu n = 14	Coreguaje n = 25	U.S. Nat. Am. n = 28
D3S1358						
14	0.1190	0.2381	0	0.0357	0	0.0893
15	0.5238	0.3571	0.3571	0.6072	0.6200	0.4821
16	0.2381	0.1429	0.0714	0.2500	0.1600	0.1964
17	0.1190	0.2619	0.5714	0.0357	0.2200	0.1250
18	0	0	0	0.0714	0	0.1071
vWA						
14	0	0	0	0.0357	0	0.0357
15	0.0714	0	0.0357	0.0714	0	0.1071
16	0.4762	0.6667	0.2857	0.4286	0.6071	0.2321
17	0.3810	0.2143	0.4643	0.2857	0.3036	0.2679
18	0.0714	0.119	0.1786	0.0357	0.0714	0.2500
19	0	0	0.0357	0.1429	0.0179	0.1071
FGA						
18	0	0	0	0	0	0.0536
19	0	0	0	0.0357	0.0385	0.1964
20	0	0	0	0	0	0.0893
21	0.1905	0.1429	0.0357	0.1786	0.0962	0.0883
22	0	0	0.0357	0.0714	0.0577	0.1250
23	0	0	0.0357	0.0714	0	0.1615
24	0.4524	0.5952	0.3214	0.3214	0.1923	0.0536
25	0.3333	0.1667	0.4286	0.1786	0.0962	0.1429
26	0	0	0.1071	0.0714	0.4038	0.0714
27	0.0238	0.0952	0.0357	0.0714	0.1153	0.0179
D8S1179						
8	0	0	0	0	0	0.0536
10	0.0714	0	0	0.1786	0.2000	0.0714
11	0.2143	0.0952	0.2857	0.1429	0.1000	0.1071
12	0.0714	0.0714	0.1071	0.1429	0.0800	0.1429
13	0.2381	0.1905	0.4643	0.2143	0.3200	0.3214
14	0.3333	0.5000	0.0714	0.2857	0.1200	0.1429
15	0.0714	0.1190	0.0714	0.0357	0.1200	0.1429
16	0	0.0238	0	0	0.0600	0.0179
D21S11						
28	0	0	0	0.1071	0.0200	0.0893
29	0.1190	0.1429	0	0.1429	0.3800	0.1964
30	0.2381	0.3571	0.3571	0.4286	0.1200	0.3214
30.2	0	0	0	0	0	0.0357
31	0.0952	0.0952	0.2857	0.0357	0	0.0714
31.2	0.1190	0	0	0	0.3200	0.0536
32	0.2143	0.2381	0.0357	0.0714	0	0
32.2	0.1429	0.1667	0.3214	0.1786	0.0800	0.1964
33	0.0238	0	0	0.0357	0	0
33.2	0	0	0	0	0.0800	0.0357
34	0.0477	0	0	0	0	0
D18S51						
10	0	0	0	0	0	0.0185
11	0	0	0	0	0	0.0185
12	0.0238	0.0476	0.0357	0.0714	0.0600	0.0926
13	0.3810	0.3571	0.1071	0.2143	0.0400	0.0741
14	0.0952	0.0238	0.2500	0.1429	0.1600	0.2037
15	0.1190	0.1429	0.2143	0.1786	0.0600	0.1296
16	0.1429	0.0952	0.0714	0.0357	0.1000	0.1296
17	0.1667	0.2381	0.3214	0.1429	0.5400	0.2222

Table 4. *Continued*

Locus/ Allele	Arsario n = 21	Kogui n = 21	Ijka n = 14	Wayuu n = 14	Coreguaje n = 25	U.S. Nat. Am. n = 28
18	0.0238	0.0952	0	0	0.0400	0.0926
19	0.0476	0	0	0.1071	0	0
20	0	0	0	0.1071	0	0.0185
D5S818						
7	0.4286	0.4286	0.3571	0.1786	0.0200	0.1429
8	0	0	0	0	0	0.0357
9	0	0	0	0.0714	0.5000	0.0893
10	0.1429	0	0.1071	0.0357	0.0200	0.0179
11	0.2857	0.4524	0.2857	0.5357	0.2600	0.4107
12	0.1429	0.1190	0.0357	0.1071	0.1000	0.1786
13	0	0	0.2143	0.0714	0.0600	0.1250
15	0	0	0	0	0.0400	0
D13S317						
8	0	0	0.0357	0	0.0400	0.0714
9	0.6190	0.6667	0.4643	0.3571	0.3000	0.0893
10	0.0714	0.0476	0.0357	0.1429	0.2000	0.1071
11	0.1190	0.2619	0.1071	0.0714	0.0200	0.2143
12	0.0952	0	0.1071	0.3214	0.1600	0.3036
13	0.0238	0	0.2143	0.0714	0.1200	0.1429
14	0.0714	0.0238	0.0357	0.0357	0.1600	0.0714
D7S820						
7	0	0	0	0.0357	0	0
8	0	0	0	0.0714	0.0200	0.1296
9	0	0.0476	0	0.0357	0	0.0741
10	0.0476	0.0476	0.0357	0.5357	0.2200	0.2963
11	0.7857	0.7143	0.7143	0.1429	0.3800	0.2407
12	0.1667	0.1905	0.2142	0.0714	0.3400	0.2037
13	0	0	0.0357	0.1071	0.0400	0.0555

Americans do not possess allele 32. Allele 32.2 is seen in all tribes at high frequencies, but highest in the Ijka (32%).

The D18S51 locus exhibits a multimodal distribution (Table 4). In the Arsario, Kogui and Wayuu tribes, allele 13 is most frequent, while in the Ijka, Coreguaje and U.S. Native Americans, allele 17 is most common. Alleles 10 and 11 occur in only 2% of U.S. Native Americans and both alleles are absent in all Colombian tribes. Allele 19 is absent in all the tribes except Wayuu and Arsario. Allele 20 is also rare, occurring only in the Wayuu (11%) and U.S. Native Americans (2%). Allele 13 in the Arsario and Kogui has a frequency above 35% and in the Wayuu it has a frequency of 21% and these values contrast with those of less than 12% for the Ijka, Coreguaje and U.S. Native American populations. Allele 14 is common to all tribes, except the Kogui, and allele 15 is common to all tribes except the Coreguaje. Allele 17 has frequencies ranging from 14% to 32% in five groups, whereas the Coreguaje possess this allele at a frequency of 54%. Allele 18 was absent in the Ijka and Wayuu, but present below 10% in the remaining tribes.

The D5S818 locus has a multimodal distribution (Table 4). The most common allele in the Arsario and Ijka is 7, where it has a frequency over 35%, but in the Kogui, Wayuu and U. S. Native Americans the most common allele is 11. The Coreguaje are distinguished by their very high frequency of allele 9 (50%) which is absent in the Arsario, Kogui and Ijka tribes, but present at low frequencies in the Wayuu (7%) and U.S. Native

Americans (9%). While other tribes display high frequencies of allele 7 (>15%), only 2% of the Coreguaje possessed this allele. Allele 15 is absent in all tribes except the Coreguaje (4%).

The D13S317 locus shows a multimodal distribution, with allele 9 the most frequent in all five Colombian tribes, ranging from 30% in the Coreguaje to over 60% in the Arsario and Kogui (Table 4). Amongst Colombians, the Wayuu show a high frequency of allele 12 (32%), double that of any other tribe. Alleles 12 and 13 are not seen in the Kogui. Allele 8 is present in the Ijka, Coreguaje and U.S. Native Americans at low frequencies, but absent in the Arsario, Kogui and Wayuu.

The D7S820 locus exhibits a unimodal distribution with the modal allele for the Arsario, Kogui, Ijka and Coreguaje tribes being 11, and its frequency extremely high (Table 4). The most common allele in the Wayuu and U.S. Native Americans, however, is 10. Allele 7 is only seen in the Wayuu. The Wayuu and U.S. Native Americans display low frequencies of alleles 8, 9 and 13, while in the remaining tribes these alleles have even lower frequencies, or are absent. The number of alleles at this locus varies considerably by tribe; the Arsario display only three, Kogui and Ijka four, Coreguaje five, and the Wayuu seven alleles.

3.1.1. R Matrix Analysis. The diagonal values in the unscaled R matrix of five Colombian tribes (Table 5) indicate that the Coreguaje exhibit the greatest gene frequency deviation from the total mean. The one positive kinship coefficient, that between Arsario and Kogui, indicates that these two tribes are more similar to each other, than the overall mean pairwise similarity. The unscaled D matrix (Table 5) shows that the Arsario and Kogui are the closest tribes, while the Coreguaje display the greatest distance from all other Colombian tribes. A plot of the first two components of this variation is represented in Fig. 2. The first component distinguishes the Coreguaje from all other tribes and the second component separates the Wayuu from all the other tribes.

A scaled R matrix, incorporating the effects of different population sizes of these five tribes, indicates that both the Ijka and Wayuu display greatest deviation from the total mean gene frequency (Table 6). The Arsario and Kogui are the most similar of pairs, and the scaled D matrix shows the lowest distance is still between these two tribes. The greatest distance is now that between the Ijka and Wayuu, while the distances between the Coreguaje and all other tribes is considerably reduced. In the plot of the variation (Fig. 3), the first component separates the Wayuu and Ijka to a great extent. The Coreguaje are separated to a lesser extent, while Arsario and Kogui plot close together.

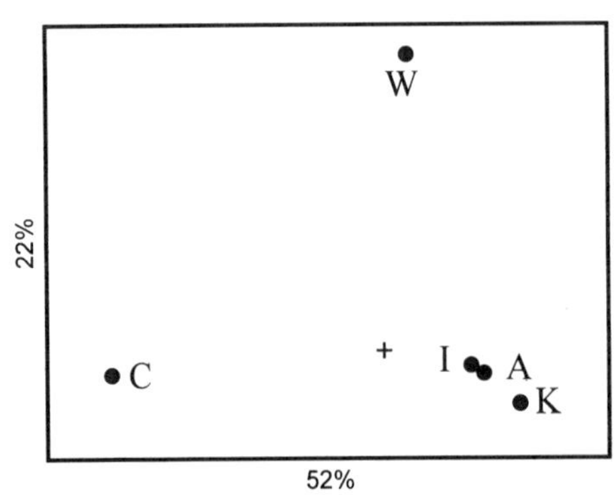

Figure 2. Plot of variation among 5 Colombian tribes (unscaled R matrix) [A = Arsario, K = Kogui, I = Ijka, W = Wayuu, C = Coreguaje].

Table 5. Unscaled R matrix and D matrix of 9 loci over 5 Colombian tribes (values 10^{-4})

Population	Arsario	Kogui	Ijka	Wayuu	Coreguaje
Arsario	567	399	−38	−204	−84
Kogui	652	883	−77	−222	−220
Ijka	1,171	1,564	528	−399	−228
Wayuu	1,438	1,789	1,790	463	−49
Coreguaje	3,041	3,627	3,289	2,865	2,305

Note: R matrix above diagonal, D matrix below diagonal.

Table 6. Scaled R matrix and D matrix of 9 loci over 5 Colombian tribes (values 10^{-4})

Population	Arsario	Kogui	Ijka	Wayuu	Coreguaje
Arsario	304	201	−36	−210	−33
Kogui	321	420	−69	−216	−80
Ijka	1,289	1,471	912	−739	−158
Wayuu	1,642	1,770	3,308	918	−36
Coreguaje	1,009	1,220	1,868	1,631	640

Note: R matrix above diagonal, D matrix below diagonal.

The second component distinguishes the Coreguaje, and to a lesser extent, the Wayuu and Ijka. When the same analysis is performed on the Colombians and the U.S. Native Americans together, the Wayuu plot closest to the U.S. Native Americans, while the Ijka, Arsario and Kogui cluster together and the Coreguaje still remain distinct from all other groups (data not shown).

In an attempt to determine the source of gene flow into these tribes we compared the present data with similar data from a well defined Australian Caucasian sample (Guarino *et al.* unpublished) and African Americans (Perkin Elmer Corporation, 1997) which were typed by the same methodology. The frequency of alleles at all nine loci in Australians were very close to those seen in a U.S. Caucasian sample of the Perkin-Elmer Corporation. The African-American sample is the only data set available for these nine microsatellites that contains African genes, but, of course, a "true" African frequency may vary quite considerably from the values seen in this sample (Perkin-Elmer does not

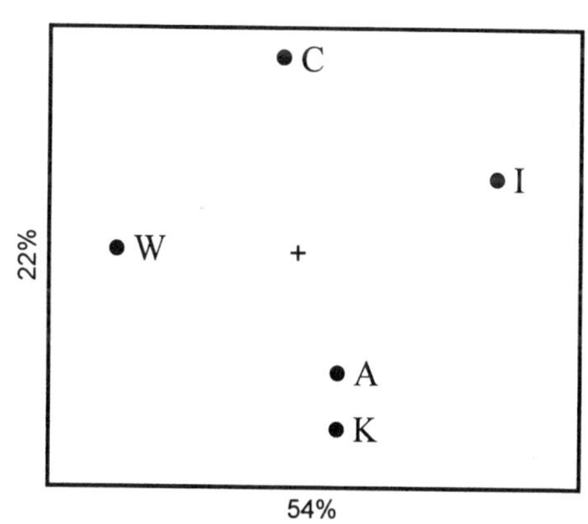

Figure 3. Plot of variation among 5 Colombian tribes (scaled R matrix).

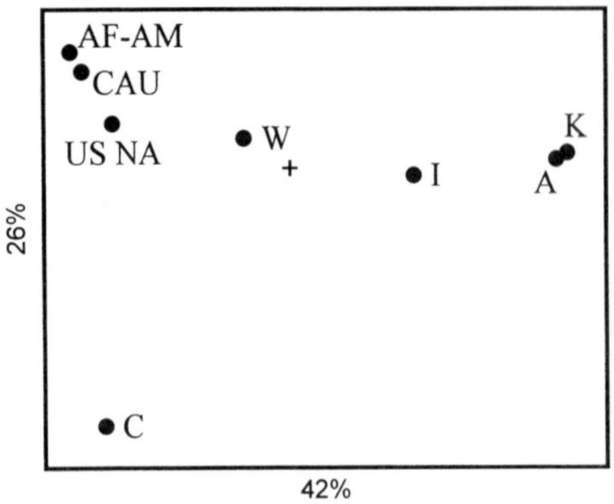

Figure 4. Plot of variation for 8 populations, including the Colombian tribes [AF-AM = African-American, CAU = Caucasian, US NA = U.S. Native American].

indicate the level of admixture in their sample). A plot of the variation (Fig. 4) shows that the two major ethnic groups are most different to the Arsario, Kogui, Coreguaje and Ijka tribes, while the Wayuu, and to a greater extent the U.S. Native Americans, plot closest. Genetic distances are largest between the Colombian tribes and either Caucasians or African-Americans, but the smallest distances are observed when comparisons of either are made to the Wayuu. The greatest distances are those between either Caucasians or African-Americans and the Coreguaje (Table 7).

3.1.2. Distance Versus Heterozygosity. The further a population is from the centroid (i.e. the greater the deviation from the mean of all the sub-populations), the more isolated it is, the more opportunities for drift and "inbreeding" or other causes of gene fixation, and, therefore, lower heterozygosity. When distance versus heterozygosity is plotted for the five Colombian populations following Harpending and Ward (1982) no regression occurs (Fig. 5). When the Coreguaje, by far the most distant population, is removed a slope exists (Fig. 6). Populations above the regression line are considered to have heterozygosity in excess of expectation.

3.1.3. Phylogenetic Analysis. The phylogenetic tree comprising the six Native American groups and the two ethnic groups is shown in Fig. 7. The tree divides into Native Americans and the two ethnic populations, with U.S. Native Americans and the Wayuu forming the branches that connect the two major groups and the other Native Americans. The Arsario and Kogui cluster is highly significant. The Ijka, Coreguaje and Wayuu tribes individually branch off in a sequential manner from the Arsario and Kogui.

Table 7. R matrix and D matrix of 9 loci over 8 populations (values 10^{-4})

Population	Arsario	Kogui	Ijka	Wayuu	Coreguaje	US NA	Caucasian	AF-AM
Arsario	1,516	1,291	748	353	196	−113	−279	−244
Kogui	559	1,625	697	344	177	−125	−272	−226
Ijka	1,193	1,404	1,173	101	240	6	−240	−253
Wayuu	1,401	1,529	1,563	591	196	4	−159	−93
Coreguaje	3,001	3,148	2,571	2,076	1,877	137	−150	−147
US NA	1,966	2,100	1,386	806	1,827	224	−32	−30
Caucasian	2,184	2,281	1,794	1,019	2,287	399	110	24
AF-AM	2,366	2,438	2,040	1,137	2,532	646	422	361

Note: R matrix above diagonal, D matrix below diagonal.

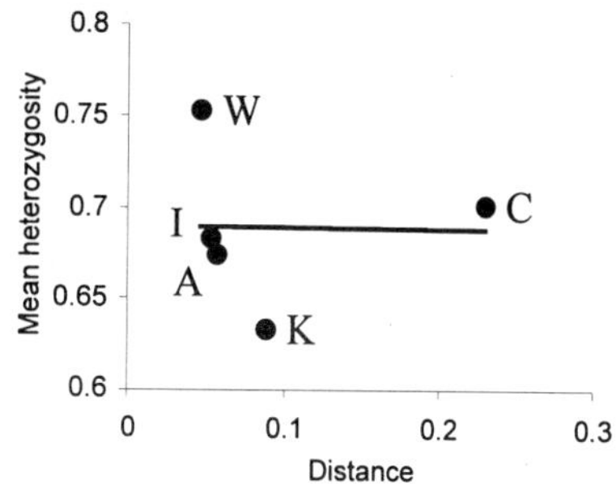

Figure 5. Heterozygosity versus distance from the centroid for 5 Colombian tribes.

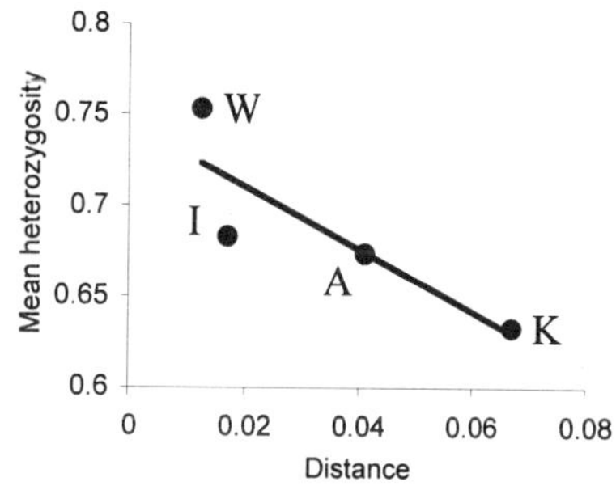

Figure 6. Heterozygosity versus distance from the centroid for 4 Colombian tribes.

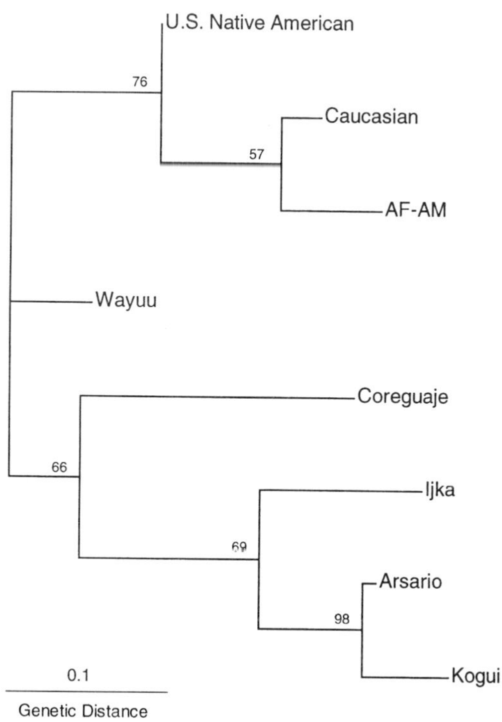

Figure 7. UPGMA tree for 8 populations based on allele frequencies of 9 microsatellite loci. Bootstrap probabilities of 100 resamplings are displayed on internal branches.

The bootstrap probabilities indicate that the clustering observed amongst the five Colombian tribes is more likely to be correct than incorrect. The U.S. Native Americans display a very low genetic distance, while the Coreguaje display the largest genetic distance of all.

3.1.4. Y-Chromosome Microsatellite Variation in Colombian Native Americans. We also scored five Y-chromosome specific microsatellites (DYS19, DYS390, DYS391, DYS392 and DYS393) on a small number of males from four of the Colombian tribes (Arsario n = 11, Kogui n = 9, Ijka n = 10, and Wayuu n = 11) and our preliminary findings are presented. In a total of 41 males we found only two Y-haplotypes that were not Native American, one in the Ijka (Caucasian) and one in the Wayuu (African). DYS19 allele 13 is the most characteristic Y-chromosome microsatellite allele in Native Americans (Santos *et al.* 1996) and this is also the case with Colombians. The largest number of haplotypes are seen in the Wayuu (8) and the smallest number in the Ijka (3). Sharing of Y-haplotypes is observed among Arsario, Kogui and Ijka but not with the Wayuu. The Arsario and Kogui share three haplotypes while the Ijka share two haplotypes with the Arsario and Kogui. Among the 39 males with Native American Y-chromosomes a total of 14 unique five-locus haplotypes were identified. These haplotypes were joined together using the median network method of Bandelt *et al.* (1995) and the result is shown in Fig. 8. All the haplotypes form a single cluster, with only one inferred haplotype, (joining haplotypes 2, 11, 12 and 13) required. The locus at which a mutational change occurred is given on the line joining the two haplotypes.

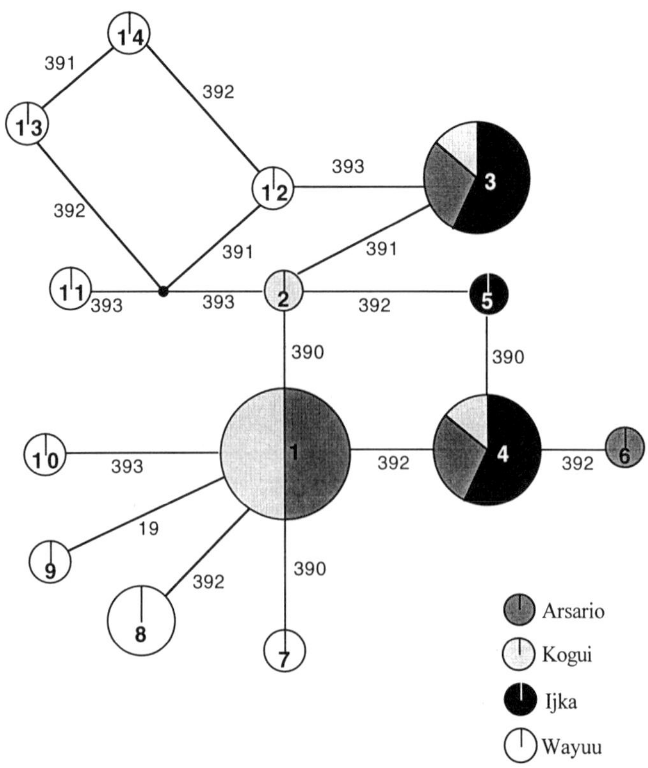

Figure 8. Network of Y-chromosome haplotypes in 4 Colombian tribes.

4. DISCUSSION

The nine autosomal microsatellites typed in this study generate an overall picture of the relationships among the five Colombian tribes that closely relates to their geographical location, ecology, and history. The Arsario and Kogui tribes are genetically very similar to each other and show no evidence of admixture. This is supported by heterozygosity, kinship and distance analyses, and also by their clustering in the phylogenetic tree (Fig. 7). Both the Arsario and Kogui belong to the Chibcha linguistic group and geographically they are neighbours, with both inhabiting the Sierra Nevada de Santa Marta. Heterozygosity levels in the Arsario and Kogui are the lowest of all Colombian tribes, and are considerably lower in comparison with mainstream Caucasian and Asian populations scored for the same microsatellites (Guarino *et al.* unpublished). These findings suggest that the two tribes are, and have been, isolated. No evidence of admixture was found, suggesting that the Arsario and Kogui are representative of non-admixed Native Colombians and possibly all Amerindians for these loci. These findings on the Arsario and Kogui are in agreement with those based on "classical" blood groups and HLA (Yunis *et al.* 1994) and DNA sequence polymorphisms (Briceno *et al.* 1996a).

The Ijka tribe also belongs to the Chibcha linguistic group, and inhabits the same mountainous elevations as the Arsario and Kogui, but lie further to the east, nearer the Guajira desert. The Ijka display many genetic similarities to the Arsario and Kogui, but are distinguished from these tribes chiefly by the presence of extra alleles at some loci. This is best seen at the FGA locus, where the Ijka possess seven in comparison to four alleles. The additional alleles in the Ijka could be due to either admixture or larger population size, but the suggestion of this, and other studies, is that the former is most likely. For example, admixture is supported in Fig. 3, where the Ijka separate from all other tribes in both components, and also by the data of Yunis *et al.* (1994), who found high frequencies of the "African" cDe (R_o) haplotype.

The Coreguaje are genetically distinct from the other Colombian tribes. This was seen in every analysis, with genetic drift the most likely explanation. The greatest distances are almost always observed between the Coreguaje and all other groups; this is also shown by the Coreguaje possessing the longest branch length (i.e. genetic distance) in the phylogenetic tree. There are many forces leading to their distinctiveness. They dwell in the Amazon forest in southern Colombia and the total population is very low, approximately 1,200 people (Briceno *et al.* 1996b), and they live in isolated settlements with minimal or no gene flow (Briceno *et al.* 1996b). This isolation and drift is evidenced from the allele frequencies: the Coreguaje possess the common Native American alleles, however, the most frequent allele in the Coreguaje is often different from that in other tribes at a number of loci (e.g. D21S11, D5S818) or alleles seen in the other tribes were absent in the Coreguaje (e.g. D21S11 *31 and *32). Also, for six of the nine loci, heterozygosity levels were low compared to the other tribes. Comparison of the position of the Coreguaje in the two plots (Fig. 2 and 3) confirms the significance of drift in explaining their distinctiveness. Our findings on the Coreguaje confirm those of Briceno *et al.* (1996b) who detected no admixture using HLA loci, and also concur with those of Salzano and Callegari-Jacques (1988), who studied the Secoya, a tribe linguistically and geographically close to the Coreguaje, and found admixture to be extremely minimal.

The Wayuu tribe appears genetically distinct from all other Colombian tribes and this is the result of a number of factors. They inhabit a harsh desert region on the Guajira peninsula in northern Colombia, which makes them subject to extreme ecological pressures. The Wayuu belong to the most numerous linguistic group in Colombia (Arawak) and their population size is also large (between 40,000 and 80,000 people) compared to other Colombian tribes (Yunis *et al.* 1994). The microsatellite data indicate that admixture is the most likely cause of the differences of the Wayuu from the other tribes. The Wayuu display the highest heterozygosity levels, suggesting that isolation is not affecting this population to the extent observed in the other tribes, and in Fig. 4 they plot closest to the U.S. Native American group, which is known to be admixed. Across most loci, alleles consistently observed in the Wayuu (e.g. D3S1358 *18; D18S51 *20) are absent in the other Colombian tribes, but are seen in both Caucasians and African-Americans. The phylogenetic tree also supports gene flow in the Wayuu, as the branch connecting Caucasians and African-Americans joins to the U.S. Native American branch, which subsequently joins the Wayuu branch. Our findings concur with those of Yunis *et al.* (1994) who observed high degrees of both Caucasian and African admixture in the Wayuu, indicated by the frequencies of the cDe (R_o) haplotype, Fy null, and A and B alleles of the ABO system. Salzano and Callegari-Jacques (1988) also reported high levels of African and Caucasian admixture in the Wayuu based on blood groups, red cell enzymes and serum protein markers.

This evidence for admixture agrees with the known history of the Wayuu. Africans were introduced to Colombia as slaves during the Spanish colonial years. After Spanish colonization ended (in the early 1800s), African populations remained in the Guajira peninsula, where the Wayuu tribe is located (Yunis *et al.* 1994). Furthermore, extensive trading between the Wayuu and peoples of both African and Dutch descent inhabiting nearby Caribbean islands also occurred and continued after colonization (Yunis *et al.* 1994).

The distance versus heterozygosity analyses (Figs. 5 and 6), show that the Wayuu tribe in particular, possess heterozygosity in excess of expectation. There are many possible sources of this excess heterozygosity, however given the history of the Colombian tribes, admixture with non-Indian populations needs to be considered. The microsatellite loci do not provide an independent estimate of admixture. However, two other studies have estimated admixture in these populations (Salzano and Callegari-Jacques 1988 and Yunis *et al.* 1994). Regression of heterozygosity on admixture indicates that admixture accounts for 80–82% of the variation in heterozygosity.

As disparate as the Coreguaje are from the other Colombian Indians the U.S. Native Americans are even more disparate. What are the possible explanations for these differences? Admixture certainly accounts for some of the differences seen between the U.S. Native Americans and the Colombians; however two of the Colombian populations (Ijka and Wayuu) are also admixed. One possible explanation is that these three groups, Coreguaje, other Colombian Indians, and U.S. Indians represent different migrations. The study of Schanfield (1992), using immunoglobulin allotypes as markers to investigate migrations of the ancestors of Native Americans into the New World, would support the position that the Coreguaje are more characteristic of the Paleo Indian migration, that the Arawak-Chibchan speaking tribes represent the next migration and that the U.S. Indian population represent an even later migration.

Although these populations all share attributes of American Indians, it is possible that they represent migrations of different time depth. This possibility adds a further complexity to understanding the pattern of diversity seen in these tribes, on top of the sampling difficulties and the effects of non-Indian admixture.

Overall, the data suggest that two of the five tribes (Wayuu and Ijka) have experienced admixture with outside groups, in particular Caucasians and Africans. Our analysis indicates that the Wayuu have experienced more gene flow than the Ijka, and this also agrees with earlier studies on these groups (Salzano and Callegari-Jacques 1988; Yunis *et al.* 1994). Unfortunately, attempts to identify the source of admixture using the microsatellites of the present study proved unsuccessful, as the "admixed" alleles in the Wayuu and Ijka are found at a relatively high and often similar frequency in both African-American and Caucasian populations. This is not the case when using a number of "classical markers" (e.g. ABO, HLA antigens) as shown by the work of Salzano and Callegari-Jacques (1988) and Yunis *et al.* (1994).

The preliminary findings on Y-chromosome variation suggest that all the Colombian males sampled are members of the same patrilineage. Also, the data confirm the genetic similarity of the three Chibcha speaking tribes, (Arsario, Kogui, and Ijka) and, in particular, the closeness between the Arsario and Kogui. Further, the absence of any "exotic" Y-haplotypes in the Arsario and Kogui confirm their 100% Amerindian gene pool. Even with a small sample, evidence of gene flow was detected in the Ijka and Wayuu.

Another interesting feature of this study is the overall similarity of the Caucasians and African-Americans at these nine loci, evidenced by their clustering (Figs. 4 and 7). It is clear that these nine microsatellites do not readily distinguish the two ethnic groups and the most likely explanation is that the shared alleles are not identical by descent, but rather a product of the high mutation rate. African and Caucasian populations are thought to have diverged more than 60,000 years ago. It has been argued that in the case of Y chromosome microsatellites, genetic "signatures" can be detected up to 50,000 years ago (de Knijff *et al.* 1997), but Ciminelli *et al.* (1995) postulated that tetranucleotide repeat alleles on the Y were particularly subject to randomisation in their study of European and Middle Eastern populations. Jorde *et al.* (1997) demonstrated that a large number of tetranucleotide loci (minimum of 36) was needed to detect a significant increased level of diversity in Africans compared to Asians and Caucasians. These authors also observed that microsatellite loci with low allele variances were more suitable for detecting these differences than loci with high variances (i.e. large allele numbers). The latter are clearly highly suitable for detecting relationships amongst recently diverged groups, such as the Native Americans of the present study. This latter characteristic has also been confirmed by Brinkmann *et al.* (1998).

For these reasons it proved impossible to identify the source of admixture in the Native Colombians of the present study. Therefore, if admixture studies are to be performed on indigenous groups (and also attempt to identify the origin of exotic alleles) then particular attention must be made to the selection of appropriate microsatellites (those with alleles distinct to each exotic group) or the investigation should score "classical markers" or DNA sequence polymorphisms in addition.

5. ACKNOWLEDGMENTS

The authors wish to thank Pontificia Universidad Javeriana and Colciencias for their financial support for part of this work. We also wish to thank Jeanette Auer and Clinton Hale for help with statistical analyses.

6. REFERENCES

AmpFlSTR Profiler Plus™ PCR Amplification Kit: User's Manual (1997) The Perkin-Elmer Corporation. Printed in the U.S.A

Balazs I, Baird M, Clyne M, *et al.* (1992) Human population genetic studies of five hypervariable DNA loci. American Journal of Human Genetics 44:182–190.

Bandelt H-J, Forster P, Sykes BC, *et al.* (1995) Mitochondrial portraits of human populations using median networks. Genetics 141:743–753.

Beckmann JS, Weber JL (1992) Survey of human and rat microsatellites. Genomics 12:627–631.

Bortolini MC, Da Silva-Junior WA, Weimer TA, *et al.* (1998) Protein and hypervariable tandem repeat diversity in eight African-derived South American populations: inferred relationships do not coincide. Human Biology 70:443–461.

Bowcock AM, Ruiz-Linares A, Tomfohrde A, *et al.* (1994) High resolution of human evolutionary trees with polymorphic microsatellites. Nature 368:455–457.

Briceno I, Gomez A, Bernal JE, *et al.* (1996a) HLA-DPB1 polymorphism in seven South American Indian tribes in Colombia. European Journal of Immunogenetics 23:235–240.

Briceno I, Bernal JE, Duran C, *et al.* (1996b) HLA antigens in Amerindian groups of two different linguistic families from Colombia. European Journal of Immunogenetics 23:21–28.

Brinkmann B, Junge A, Meyer E, *et al.* (1998) Population genetic diversity in relation to microsatellite heterogeneity. Human Mutation 11:135–144.

Chikhi L, Destro-Bisol G, Pascali V, *et al.* (1998) Clinal variation in the nuclear DNA of Europeans. Human Biology 70:643–657.

Ciminelli BM, Pompei F, Malaspina P, *et al.* (1995) Recurrent simple tandem repeat mutations during human Y-chromosome radiation in Caucasian subpopulations. Journal of Molecular Evolution 41:966–973.

Crawford MH (1998) The origins of Native Americans: evidence from anthropological genetics. Cambridge University Press: Cambridge, UK.

Deka R, Shriver MD, Yu LM, *et al.* (1995) Intra- and inter population diversity at short tandem repeat loci in diverse populations of the world. Electrophoresis 16:1659–1664.

de Knijff P, Kayser M, Caglia A, *et al.* (1997) Chromosome Y microsatellites: population genetic and evolutionary aspects. International Journal of Legal Medicine 110:134–140.

Edwards A, Civitello A, Hammond HA, *et al.* (1991) DNA typing and genetic mapping with trimeric and tetrameric tandem repeats. American Journal of Human Genetics 49:746–756.

Edwards A, Hammond HA, Jin L, *et al.* (1992) Genetic variation at five trimeric and tetrameric tandem repeat loci in four human population groups. Genomics 12:241–253.

Felsenstein J (1993) PHYLIP—Phylogeny Inference Package (Version 3.5c). Distributed by the author. Department of Genetics, University of Washington, Seattle.

Hanratty DM, Meditz SW (1990) Colombia: a country study. Library of congress cataloging-in-publication data.

Harpending HC, Jenkins T (1973) Genetic distance among Southern African populations. In: Crawford MH, Workman PL (eds) Methods and theories of anthropological genetics. University of New Mexico Press: Abuquerue pp 177–199.

Harpending HC, Ward R (1982) Chemical systematics and human populations. In: Nitecki M (ed) Biochemical aspects of evolutionary biology. University of Chicago Press: Chicago pp 213–256.

Jeffreys AJ, Royle NJ, Wilson V, *et al.* (1988) Spontaneous mutation rates to new length alleles at tandem-repetitive hypervariable loci in human DNA. Nature 332:278–281.

Jorde LB, Rogers AR, Bamshad M, *et al.* (1997) Microsatellite diversity and the demographic history of modern humans. Proceedings of the National Academy of Science USA 94:3100–3103.

Kimpton CP, Gill P, Walton A, *et al.* (1993) Automated DNA profiling employing multiplex amplification of short tandem repeat Loci. PCR Methods and Applications 3:13–22.

Kline HF (1983) Colombia: portrait of unity and diversity. Westview press, Boulder, Colorado.

Relethford JH (1996) Gentic drift can obscure population history: problem and solution. Human Biology 68:29–44.

Salzano FM, Callegari-Jacques SM (1988) South American Indians: a case study in evolution. Clarendon Press, Oxford.

Sambrook J, Fritsch EF, Maniatis T (1989) Molecular cloning: a laboratory manual. Cold Spring Harbour, NY: Cold Spring Harbour Laboratory Press.

Santos FR, Gerelsaikhan T, Munkhtuja B, *et al.* (1996) Geographic differences in the allele frequencies of the human Y-linked tetranucleotide polymorphism DYS19. Human Genetics 97:309–313.

Schanfield MS (1992) Immunoglobulin allotypes (GM and KM) indicate multiple founding populations of Native Americans: Evidence of at least four migrations to the new world. Human Biology 64:381–402.

Szathmary EJE (1993) mtDNA and the peopling of the Americas. American Journal of Human Genetics 53:793–799.

Torroni A, Schurr TG, Cabell MF, *et al.* (1993a) Asian affinities and continental radiation of the four founding native American mtDNAs. American Journal of Human Genetics 53:563–590.

Torroni A, Sukernik RI, Schurr TG, *et al.* (1993b) mtDNA variation of Aboriginal Siberians reveals distinct genetic affinities with Native Americans. American Journal of Human Genetics 53:591–608.

Wyman AR, White R (1980) A highly polymorphic locus in human DNA. Proceedings of the National Academy of Science USA 77:6754–6758.

Yunis JJ, Ossa H, Salazar M, *et al.* (1994) Major histocompatibility complex class II alleles and haplotypes and blood groups of four Amerindian tribes of Northern Colombia. Human Immunolgy 41:248–258.

Zhivotovsky LA, Feldman MW (1995) Microsatellite variability and genetic distances. Proceedings of the National Academy of Science USA 92:11549–11552.

4

TRINUCLEOTIDE REPEATS, GENETIC INSTABILITY AND VARIATION IN THE HUMAN GENOME

Ranjan Deka[1] and Ranajit Chakraborty[2]

[1]Department of Environmental Health
University of Cincinnati, Cincinnati, Ohio
[2]Human Genetics Center
University of Texas Health Science Center
Houston, Texas

ABSTRACT

Trinucleotide repeats have been implicated in the genesis of over a dozen human diseases. These repeats are normally polymorphic in human populations. However, disease manifests when the repeat arrays become longer than a certain threshold. The expanded repeats are meiotically unstable. Recently, two other trinucleotide repeat loci have been reported with very large and unstable alleles in the general population without being associated with any known disease. These findings imply that the process of instability is not exclusively a disease-associated phenomenon. In this chapter, we discuss some of the population genetic properties of these loci, with a general emphasis on questions relating to maintenance of the expanded alleles in human populations.

1. INTRODUCTION

The human genome contains thousands of highly polymorphic loci, which often occur as tandem repeats of short and simple DNA sequences. The first of these loci, now described as "microsatellites", or "short tandem repeats" (STRs), were reported in 1989 by two groups of investigators (Litt and Luty 1989; Weber and May 1989). The polymorphism was shown to be associated with tandem arrays of simple "CA" sequences. Because of the involvement of two nucleotides, they also came to be known as "dinucleotide" or "CA repeats". Since 1989, thousands of dinucleotide repeat loci have been characterized in the human genome with such abundance that on average one microsatellite is found every 30-kb in the genome. Apart from the dinucleotide repeats, two other types of microsatellite loci were also shown to be highly abundant and polymorphic in

Genomic Diversity: Applications in Human Population Genetics
edited by Papiha, Deka, and Chakraborty, Kluwer Academic / Plenum Publishers, New York, 1999.

the genome, viz., the trinucleotide and the tetranucleotide repeats. Their associated hypervariability, defined by the number of alleles and heterozygosity, has been the key in the success of the human genome project. Recent studies have demonstrated that microsatellites are also powerful tools in tracing evolutionary relationships and demographic histories of human populations (Bowcock *et al.* 1994; Jorde *et al.* 1997; Kimmel *et al.* 1998; Shriver *et al.* 1997) and in the determination of parentage and relatedness of individuals (Chakraborty and Jin 1993).

Within a relatively short period of time, an enormous number of microsatellite loci have been characterized and an explosive growth in the accumulation of population data has occurred. However, the question remained open whether microsatellites have any important biological function. Several lines of evidence indicated that in general microsatellites are selectively neutral and thus without any associated function. The number of such loci is enormously large with large numbers of alleles and high heterozygosity. Biological function would intuitively result in reduced variability and constraint on repeat number. Secondly, allele frequency distributions at most microsatellite loci do not show deviation from the expectations of the stepwise mutation model (Valdes *et al.* 1993; Shriver *et al.* 1993; Di Rienzo *et al.* 1994). In addition, as noted before, microsatellites provide reliable evolutionary histories of human populations. Were allelic distributions at these loci constrained by strong selection, these observations would not have been expected. Apparently, there is a compelling argument that microsatellites are generally unselected and they evolve in a neutral fashion. However, this does not mean that all microsatellite loci have no function nor are they affected by selection. A number of studies have indicated that CA sequences can enhance gene expression (Hamada *et al.* 1984) and promote homologous meiotic recombination (Treco and Arnheim 1986). Some microsatellites are also binding sites for specific nuclear proteins (Richards *et al.* 1993). Evolutionary conservation of these repeats can also be interpreted that they have functional roles (Stallings *et al.* 1991).

2. TRINUCLEOTIDE REPEATS IN HUMAN DISEASES

Regardless of whether average microsatellite loci have any biological function, a series of surprising discoveries beginning in 1991 provided evidence that at least some microsatellite elements have dramatic biological effects. A new class of mutation involving trinucleotide repeats was implicated in a number of human genetic diseases. The first of these, in which a trinucleotide repeat was found to be involved, is the fragile-X syndrome. Inheritance of fragile-X syndrome was an enigma for geneticists: it is an unusual form of X-linked mental disorder, because one-third of carrier females are affected and about one-fifth of the males with the fragile-X chromosome are normal. These phenotypically normal males, called the "normal transmitting males", transmit the fragile-X chromosome to their daughters. These daughters by themselves are not affected, but have mentally retarded children. Analysis of affected families showed an unusual pattern of segregation of the disease; brothers of normal transmitting males have a lower risk (9%), while the risk in grandsons (40%) and great grand sons (50%) is considerably higher (Sherman *et al.* 1984, 1985). This peculiar finding, that the risk of being affected with fragile-X mental retardation depends on the position of the individual in the pedigree and that the risk increases in subsequent generation came to be known as the "Sherman paradox". In 1991, DNA that give rise fragile-X syndrome was isolated (Verkerk *et al.* 1991; Kremer *et al.* 1991; Fu *et al.* 1991) and a new gene named as FMR1 (Fragile-X-mental reatrdation-1) was identified. DNA sequencing revealed that a normally polymorphic "CCG" trinucleotide repeat located in the 5′ untranslated region of the gene is massively expanded in the affected individuals. It was found that this repeat is polymorphic in the general population with alleles of 6 to 54 repeats. However, among the normal transmitting males and their daughters, the copy number of the repeat is moderately expanded, with 52 to ~200 CCGs, which were termed as pre-mutation alleles; while among the affected males, the repeat array is massively expanded, with 200 to more than 1,000 triplets, called the full mutation alleles. Pedigree analysis revealed that the normal sized alleles are meiotically stable. But, the pre-mutation alleles show meiotic instability and are prone to massive expansion during transmission resulting in the disease phenotype. Interestingly, expansion from pre-mutation to full mutation almost always occurs in female meiosis, explaining why daughters of normal transmitting males are not affected, while these daughters are likely to have affected sons. These findings provided a resolution of the Sherman paradox (Fu *et al.* 1991).

Within a year of the above discovery, expanded trinucleotide repeats were implicated in two other diseases. An expanded CAG-repeat located in the first exon of the androgen receptor gene was shown to cause X-linked spinal and bulbar muscular atrophy (SBMA), also known as Kennedy's disease (LaSpada *et al.* 1991). The CAG repeat encodes a long tract of glutamine residues with 13 to 30 repeats in normal individuals, but in affected individuals the repeat array almost doubles, with 40 to 60 repeats. Meiotic instability of the expanded repeat was observed. In 1992, the molecular defect causing myotonic dystrophy (DM) was identified. A normally polymorphic CTG repeat located in the 3′ untranslated region of the myotonin protein kinase gene on chromosome 19q13.3 was found to be unstable and massively expanded in DM patients (Brook *et al.* 1992; Fu *et al.* 1992; Mahadevan *et al.* 1992). Normal individuals have 5–35 CTG repeats; patients with fewer symptoms have 50–100 repeats; whereas, patients with severe symptoms have 100 to ~2,000 copies of the repeat. The expanded alleles are prone to progressive expansion. Although expansion of the repeat occurs in both male and female meiosis, congenital form of DM is exclusively maternally transmitted. Severity of DM had been shown to progress through generations. It was demonstrated that this was the result of an increasing expansion of the CTG repeat through succeeding generations. With this observation, DM became the hallmark of the phenomenon of "anticipation"!

It was evident that a new class of mutation exists in the human genome. The findings were dramatic and unexpected. That genomic instability can progress through generations appeared to defy the classic Mendelian principles. Even to the leading and respectful geneticists like Penrose, the concept of anticipation was barely acceptable, who considered it to be due to ascertainment bias (Penrose 1948). To highlight these developments, the July 1992 issue of the *American Journal of Human Genetics* published an invited editorial by Grant Sutherland and Robert Richards with the title "Anticipation legitimized: unstable DNA to the rescue". They named these mutations "dynamic" as opposed to the hitherto more acceptable concept of "static" mutation (Sutherland and Richards 1995).

It soon became apparent that instability is not limited to a few segments of the genome. Thus far, expanded trinucleotide repeats have been shown to be the causes of over a dozen human diseases. Among them are the Huntington's disease, HD (The Huntington's Disease Collaborative Research Group 1993), dentatorubral pallidoluysian atrophy, DRPLA (Koide *et al.* 1994), several forms of spinocerebellar ataxias, SCA1, SCA2, SCA3, SCA6, SCA7 and SCA8 (Orr *et al.* 1993; Pulst *et al.* 1996; Kawaguchi *et al.* 1994; Zhuchenko *et al.* 1997; David *et al.* 1997; Koob *et al.* 1999), a second and

Table 1. Trinucleotide-repeat associated diseases

Repeat Motif	Diseases
CCG	FRAXA, FRAXE
CTG	DM, SCA8
CAG	SBMA, HD, DRPLA, HD, SCA1, SCA2, SCA3, SCA6, SCA7
GAA	FRDA

relatively milder form of fragile X mental retardation, FRAXE (Knight *et al.* 1993), and Friedreich's ataxia, FRDA (Campuzano *et al.* 1996). These diseases with trinucleotide repeat expansion share a number of common features, viz., the repeat is polymorphic in normal individuals; the repeat becomes unstable when the repeat array becomes longer than a certain threshold; anticipation and parental sex bias in transmission is noticed in several of these diseases. Notably, three repeat motifs are predominantly associated with these diseases, viz., CAG, CCG and CTG (Table 1).

3. POPULATION GENETICS

The discovery of dynamic mutations posed new challenges to molecular biologists and geneticists in deciphering a host of unknown biological facts about these repeats. In the domain of population biology as well, a new challenge came. Note that these repeats are normally polymorphic in humans. In this context, several important population genetic questions arise and need to be addressed. For example, how variable are human populations with respect to the distribution of normal alleles at these loci? The biological basis of repeat expansion is believed to be caused by dynamic instability of the expanded repeats. But once expanded, these alleles impose a reduction in fitness of the affected individuals. How then are the expanded alleles maintained in human populations through time? Are some of the normal alleles, particularly the larger normal alleles, predisposed to expansion compensating for the loss of the expanded alleles?

3.1. Distribution of CTG Repeats at the DM Locus in Global Populations

We will address some of the issues raised above by looking at the distribution of normal CTG-repeat alleles at the DM locus in several worldwide populations. Myotonic dystrophy is a multisystemic disorder characterized by progressive weakness and wasting of distal muscles with cardiac involvement (Harper 1989). Epidemiological studies have shown that prevalence of DM varies considerably across the continents; from about 1 in 8,000 among populations of west European descent (Harper 1989) to about 1 in 18,000 among the Japanese (Ashizawa and Epstein 1991). Among Africans, DM is extremely rare; to date only one family from Nigeria has been reported (Dada 1973). We surveyed 15 ethnically diverse populations from five major human groups, namely Caucasian, Asian, African, American Indian and Pacific Islander (details of this work have been published elsewhere, Deka *et al.* 1996). In all, twenty-seven alleles were detected in the total sample. Allele frequencies, grouped in six different classes, are shown in Table 2. Although there is considerable variation among populations in the frequencies of the CTG repeats, $(CTG)_5$ is the smallest and the most common allele in the majority of the populations, and combined frequency of the $(CTG)_{11-14}$ repeats is the most prevalent of

Table 2. Distribution of CTG repeats at the DM locus in 15 human populations

Populations (no. of chromosomes)		Range of Repeat Size Min	Max	No. of Alleles	Repeat Size Class and their Frequencies ≤5	6–10	11–14	15–18	19–30	≥31
Caucasian	CEPH (158)	5	28	14	0.399	0.038	0.430	0.057	0.076	—
	German (104)	5	27	15	0.327	—	0.442	0.019	0.212	—
	Brazilian White (100)	5	25	15	0.390	0.050	0.460	0.030	0.070	—
	Indian Brahmin (72)	5	27	12	0.333	0.042	0.472	0.111	0.042	—
Asian	Chinese (102)	5	23	12	0.294	0.039	0.569	0.088	0.010	—
Mongoloid	Japanese (176)	5	28	15	0.176	0.017	0.656	0.074	0.074	—
	Kachari (86)	5	26	10	0.302	0.128	0.523	0.023	0.023	—
African	Benin (100)	5	18	11	0.320	0.140	0.520	0.020	—	—
	Sokoto (106)	5	19	11	0.264	0.198	0.500	0.028	0.009	—
	Brazilian Black (84)	5	28	13	0.191	0.083	0.655	0.012	0.060	—
American	Dogrib (100)	5	34	9	0.310	0.020	0.620	—	0.020	0.030
Indian	Cabecar (104)	11	13	3	—	—	1.00	—	—	—
	Pehuenche Indian (92)	5	27	9	0.065	0.087	0.761	0.022	0.065	—
Pacific Islander	New Guinea Highlander (100)	9	19	10	—	0.130	0.810	0.040	0.020	—
	Samoa (198)	4	25	9	0.338	0.036	0.484	0.132	0.010	—

all classes of CTG repeats in all populations, irrespective of their origin. Interestingly, a $(CTG)_4$ allele was found in a Samoan individual, which is perhaps the smallest known allele at this locus. With the exception of Brazilian White ($\chi^2 = 3.77$, $P = 0.04$) and New Guinea Highlander ($\chi^2 = 7.68$, $P < 0.001$), all of the studied populations showed conformity with Hardy-Weinberg expectations. A reduced diversity was, however, observed in a few populations. The Cabecar from Costa Rica have only 3 alleles, all of which are in the class of $(CTG)_{11-14}$ repeats. $(CTG)_5$ is absent in New Guinea Highlanders and present in relatively low frequency among the Pehuenche Indians from Chile. These two populations are also characterized by high frequencies of alleles in the class of $(CTG)_{11-14}$ repeats, 76% and 81%, respectively. The most likely explanation for these departures from the norm is the effect of genetic drift on these populations, whose effective population sizes are much smaller.

3.2. Distribution of $(CTG)_{\geq 19}$ Alleles and Their Implications

One of the questions that we raised above relates directly to the maintenance of DM in human populations. With reduced fitness of the severely affected individuals, DM alleles should have been abolished from the human gene pool. However, this is not the case; DM is maintained in human populations for thousands of years. It was postulated that normal alleles of $(CTG)_{\geq 19}$ repeats are predisposed to expansion and constitute a reservoir for recurrent mutation to the premutation size (Imbert *et al.* 1993; Neville *et al.* 1994), which then become expanded to the disease-causing size. Such a process would compensate the disease alleles, lost due to the negative effect of anticipation. Consequently, the frequency of large normal alleles (≥19 repeats) should then be an indicator of prevalence of DM in a population. As noted above, west Europeans have the highest prevalence of DM followed by the Japanese and with extreme rarity in Africans. This prevalence rate parallels the observed frequencies of $(CTG)_{\geq 19}$ alleles in our examined populations, from a frequency of 21% among Germans to 7.4% among the Japanese to virtual absence among the native African populations. The relatively higher frequency (6%) of these alleles in Brazilian Blacks is attributable to Caucasian admixture in that country. Although, closer to Japanese, the Chinese population has a very low frequency (1%) of the large alleles. This is noteworthy in view of reported rarity of DM in China (Ashizawa and Epstein 1991). There is no epidemiological data on DM prevalence among the American Indians and Pacific Islanders. However, the Dogrib Indians have some of the very large alleles (32, 33, 34 repeats), which were not observed in any other population. The Pehuenche Indians also have a comparatively higher frequency (6.5%) of ≥19 repeat alleles. It would be of interest to screen these populations for presence of DM. Such patterns of CTG allelic distributions correlating with disease prevalence have been reported by other investigators as well (Davies *et al.* 1992; Goldman *et al.* 1994; Watkins *et al.* 1995; Rubinsztein *et al.* 1994; Zerylnick *et al.* 1995). These observations are analogous to the ones at another trinucleotide repeat, the DRPLA locus (Burke *et al.* 1994; Deka *et al.* 1995). Frequencies of large normal alleles in the Japanese population are significantly elevated, among whom DRPLA is more common, than among Africans or Europeans, where the disease is rare.

3.3. Segregation Distortion at the DM Locus

If indeed, the large normal alleles (≥19 repeats) constitute the reservoir for recurrent mutations to the disease causing range, a second question arises, by what mechanisms are these predisposed alleles recruited? In an investigation of this question, two studies (Carey *et al.* 1994; Generalli *et al.* 1994) advanced a "meiotic drive" hypothesis with a finding that heterozygous individuals transmit alleles with unequal probabilities. They observed that large normal alleles (≥19 repeats) are preferentially transmitted by healthy fathers to their children. These alleles would eventually replenish the lost mutant alleles and maintain DM in human populations at a constant frequency. In these studies, however, segregation distortion was estimated for only the selected class of larger alleles.

In a detailed study (Chakraborty *et al.* 1996), we asked a more general question, whether heterozygous parents, irrespective of any allele size, transmit their alleles with equal probabilities? We analyzed 726 meioses in 95 nuclear families from 40 CEPH reference families. The data showed a preferential transmission of longer alleles (irrespective of their size) from heterozygous mothers, whereas males transmitted both short and long alleles with equal probabilities. The point estimate of the probability of transmitting the longer allele in females was 0.563 ± 0.03. A rounded value of 0.55 can be considered representative of this figure. This yields segregation distortion of $\varepsilon = 0.05$.

3.4. Balance between Segregation Distortion and Other Evolutionary Forces

With the assumption of segregation distortion over the entire range of allelic distribution, a continuous shift of alleles from short to longer sizes is expected during transmission from parents to children. This raises another question: What kind of balance between the shift caused by segregation distortion and other evolutionary forces explain the observed steady-state allele frequency distributions of normal-size CTG repeats? It should also be noted that, in general, population data show a good fit to the Hardy-Weinberg expectations. Three other basic evolutionary forces, viz., mutation, genetic drift and selection can balance the effect of segregation distortion. Comparisons of the relative strengths of these forces suggest the following. First, mutation rate at the DM locus, as well as at average microsatellite loci, is 10^{-3} per gamete per generation

(Weber and Wong 1993). Thus mutation rate is a much weaker force implying that mutation can not balance segregation distortion of the estimated intensity ($\varepsilon = 0.05$) and maintain the steady-state distribution of normal alleles. Second, a similar conclusion emerges with regard to the relationship with genetic drift. Following approximations similar to those described in Nagylaki (1990), we have shown earlier (Polanski *et al.* 1998) that our estimated intensity of segregation distortion also dominates genetic drift in populations of reasonable sizes (10^3–10^4), which would then cause the extinction of all alleles excepting the longest normal allele. However, observed distribution of CTG alleles in human populations contradicts this inference. Thus, the only remaining possibility is a balance between segregation distortion and selection. Although there is no direct evidence that individuals with normal-size alleles differ in their reproductive fitness, we have argued that it is necessary to introduce such fitness differences to explain the observed allelic distributions (Polanski *et al.* 1998). A highly mathematical model developed in the Polanski study demonstrated that a weak segregation distortion together with selection of comparable strength, within the normal size range, could maintain an equilibrium distribution of allele frequencies as observed in human populations.

4. EXPANDED TRINUCLEOTIDE REPEATS IN GENERAL POPULATIONS

Trinucleotide repeats associated with diseases are moderately to highly polymorphic in normal individuals. Disease results when the repeat array becomes longer than a certain threshold. The normal alleles, however, have not been seen to exceed 55 repeats at most, e.g., in FRAXA. In the case of CAG-repeat associated diseases, the normal allele is not longer than 40 repeats. Recently, two polymorphic CAG/CTG repeat loci, ERDA1 (Nakamoto *et al.* 1997; Ikeuchi *et al.* 1998) and SEF2-1 (Breschel *et al.* 1997), were reported with massively expanded alleles in the general populations. Neither of these two loci is shown to be associated with any disease. The ERDA1 locus is located on human chromosome 17q21.3, with alleles as large as 50 to 90 CAG repeats. The SEF2-1 locus is located on chromosome 18q21.1 and has much larger alleles of sizes 53 to 2,100 CTG repeats. Both these loci also have alleles in the range of normal repeat sizes observed in the disease-associated loci, with 10 ~ 40 repeats. Because of this rather unusual finding of expanded trinucleotide repeats in normal individuals, we studied allelic distribution at the ERDA1 locus in five human populations and characterized mutational patterns in a large number of families (details of this study is given in Deka *et al.* 1999). We observed that large alleles (≥40 repeats) are present in polymorphic frequencies in all populations, but are more common in Asian populations, intermediate in Europeans and less frequent in Africans (Table 3). Notably, the Nigerians have an intermediate class of alleles in the range of 27 to 40 repeats, which are almost absent in any other populations.

Since expanded alleles at disease associated trinucleotide repeats show inter-generational instability, it ought to be investigated if the large alleles at the ERDA1 locus are meiotically unstable. We examined a total of 731 meioses from 38 multigenerational CEPH pedigrees and a collection of 43 Southwestern Chinese families. The data in Table 4 show a total of 45 mutations (mutation rate of 6.2 ± 0.9% per generation). Of these mutations, there was only one that involved an allele smaller than 40 repeats, a maternal 17 repeat allele had a single-repeat contraction. This figure of one mutation in 636 meioses (all alleles of <40 repeats) with a mutation rate of 0.0016 per generation is consistent with the rate of mutations at other microsatellite loci (Weber and Wong 1993). In

Table 3. Distribution of large (≥40 repeats) CAG alleles at the ERDA1 locus in five human populations

Populations (No. of chromosomes)	Range of Repeat Size Min	Range of Repeat Size Max	No. of Alleles	Repeat Size Class and their Frequencies ≤10	10	11–24	25–39	40–60	≥61
Chinese (254)	10	93	42	—	0.358	0.366	—	0.039	0.236
New Guinea Highlander (98)	10	88	20	—	0.500	0.225	—	0.020	0.255
German (98)	9	73	18	0.020	0.316	0.490	—	0.112	0.061
CEPH (226)	7	75	28	0.013	0.288	0.579	0.004	0.062	0.053
Nigerian (102)	10	90	23	—	0.216	0.382	0.343	0.020	0.039

contrast, alleles in the size of ≥40 repeats are significantly unstable with a mutation rate of 46.3 ± 5.1%. In fact, within this class of large alleles, mutation rates also showed allele size dependency, with mutation rates reaching almost 55% for alleles of ≥40 repeats. In the total data, although paternal mutations appear to be more common (7.8 ± 1.4%) than maternal mutations (4.6 ± 1.1%), the difference is not statistically significant (P = 0.09); there is virtually no difference in the class of large alleles. Interestingly, however, mutation patterns differed depending on parental sex. Of the 28 paternal mutations, 26 (93%) resulted in contractions; whereas, of the 16 maternal mutations (≥40 repeats) 12 (75%) resulted in expansions.

Microsatellite repeat stability depends on the length of the perfect repeat array (Weber 1990). In two of the trinucleotide repeat disorders, SCA1 and FRAXA, sequence interruptions within the repeat array have been shown to confer stability to the normal size alleles (Chung *et al.* 1993; Eichler *et al.* 1994). In the context of allelic distributions at the ERDA1 locus, we observed that large alleles are comparatively rarer among the Nigerians (6% as opposed to 12 to 27% among the other populations, see Table 3), and interestingly enough, they also have a set of alleles that are intermediate in size (27 to 40 repeats), which are absent in other populations. The question then arises whether imperfections are present in these intermediate sized alleles. Such imperfections would then provide a stability preventing expansion to the expanded range, and consequently has resulted in the lower prevalence of large alleles in the Nigerian population. We sequenced 18 small alleles (10 to 26 repeats) drawn from various populations, and five intermediate (30 to 39 repeats) sized alleles from the Nigerians. The small alleles were found to be consisting of pure stretches of CAG repeats, whereas the intermediate sized alleles have CAT and CAC interruptions within the CAG repeat array (Table 5). Our mutation data suggest that the small alleles are inherently stable. Interruptions in the intermediate sized alleles probably preserve their stability as well. A few small and large alleles were also sequenced earlier (Nakamoto *et al.* 1997; Ikeuchi *et al.* 1998), but sequence interruptions had not

Table 4. Mutation at the ERDA1 locus

	Transmissions (No.)	No. of Mutations	Mutation Rate (%)
All Alleles	Paternal (360)	28	7.8 ± 1.4
	Maternal (371)	17	4.6 ± 1.1
	Total (731)	45	6.2 ± 0.9
≥40 Repeats	Paternal (61)	28	45.9 ± 6.4
	Maternal (34)	16	47.1 ± 8.6
	Total (95)	44	46.3 ± 5.1

Table 5. Sequences of ERDA1 alleles

No. of Repeats	Sequence
<27	AAAG$(CAG)_n$AAAAAG
30–32	AAAG$(CAG)_9$CAT$(CAG)_n$AAAAAG
38–39	AAAG$(CAG)_{13}$CAT(CAG)CAC$(CAG)_n$AAAAAG

been detected. Nakamoto and colleagues had, however, reported a single nucleotide polymorphism immediately flanking the CAG repeat array at the 3′ end, with an "A" in short alleles and a "C" in long alleles in individuals of Japanese and European ancestry. In all of our sequences of short and intermediate alleles, we found an "A" residue. This would suggest that the intermediate alleles in the Nigerians are not the progenitors of the large alleles in non-African populations. At this time, it is not known if the "A" to "C" substitution is responsible for the instability of the large alleles. Flanking sequence polymorphism has been shown to regulate minisatellite mutation rates (Monckton *et al.* 1994). It may, therefore, be not surprising if such mechanisms also regulate mutational processes at microsatellite loci.

All of the above observations at the ERDA1 locus mimic the phenomena of anticipation and dynamic mutation, features related to disease associated trinucleotide repeats. Although, the ERDA1 repeat is shown to be transcribed in human brain and testis, it is not yet known if the CAG repeat is located within an open reading frame (Ikeuchi *et al.* 1998). It is not even shown to be associated with any disease phenotype. In this case, it would imply that the process of instability of repeat alleles is not exclusively a disease-associated phenomenon. Alternatively these repeat sequences are associated with yet unknown disease phenotypes. In either case, we have yet to understand the implications of the dynamic nature of the genome.

5. CONCLUDING WORDS

Until the beginning of the present decade, dynamic mutation, anticipation, genomic instability were words somewhat alien to the scientific community involved in human genetic research. However, with the finding of expanded trinucleotide repeats involving human diseases, the notions changed. It is no longer a surprise when a new repeat locus is reported being associated with disease. The expectation is that many more such loci will be found in the genome. The latest finding of expanded and unstable trinucleotides in the general populations without any disease association further dramatizes these observations. Many questions surrounding genomic instabilities and their implications on human genome diversity are, however, yet to be answered.

6. ACKNOWLEDGMENTS

A number of colleagues have contributed to our work on population biology of trinucleotide repeats, among whom, we owe our gratitude especially to Drs. Li Jin, Marek Kimmel and Mark Shriver. This work is supported by US Public Health Service Research grants GM 45861 and GM 41399.

7. REFERENCES

Ashizawa T, Epstein HF (1991) Ethnic distribution of the myotonic dystrophy gene. Lancet 338:642–643.

Bowcock AM, Ruiz-Linares A, Tomfohrde J, Minch E, Kidd JR, Cavalli-Sforza LL (1994) High resolution evolutionary trees with polymorphic microsatellites. Nature 368:455–457.

Breschel TS, McInnis MG, Margolis RL, Sirugo G, Corneliussen B, Simpson SG, McMahon FJ *et al.* (1997) A novel, heritable, expanding CTG repeat in an intron of the SEF2-1 gene on chromosome 18q21.1. Hum Mol Genet 11:1855–1863.

Brook JD, McCurrah ME, Harley HG, Buckler AJ, Church D, Aburatani H, Hunter K *et al.* (1992) Molecular basis of myotonic dystrophy: expansion of trinucleotide (CTG) repeat at the 3′ end of a transcript encoding a protein kinase family member. Cell 68:799–808.

Burke JR, Ikeuchi T, Koide R, Tsuji S, Yamada M, Pericak-Vance MA, Vance JM (1994) Dentatorubral-pallidoluysian atrophy and Haw River syndrome. Lancet 344:1711–1712.

Campuzano V, Montermini L, Moltò MD, Pianese L, Cossée M, Cavalcanti F, Monros E *et al.* (1996) Friedreich's ataxia: autosomal recessive disease caused by an intronic GAA triplet repeat expansion. Science 271:1423–1427.

Carey N, Johnson K, Nokelainen P, Peltonen L, Savontous M-L, Juvonen V, Anvret M *et al.* (1994) Meiotic drive at the myotonic dystrophy locus? Nature Genet 6:117–118.

Chakraborty R, Jin L (1993) Determination of relatedness between individuals by DNA fingerprinting. Hum Biol 65:875–895.

Chakraborty R, Stivers DN, Deka R, Yu LM, Shriver MD, Ferrell RE (1996) Segregation distortion of the CTG repeats at the myotonic dystrophy locus. Am J Hum Genet 59:109–118.

Chung M-Y, Ranum LPW, Duvick LA, Servadio A, Zoghvi H, Orr HT (1993) Evidence for a mechanism predisposing to intergenerational CAG repeat instability in spinocerebellar ataxia type I. Nature Genet 5:254–258.

Dada TO (1973) Dystrophia myotonica in Nigerian family. East Afr Med J 50:213–228.

David G, Abbas N, Stevanin G, Dürr A, Yvert G, Cancel G, Wever C *et al.* (1997) Cloning of the *SCA7* gene reveals a highly unstable CAG repeat expansion. Nature Genet 17:65–70.

Davies J, Yamagata H, Shelbourne P, Buxton J, Ogihara T, Nokelainen P, Nakagawa R *et al.* (1992) Comparison of the myotonic dystrophy associated CTG repeat in European and Japanese populations. J Med Genet 29:766–769.

Deka R, Miki T, Yin S-J, McGarvey ST, Shriver MD *et al.* (1995) Normal CAG repeat variation at the DRPLA locus in world populations. Am J Hum Genet 57:508–511.

Deka R, Majumder PP, Shriver MD, Stivers DN, Zhong Y *et al.* (1996) Distribution and evolution of CTG repeats at the myotonin protein kinase gene in human populations. Genome Res 6:142–154.

Deka R, Guangyun S, Wiest J, Smelser D, Chunhua S, Zhong Y, Chakraborty R (1999) Patterns of instability of expanded CAG repeats at the ERDA1 locus in general populations. Am J Hum Genet 65:192–198.

Di Rienzo A, Peterson AC, Garza JC, Valdes AM, Slatkin M *et al.* (1994) Mutational processes of simple-sequence repeat loci in human populations. Proc Natl Acad Sci USA 91:3166–3170.

Eichler EE, Holden JJA, Popovich BW, Reiss AL, Snow K, Thibodeau SN, Richards CS *et al.* (1994) Length of uninterrupted CGG repeats determines stability in the FMR1 gene. Nature Genet 8:88–94.

Fu Y-H, Kuhl DPA, Pizzuti A, Pieretti M, Sutcliffe JS, Richards S, Verkerk AJMH *et al.* (1991) Variation of the CGG repeat at the fragile X site results in genetic instability: resolution of the Sherman paradox. Cell 67:1047–1058.

Fu Y-H, Pizzuti A, Fenwick Jr. RG, King J, Rajnarayan S, Dunne PW, Dubel J *et al.* (1992) An unstable triplet repeat in a gene related to myotonic muscular dystrophy. Science 255:1256–1258.

Gennarelli M, Dallapiccola B, Baiget M, Martorell I, Novelli G (1994) Meiotic drive at the myotonic dystrophy locus. J Med Genet 31:980.

Goldman A, Ramsay M, Jenkins T (1994) Absence of myotonic dystrophy in southern Africa Negroids is associated with a significantly lower number of CTG trinucleotide repeats. J Med Genet 31:37–40.

Hamada H, Seidman M, Howard BH, Gorman CM (1984) Enhance gene expression by poly(dT-dG).(dC-dA) sequence. Mol Cell Biol 4:2622–2630.

Harper PS (1989) Myotonic Dystrophy. Saunders, London.

Ikeuchi T, Kazuhiro S, Takano H, Sasaki H, Tashiro K, Cancel G, Brice A *et al.* (1998) A novel long and unstable CAG/CTG trinucleotide repeat on chromosome 17q. Genomics 49:321–326.

Imbert G, Kretz C, Johnson K, Mandel J-L (1993) Origin of expansion mutation in myotonic dystrophy. Nature Genet 4:72–76.

Jorde LB, Rogers AR, Bamshad M, Watkins WS, Krakowiak P, Sung S, Kere J *et al.* (1997) Microsatellite diversity and the demographic history of modern humans. Proc Natl Acad Sci USA 94:3100–3103.

Kawaguchi Y, Okamota T, Taniwaki M, Aizawa M, Inoue M, Katayama S, Kawakami H *et al.* (1994) CAG expansions in a novel gene for Machado-Joseph disease at chromosome 14q32.1. Nature Genet 8:221–228.

Kimmel M, Chakraborty R, King JP, Bamshad M, Watkins WS, Jorde LB (1998) Signature of population expansion in microsatellite repeat data. Genetics 148:1921–1930.

Knight SJL, Flannery AV, Hirst MC, Campbell L, Christodoulou Z, Phelps SR, Pointon J *et al.* (1993) Trinucleotide repeat amplification and hypermethylation of a CpG island in FRAXE mental retardation. Cell 74:127–134.

Koide R, Ikeuchi T, Onodera O, Tanaka H, Kondo R, Ishikawa A, Hayashi T *et al.* (1994) Unstable expansion of CAG repeat in hereditary dentatorubral-pallidoluysian atrophy (DRPLA). Nature Genet 6:9–13.

Koob MD, Moseley ML, Schut LJ, Benzow KA, Bird TD, Day JW, Ranum LPW (1999) An untranslated CTG expansion causes a novel form of spinocerebellar ataxia (SCA8). Nature Genet 21:379–384.

Kremer EJ, Pritchard M, Lynch M, Yu S, Holman K, Baker E, Warren ST *et al.* (1991) Mapping of DNA instability at the fragile X to a trinucleotide repeat sequence p(CCG)n. Science 252:1711–1714.

LaSpada AR, Wilson AM, Lubahn DB, Harding AE, Fishbeck KH (1991) Androgen receptor gene mutations in X-linked spinal and bulbar muscular atrophy. Nature 352:77–79.

Litt M, Luty JA (1989) A hypervariable microsatellite revealed by in vitro amplification of a dinucleotide repeat within the cardiac muscle actin gene. Am J Hum Genet 44:388–396.

Mahadevan M, Tsilfidis C, Sabourin L, Shutler G, Amemiya C, Jansen G, Neville C *et al.* (1992) Myotonic dystrophy mutation: an unstable CTG repeat in the 3′ untranslated region of the gene. Science 255:1253–1255.

Monckton DG, Neumann R, Guram T, Fretwell N, Tamaki K, MacLeod A, Jeffreys AJ (1994) Minisatellite mutation rate variation associated with a flanking DNA sequence polymorphism. Nature Genet 8:162 170.

Nagylaki T (1990) Models and approximations for random genetic drift. Theor Pop Biol 37:192–212.

Nakamoto M, Takebayashi H, Kawaguchi Y, Narumiya S, Taniwaki M, Nakamura Y, Ishikawa Y *et al.* (1997) A CAG/CTG expansion in the normal population. Nature Genet 17:385–386.

Neville CE, Mahadevan MS, Barcelo JM, Korneluk RG (1994) High resolution genetic analysis suggests one ancestral predisposing haplotype for the origin of the myotonic dystrophy mutation. Hum Mol Genet 3:45–57.

Orr HT, Chung M, Banfi S, Kwiatkowski TJ Jr, Servadio A, Beaudet AL, McCall AE *et al.* (1993) Expansion of an unstable trinucleotide CAG repeat in spinocerebellar ataxia type 1. Nature Genet 4:221–226.

Penrose LS (1948) The problem of anticipation in pedigrees of dystrophia myotonica. Ann Eugenics 14:125–132.

Polanski A, Chakraborty R, Kimmel M, Deka R (1998) Dynamic balance of segregation distortion and selection maintains normal allele sizes at the myotonic dystrophy locus. Math Biosci 147:93–112.

Pulst S-M, Nechiporuk A, Nechiporuk T, Gispert S, Chen X-N, Lopes-Cendes I, Pearlman S *et al.* (1996) Moderate expansion of a normally biallelic trinucleotide repeat in spinocerebellar ataxia type 2. Nature Genet 14:269–276.

Richards RI, Holman K, Yu S, Sutherland GR (1993) Fragile X syndrome unstable element, p(CCG)n, and other simple tandem repeat sequences are binding sites for specific nuclear proteins. Hum Mol Genet 2:1429–1435.

Rubinsztein DC, Leggo J, Amos W, Barton DE, Ferguson-Smith MA (1994) Myotonic dystrophy CTG repeats and the associated insertion/deletion polymorphism in human and primate populations. Hum Mol Genet 3:2031–2035.

Sherman SL, Morton NE, Jacobs PA, Turner G (1984) The marker (X) syndrome: a cytogenetic and genetic analysis. Ann Hum Genet 48:21–37.

Sherman SL, Jacobs PA, Morton NE, Froster-Iskenius U, Howard-Peebles PN, Nielson KB, Partington NW *et al.* (1985) Further segregation analysis of the fragile X syndrome with special reference to transmitting males. Hum Genet 69:3289–3299.

Shriver MD, Jin L, Chakraborty R, Boerwinkle E (1993) VNTR allele frequency distributions under the stepwise mutation model: a computer simulation approach. Genetics 134:983–993.

Shriver MD, Jin L, Ferrell RE, Deka R (1997) Microsatellite data support an early population expansion in Africa. Genome Res 7:586–591.

Stallings RL, Ford AF, Nelson D, Torney DC, Hildebrand CE, Moyzis RK (1991) Evolution and distribution of (GT)n repetitive sequences in mammalian genomes. Genomics 10:807–815.

Sutherland GR, Richards RI (1992) Anticipation legitimized: unstable DNA to the rescue. Am J Hum Genet 51:7–9.

Sutherland GR, Richards RI (1995) Simple tandem DNA repeats and human genetic disease. Proc Natl Acad Sci USA 92:3636–3641.

The Huntington's Disease Collaborative Research Group (1993) A novel gene containing a trinucleotide repeat that is expanded and unstable on Huntington's disease chromosomes. Cell 72:971–983.

Treco D, Arnheim N (1986) The evolutionarily conserved repetitive sequence d(TG.AC)n promotes reciprocal exchange and generates unusual recombinant tetrads during yeast meiosis. Mol Cell Biol 6:3934–3947.

Valdes AM, Slatkin M, Freimer NB (1993) Allele frequencies at microsatellite loci: the stepwise mutation model revisited. Genetics 133:737–749.

Verkerk AJMH, Pieretti M, Sutcliffe JS, Fu Y-H, Kuhl DPA, Pizutti A, Reiner O *et al.* (1991) Identification of a gene (FMR-1) containing CGG repeat coincident with a breakpoint cluster region exhibiting length variation in fragile X syndrome. Cell 65:905–914.

Watkins WS, Bamshad M, Jorde LB (1995) Population genetics of trinucleotide repeat polymorphisms. Hum Mol Genet 4:1485–1491.

Weber JL (1990) Informativeness of human (dC-dA)n.(dG-dT) polymorphisms. Genomics 7:524–530.

Weber JL, May PE (1989) Abundant class of human DNA polymorphisms which can be typed using the polymerase chain reaction. Am J Hum Genet 44:397–401.

Weber JL, Wong C (1993) Mutation of human short tandem repeats. Hum Mol Genet 8:1123–1128.

Zerylnick C, Torroni A, Sherman SL, Warren ST (1995) Normal variation at the myotonic dystrophy locus in global human populations. Am J Hum Genet 56:123–130.

Zhuchenko O, Bailey J, Bonnen P, Ashizawa T, Stockton DW, Amos C, Dobyns WB *et al.* (1997) Autosomal dominant cerebellar ataxia (SCA6) associated with small polyglutamine expansions in the α_{1A}-volatge-dependent calcium channel. Nature Genet 15:62–69.

5

Y-CHROMOSOMAL DNA MARKERS

Chris Tyler-Smith

CRC Chromosome Molecular Biology Group
Department of Biochemistry
University of Oxford
South Parks Road, Oxford OX1 3QU
UK

ABSTRACT

Many Y-chromosomal DNA polymorphisms are now available. The most widely used are biallelic markers such as base substitutions, and microsatellites. This chapter discusses some of the issues and complications raised by the use of these markers, including recurrent mutation, duplication of loci, sample reliability and nomenclature.

1. INTRODUCTION

The first Y-chromosomal DNA polymorphisms were described in 1985 (Casanova *et al.* 1985; Lucotte and Ngo 1985), so by the year 2000 the field will be 15 years old. During this time, it has developed from an obscure specialist area involving only a few labs to a routine part of molecular anthropology. Such progress has been possible because of technical advances, especially the development of improved markers. The early markers could only be scored using hybridisation, and the results were often complex and difficult to interpret. They have been replaced by (or in a few cases converted into) markers that can be scored using PCR and interpreted easily. The attraction of Y markers is that they allow paternal lineages to be identified and thus provide a male perspective on anthropological questions. Their uses have been reviewed elsewhere (Jobling and Tyler-Smith 1995; Mitchell and Hammer 1996; Hammer and Zegura 1996; Jobling *et al.* 1997) and are covered in other chapters in this volume.

The ultimate "marker" for the Y chromosome for the 21st century would be the DNA sequence of the entire chromosome. However, the technology for acquiring data on this scale is not yet available, and is not likely to become available in the near future. The sequence variation on the Y chromosome is low (Hammer 1995; Whitfield *et al.* 1995), so, unlike analysis of the mitochondrial DNA control regions, re-sequencing on the scale that is practical does not provide sufficient information to address many anthropological questions. Therefore the strategy usually adopted is to score previously identified Y polymorphisms on the new samples being investigated. The most widely used polymorphisms are biallelic markers and microsatellites. The sole known Y-specific minisatellite, MSY1 (Jobling *et al.* 1998a) and the alphoid heteroduplex system (Santos *et al.* 1995) are more complex markers that can be scored using PCR, but will not be considered in detail here. This chapter is aimed at non-specialists who wish to begin to analyse Y markers in their populations. It assumes that the analysis will be carried out on a moderate scale using the technologies available to small labs at the end of the 20th century, and will discuss some of the issues associated with using the current Y markers that regular scientific papers rarely provide the space to address.

2. BIALLELIC MARKERS

Some of the Y markers available have low mutation rates and thus identify stable lineages. Examples of these markers are base substitutions, small insertions and deletions, and retroposon insertions. There is no perfect collective word for them, but the best available is "biallelic markers". This is because their low mutation rates mean that there have seldom been recurrent mutations, so usually just two alleles are found. Undoubtedly, if a large number of these markers are examined in a lot of individuals, examples with more than two alleles will be found. Conversely, it is already clear that some markers with two alleles can have quite high mutation rates: for example, the presence or absence of the 50f2/C locus (Jobling *et al.* 1996). Alternative names such as SNPs (single nucleotide polymorphisms) and UEPs (unique event polymorphisms) are sometimes used to describe similar sets of polymorphisms.

2.1. Ancestral State

The ancestral state of a stable marker can be inferred by sequencing its homologue from an outgroup species. Apes, particularly chimpanzees, are the usual choice, but there is no consensus about the number of species/subspecies or individuals that need to be examined. If all the apes tested show the same state of the marker, and this corresponds to one of the human alleles, it can be assumed to be the ancestral state. Any other finding will introduce uncertainty. In practice, the greatest problems are encountered with repeated sequences such as 92R7 (Mathias *et al.* 1994) or *DYS257* (Hammer *et al.* 1998), where it may not be obvious which ape locus corresponds to the polymorphic human locus. Thus the ancestral state of some biallelic markers will be ambiguous. Such ambiguity can be resolved to most people's satisfaction by examining haplotypes containing these markers in combination with other biallelic markers whose ancestral states are clear. In the future, when the number of markers available is not limiting, it may no longer be necessary to use markers with ambiguous ancestral states.

2.2. Recurrent Mutation

Since the low mutation rate and stability of biallelic markers such as base substitutions have been emphasised above, it is a little disconcerting to find that one of the very first base substitutions to be identified, an A to G transition 1,532 bp upstream of the *SRY* gene, has reverted. This polymorphism was detected during the sequencing of an *SRY*-containing cosmid (Whitfield *et al.* 1995), where the A was present in a San individual, and in a screen of the *SRY* promoter region (Kwok *et al.* 1996) where the A was found in 16 individuals, mainly from the Indian subcontinent. The ancestral state was A, now restricted to Africa; it mutated to G which is now the most widespread allele, and then back to an A which is found in parts of Asia and Europe. Such a recurrent mutation creates a closed loop in haplotype trees (Fig. 1); the loop can be resolved if additional suitable markers (e.g. M9) are used. This recurrent mutation does raise the question of whether it is a chance event, whether the mutation rates at some positions on the chromosome are high, or whether recombination between Y chromosomes is occasionally possible. Of course, all base substitutions will be recurrent if sufficient individuals are screened: a mutation that occurs with a frequency of 1 chromosome in 10^8 will recur ten times in a generation of 10^9 males, but these ten individuals are unlikely to be sampled.

True recurrent mutation must be distinguished from apparent recurrent mutation. If a marker is scored by the presence or absence of a restriction enzyme site, cleavage can usually be abolished by a base change at any position in the site, so the absence of cleavage does not always imply the same base change. Sequencing should always be used to investigate apparent recurrent mutations, and especially those due to loss of a restriction site.

2.3. Nomenclature

Different labs often use different haplotype nomenclatures, and these can change from publication to publication. While scientifically trivial, this matter already impedes communication, and could get much worse in the future. The nomenclatures currently in use by the Arizona, Stanford and Oxford/Leicester labs are compared in Table 1 and Fig. 1.

What should be done in the future? mtDNA provides a useful precedent. Here, groups of haplotypes, originally identified by RFLPs (Torroni *et al.* 1992), are designated "haplogroups" and given capital letters A, B, C etc. In the most recent discussion of mtDNA nomenclature (Richards *et al.* 1998), these are then subdivided by numbers and letters, to produce a hierarchy of names such as A1, A2, then A2a, A2b etc. Richards and colleagues detail ways in which this system can absorb changing knowledge.

While it seems premature to propose here a nomenclature that will be universally accepted in the field, there are plans to address this problem. The YCC collection of worldwide DNA samples has been distributed by Mike Hammer to laboratories known to be developing new markers, and it is hoped that all markers currently available will be scored on these samples. Details of the YCC collection are available from the web site http://www.arl.arizona.edu/lmse.ycc.html When scoring of these samples is complete, many, perhaps most, of the major haplogroups can be identified and the basis for a stable nomenclature established.

Table 1. Marker nomenclature complications

Markers with more than one name	
SRY-1532 (Pandya 1998)[1]	10831 (Hammer *et al.* 1998)
SRY-8299 (Pandya 1998)	4064 (Hammer *et al.* 1998)
Independent markers that identify similar subsets of Y chromosomes	
sY81 (M2) (Underhill *et al.* 1997)	PN1 (Hammer *et al.* 1998)
M911 and 92R7 (Pandya 1998)	*DYS257* (Hammer *et al.* 1998)
SRY-1532 (Pandya 1998)	M17 (Underhill *et al.* 1997)

[1]The reference is to the haplotype description (see Fig. 1).

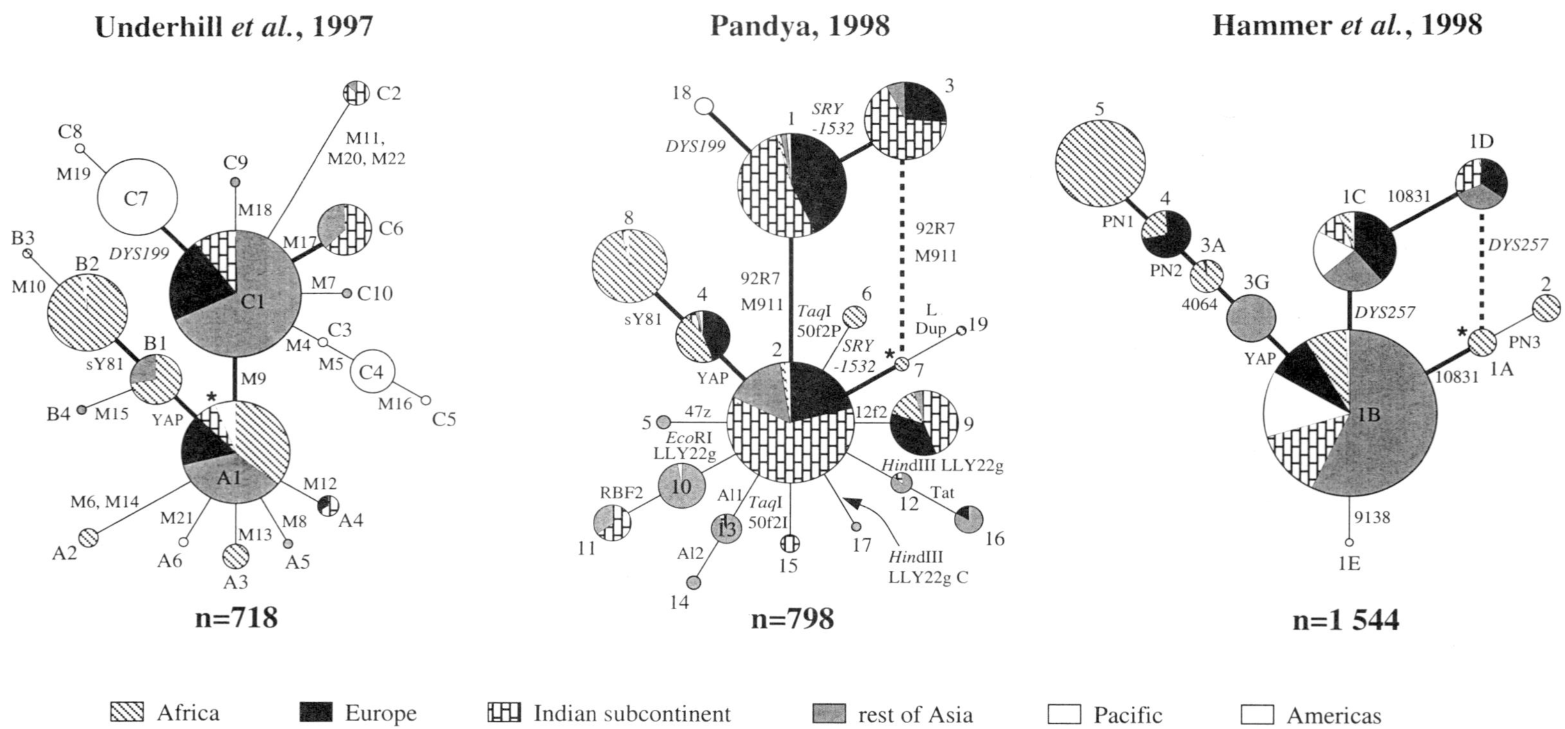

Figure 1. Marker and haplogroup nomenclatures used in three surveys of worldwide Y chromosome samples. Marker names are indicated on the lines separating haplotypes; haplotypes are represented by circles with an area proportional to their relative frequency in the sample, and shaded according to their geographical origin. The root haplotype is marked with an asterisk. Some correspondences between the three surveys (abbreviated as U, P and H) can be identified, either because the same marker was used, or because different markers have been found to identify similar sets of chromosomes. These are indicated by thick lines. YAP was used by all three groups, and sY81 (U, P) identifies a similar set of chromosomes to PN1 (H; Hammer *et al.* 1997). 92R7 and M911 (P) identify a similar set of chromosomes to *DYS257* (H; Jobling *et al.* 1998b; M. Hammer, personal communication); this set is a large subset of those identified by M9 (U). *DYS199* was used by both U and P; while *SRY-1532* (P) and 10831 (H) are different names for the same marker. *SRY-1532* haplogroup 3 (P) contains a similar set of chromosomes to M17 haplogroup C6 (U); Raheel Qamar, unpublished observations. Other similarities in haplogroup positions between the three diagrams do not necessarily imply correspondences.

3. MICROSATELLITES

3.1. Sizing

Microsatellite alleles are usually scored by measuring the size of a PCR product. The results may be presented as the fragment size (e.g. 190 bp for *DYS19*), a code for the fragment size (e.g. B for this *DYS19* allele), or as the number of copies of the microsatellite repeat unit known, from sequencing a small number of alleles, to be present in each sized fragment (e.g. 14 in this case). The last nomenclature is the best because it reflects the most important biological property of the DNA and does not change if the primers are redesigned or the locus is rediscovered and renamed (as *DYS394* in this case).

Although microsatellite scoring appears simple, it may be less reliable than is generally appreciated. A fragment scored as 369 bp in one lab may be scored as 373 bp in another. Consequently, allele sizes determined in different labs should not be compared unless either (1) standard samples have been scored, (2) one or more alleles have been sequenced, or (3) absolute size is not important (when, for example, size variation alone is being used). Ambiguous alleles should never be assigned to allele classes on the basis of the frequency of the classes in other population samples, since allele frequencies vary greatly between populations. It would be a great benefit to the field if referees, journal editors and database managers would require compliance to points (1) or (2) before allowing publication or inclusion of results.

3.2. Duplications

Several microsatellite loci are known to be duplicated (e.g. YCAII, YCAIII, *DYS385*, *DYS389*), but it is generally assumed that the more commonly used loci are single-copy. How reliable is this assumption, and does it matter? Simple duplications would produce two alleles of the same size, leading only to a more intense band or higher peak, which would not usually be detected. Only when duplicated alleles have diverged would they be noticed, and even here, inspection of the raw data, rather than the output of an interpretative program, may be needed. Examples of duplications and triplications of loci such as *DYS19* and *DYS390*, which are usually single-copy, have been reported (e.g. Santos *et al.* 1996; Kayser *et al.* 1997). These duplications and triplications followed by divergence may be present in as many as 0.1% of the individuals sampled, and duplications or triplications without divergence may be more frequent.

Duplicated loci will begin to evolve separately, and they can undergo molecular processes such as gene conversion of one copy by the other, which do not occur in single-copy loci. The additional copies may subsequently be lost, so a locus that is currently single-copy may have had an evolutionary history as a duplicated locus. It is possible to imagine a scenario where some Y chromosomes delete one copy of a duplicated region and the remaining Y chromosomes delete the other copy. In such a case, the coalescence time of this region would be the time of duplication, which could be older than the coalescence time of the Y chromosome. These possibilities must be taken into account and may increase the uncertainty of conclusions about evolutionary history.

Although duplications of microsatellites are most readily detected as polymorphisms in the population, because the duplicated microsatellites soon start to diverge, the duplication is not confined to the microsatellite locus: adjacent sequences are also duplicated. Sporadic duplications of regions lacking microsatellites are likely to occur as frequently. One documented example is the duplication of 50f2/C and surrounding sequences in the distal Yq euchromatin (Jobling *et al.* 1996). Even important functional genes such as *SRY*, which are single copy in humans, are duplicated in some rodents (Lundrigan and Tucker 1997), and a period of evolution as a multi-copy sequence cannot be ruled out for any region.

3.3. Mutation Rates

Microsatellite mutation rates are expected to vary according to the unit size and sequence, the allele length, the homogeneity of the repeat units and the flanking sequences. These factors are widely appreciated and have been discussed elsewhere (e.g. Brinkmann *et al.* 1998). Here, I would like to draw attention to one finding that deserves further consideration. Three studies reporting microsatellite mutations (Heyer *et al.* 1997; Kayser *et al.* 1997; Foster *et al.* 1998) together identify eleven mutations, but among this very small number, two, in *DYS389II* and *DYS390*, occur in the same father-to-son transmission where paternity was established beyond reasonable doubt (Kayser *et al.* 1997). Do the mutations occur at random, or is the frequency higher in some individuals because of their genotype or environment? Variation in the rate would add uncertainty to calculations that rely on microsatellite mutation rate for calibration.

4. HAPLOTYPES

4.1. Choice of Marker

Ideally, all markers would be scored on a set of samples. In practice, this will not happen because resources of time and/or materials will be limiting. Thus it will be necessary to choose the subset of the markers available that provides the maximum amount of information. This choice is likely to become more important and less arbitrary as more markers become available. What criteria should form the basis for such a choice? These will depend on the question that is being addressed. Sometimes an unbiased measure of Y-chromosomal variation may be needed, and a set of microsatellites may approximate to this. Alternatively, the aim may be to identify the phylogenetic position of each chromosome on the Y haplotype tree as precisely as possible. One strategy for achieving this is to start by screening all samples with a small number of biallelic markers that define deep branches in the tree. Once the branch that a particular sample lies on has been identified, subsequent analysis of this sample can be restricted to markers that lie on this branch. Recurrent mutations would sometimes be missed by such an approach, but this should not usually matter. In this lab, we commonly screen all samples with *SRY-1532*, YAP, M9 and 92R7. Subsequently, appropriate additional biallelic markers are used: for example, YAP+ samples are screened with sY81. It should soon be possible to assemble an initial screening set of about ten markers that can be used worldwide to identify the deep branches, perhaps as one or two multiplex PCR reactions.

A second strategy is to select biallelic markers on the basis of geography or hypotheses about population history. For example, it would be reasonable to test samples from northern Asia with Tat (Zerjal *et al.* 1997) since the C allele is known to be widespread in this region. Similarly, a population that was thought to have migrated from this area could be tested with the same marker. This strategy is most useful when there is a simple, well-defined hypothesis to test.

A third strategy is to infer the position of the chromosome on the haplotype tree indirectly using more variable markers that are polymorphic in all populations. The alphoid heteroduplex system provides, in effect, a natural multiplex reaction that allows a chromosome to be assigned to a limited set of haplogroups. MSY1, or a collection of five to ten microsatellite loci, also provide a haplotype that can often be used to infer the biallelic marker states of the chromosome. Such markers have the advantage that they can be used in all populations.

The amount of DNA available from particularly interesting samples is sometimes very small. If no information or expectations were available about the Y haplotype and it was only possible to carry out, for example, five PCR reactions on a sample, which five would provide the most information? At present, the best choices might be YAP and M9, which each identify common and widespread subsets of chromosomes, MSY1, and two microsatellite multiplexes (Kayser *et al.* 1997).

4.2. Presentation of Haplotype Information

For the Y chromosome, the most important information is contained in the haplotype, unlike traditional unlinked autosomal markers where marker frequencies alone are often used. Complete haplotypes should therefore always be presented in publications or as supplementary information.

5. SAMPLE RELIABILITY

Much of the molecular anthropologist's work requires the analysis of DNA samples of known origin. If the identification of the samples used is wrong, the conclusions drawn will be wrong. How reliable is the identification of the samples commonly used in labs? The Y chromosome provides one simple test: it should be present in males and absent from females. Our experience is that collections of male DNAs commonly contain samples that lack all Y markers. This can be found even when cell lines are purchased from respected sources such as the NIGMS Cell Repository. The overall frequency of these Y-negative "male" samples may be as high as 1%, much higher than the frequency of Y-negative XX males in the population, which is less than 0.01% (de la Chapelle 1981). If a characteristic as obvious as sex is wrongly reported, are other characteristics more reliable? I think that we must assume that a low but detectable level of sample misidentification is widespread. Consequently, we should not attach great significance to rare samples with surprising haplotypes unless their authenticity can be traced back to the original individual.

A second problem is the inadvertent "duplication" of samples. The number of DNAs from some populations available to molecular anthropologists is small. Labs interested in such a population may make great efforts to obtain all available DNAs and may receive the same samples from more than one source, sometimes with different identification codes. The problem can be avoided if the original sample identification code is always retained.

6. FINAL WORDS

Techniques for finding and scoring polymorphisms have been one of the driving forces in molecular anthropology during the last few decades. The experimental technique that would perhaps make the largest difference to molecular anthropology in the 21st century would be the ability to carry out re-sequencing cheaply and on a very large scale. In the immediate future, a large number of methods for finding and scoring polymorphisms suitable for use by small labs is now available, and promises to provide a wealth of data on worldwide Y-chromosomal variation.

7. ACKNOWLEDGMENTS

I thank Mark Jobling and Fabricio Santos for their comments on this manuscript, Raheel Qamar for information on the comparison of *SRY-1532* and M17, and the CRC for support.

8. REFERENCES

Brinkmann, B, Klintschar, M, Neuhuber, F, Huhne, J, Rolf, B (1998) Mutation rate in human microsatellites: influence of the structure and length of the tandem repeat. *Am. J. Hum. Genet.* **62**, 1408–1415.

Casanova, M, Leroy, P, Boucekkine, C, Weissenbach, J, Bishop, C, Fellous, M, Purrello, M, Fiori, G, Siniscalco, M (1985) A human Y-linked polymorphism and its potential for estimating evolutionary distance. *Science* **230**, 1403–1406.

de la Chapelle, A (1981) Etiology of maleness in XX men. *Hum. Genet.* **58**, 105–116.

Foster, EA, Jobling, MA, Taylor, PG, Donnelly, P, de Knijff, P, Mieremet, R, Zerjal, T, Tyler-Smith, C (1998) Jefferson fathered slave's last child. *Nature* **396**, 27–28.

Hammer, MF (1995) A recent common ancestry for human Y chromosomes. *Nature* **378**, 376–378.

Hammer, MF, Zegura, SL (1996) The role of the Y chromosome in human evolutionary studies. *Evol. Anthropol.* **5**, 116–134.

Hammer, MF, Spurdle, AB, Karafet, T, Bonner, MR, Wood, ET, Novelletto, A, Malaspina, P, Mitchell, RJ, Horai, S, Jenkins, T, Zegura, SL (1997) The geographic distribution of human Y chromosome variation. *Genetics* **145**, 787–805.

Hammer, MF, Karafet, T, Rasanayagam, A, Wood, ET, Altheide, TK, Jenkins, T, Griffiths, RC, Templeton, AR, Zegura, SL (1998) Out of Africa and back again: nested cladisitc analysis of human Y chromosome variation. *Mol. Biol. Evol.* **15**, 427–441.

Heyer, E, Puymirat, J, Dieltjes, P, Bakker, E, de Knijff, P (1997) Estimation of Y chromosome specific microsatellite mutation frequencies using deep rooting pedigrees. *Hum. Mol. Genet.* **6**, 799–803.

Jobling, MA, Tyler-Smith, C (1995) Fathers and sons: the Y chromosome and human evolution. *Trends Genet.* **11**, 449–456.

Jobling, MA, Samara, V, Pandya, A, Fretwell, N, Bernasconi, B, Mitchell, RJ, Gerelsaikhan, T, Dashnyam, B, Sajantila, A, Salo, PJ, Nakahori, Y, Disteche, CM, Thangaraj, K, Singh, L, Crawford, MH, Tyler-Smith, C (1996) Recurrent duplication and deletion polymorphisms on the long arm of the Y chromosome in normal males. *Hum. Mol. Genet.* **5**, 1767–1775.

Jobling, MA, Pandya, A, Tyler-Smith, C (1997) The Y chromosome in forensic analysis and paternity testing. *Int. J. Legal Med.* **110**, 118–124.

Jobling, MA, Bouzekri, N, Taylor, PG (1998a) Hypervariable digital DNA codes for human paternal lineages: MVR-PCR at the Y-specific minisatellite, MSY1 (DYF155S1). *Hum. Mol. Genet.* **7**, 643–653.

Jobling, MA, Williams, G, Schiebel, K, Pandya, A, McElreavey, K, Salas, L, Rappold, GA, Affara, NA, Tyler-Smith, C (1998b) A selective difference between human Y-chromosomal DNA haplotypes. *Curr. Biol.* **8**, 1391–1394.

Kayser, M, Caglia, A, Corach, D, Fretwell, N, Gehrig, C, Graziosi, G, Heidorn, F, Herrmann, S, Herzog, B, Hidding, M, Honda, K, Jobling, M, Krawczak, M, Leim, K, Meuser, S, Meyer, E, Oesterreich, W, Pandya, A, Parson, W, Penacino, G, Perez-Lezaun, A, Piccinini, A, Prinz, M, Schmitt, C, Schneider, PM, Szibor, R, Teifel-Greding, J, Weichhold, G, de Knijff, P, Roewer, L (1997) Evaluation of Y-chromosomal STRs: a multicenter study. *Int. J. Legal Med.* **110**, 125–133.

Kwok, C, Tyler-Smith, C, Mendonca, BB, Hughes, I, Berkovitz, GD, Goodfellow, PN, Hawkins, JR (1996) Mutation analysis of the 2 kb 5′ to *SRY* in XY females and XY intersex subjects. *J. Med. Genet.* **33**, 465–468.

Lucotte, G, Ngo, NY (1985) p49f, a highly polymorphic probe, that detects TaqI RFLPs on the human Y chromosome. *Nucl. Acids Res.* **13**, 8285.

Lundrigan, BL, Tucker, PK (1997) Evidence for multiple copies of the male sex-determining locus, Sry, in African murine rodents. *J. Mol. Evol.* **45**, 60–65.

Mathias, N, Bayes, M, Tyler-Smith, C (1994) Highly informative compound haplotypes for the human Y chromosome. *Hum. Mol. Genet.* **3**, 115–123.

Mitchell, RJ, Hammer, MF (1996) Human evolution and the Y chromosome. *Curr. Op. Gen. Dev.* **6**, 737–742.

Pandya, A (1998) *Human Y-chromosomal DNA variation.* DPhil thesis, Oxford University.

Richards, MB, Macaulay, VA, Bandelt, H-J, Sykes, BC (1998) Phylogeography of mitochondrial DNA in western Europe. *Ann. Hum. Genet.* **62**, 241–260.

Santos, FR, Pena, SDJ, Tyler-Smith, C (1995) PCR haplotypes for the human Y chromosome based on alphoid satellite DNA variants and heteroduplex analysis. *Gene* **165**, 191–198.

Santos, FR, Gerelsaikhan, T, Munkhtuja, B, Oyunsuren, T, Epplen, JT, Pena, SDJ (1996) Geographic differences in the allele frequencies of the human Y-linked tetranucleotide polymorphism *DYS19*. *Hum. Genet.* **97**, 309–313.

Underhill, PA, Jin, L, Lin, AA, Mehdi, SQ, Jenkins, T, Vollrath, D, Davis, RW, Cavalli-Sforza, LL, Oefner, PJ (1997) Detection of numerous Y chromosome biallelic polymorphisms by denaturing high-performance liquid chromatography. *Genome Res.* **7**, 996–1005.

Whitfield, LS, Sulston, JE, Goodfellow, PN (1995) Sequence variation of the human Y chromosome. *Nature* **378**, 379–380.

Zerjal, T, Dashnyam, B, Pandya, A, Kayser, M, Roewer, L, Santos, FR, Fretwell, N, Jobling, MA, Harihara, S, Shimizu, K, Semjidmaa, D, Sajantila, A, Salo, P, Crawford, MH, Ginter, EK, Evgrafov, OV, Tyler-Smith, C (1997) Genetic relationships of Asians and Northern Europeans, revealed by Y-chromosomal DNA analysis. *Am. J. Hum. Genet.* **60**, 1174–1183.

ON THE GENETIC ORIGIN OF THE TURKS

Study of Six Y-Chromosomal Short Tandem Repeats

Burkhard Rolf[1], Arne Röhl[2], Peter Forster[1,2], and Bernd Brinkmann[1]

[1]Institut für Rechtsmedizin
Westfalische Wilhelms-Universitat Munster
Von-Esmarch-Straße 62, D-48129 Münster
Germany
[2]Mathematisches Seminar
Universität Hamburg
Bundesstraße 55, D-20146 Hamburg
Germany

ABSTRACT

Although language and history bear testimony to the migration of Asians to Turkey in mediaeval times, genetic evidence for this event has been elusive. Here we analyse six Y-chromosomal short tandem repeats (STRs) DYS19, DXYs156-Y, DYS390, DYS391, DYS392 and DYS393 for 298 Turks, Germans, Japanese and Chinese using the new median-joining (MJ) phylogenetic network method specifically designed for population DNA genealogies. The Y-Chromosomal results agree with analysis of the mtDNA data for 1,210 Eurasians, indicating about 10% of East Asian genetic input into the population of Turkey.

1. INTRODUCTION

The Altaic language family extends from northeastern and central Asia to modern Turkey, forming a wedge between the Indoeuropean languages of Europe to the north and the Middle East to the South. The reason for the elongated geographic distribution to the west is known from historical sources: migrating Turkic speakers gained control of Asia minor (present Turkey) imposing their Altaic language onto the native, mainly Indoeuropean speaking population. Given these lingustic and historical facts, it is perhaps surprising that genetic evidence for east Asian admixture into the population of

Genomic Diversity: Applications in Human Population Genetics
edited by Papiha, Deka, and Chakraborty, Kluwer Academic / Plenum Publishers, New York, 1999.

Turkey is elusive. Cavalli-Sforza *et al.* (1994) have found that allele frequency data classify the Turks as Caucasoid and that a more sensitive genetic tool seems to be required to detect Asian input. To this end, we have chosen Y-STR loci and mitochondrial DNA (mtDNA) variation as markers for migratory events. Both genetic systems are characterised by uniparental inheritance and a high mutation rate, affording a maximum opportunity for genetic drift after population separations. In addition, both molecules are non-recombining, permitting a phylogeographic analysis of DNA variants, which is a more delicate type of analysis than distance based summary statistics (Richards *et al.* 1997).

2. SUBJECTS AND METHODS

Blood samples were obtained from 80 Germans from Westphalia and surroundings (10 of these had Slavic, i.e. east German or Polish surnames, Ger 3, 15, 24, 33, 48, 54, 56, 59, 71, and 75), 86 Turks from the region around Adana, 2 further Turks (Tur87 and 88) living in Germany, 72 Han Chinese from Shenyang and 58 Japanese from Shiga area. Primer sequences, PCR conditions and allele nomenclature for Y-chromosome typing of the STR loci DYS19, DYS390, DYS391, DYS392 and DYS393 were taken from Kayser *et al.* 1997. Primer sequences, amplification conditions and allele nomenclature for DXYS156 were taken from Chen *et al.* (1994) and Karafet *et al.* (1998). The DXYS156 primers amplify both Y-specific and an X-specific fragment, however the X- and Y-chromosome fragments are distinguishable as only allele lengths 6 to 10 are present in females and 11 to 14 in males. A single exception is a German male with 7 and 10 repeats, presumably corresponding to the X and the Y locus, respectively. A total of 226 females were types (not shown) in four populations to confirm the absence of allele lengths >10 on the X chromosome (Karafet *et al.* 1998).

Construction of the DNA genealogies of closely related sequences in a population sample usually requires phylogenetic network approaches to cope with the multitude of (almost) equally plausible trees incurred by homoplasy, whereas standard tree-building algorithms are only suitable for either less homoplasious data or interspecies data. Two different types of implemented phylogenetic network algorithms are currently available: the reduced median network algorithm (Bandelt *et al.* 1995) and the median joining network algorithm (Bandelt *et al.* submitted). Both methods have been extensively studied on mtDNA data (Richards *et al.* 1996, Forster *et al.* 1996, Watson *et al.* 1997, Bandelt *et al.* 1998). Preliminary trials (not shown) however demonstrated that the rate of homoplasy is so high in Y-STRs that the network algorithms, applied singly on the whole data set, result in complex networks which include too many implausible trees. This observation is in agreement with others who either exclude STR data from their phylogenetic analysis (Hammer *et al.* 1997) or restrict the analysis to a subset of the data defined by a point mutation (Zerjal *et al.* 1997). To partially overcome the effect of character conflicts generated by homoplasy, we have devised a heuristic strategy for weeding out spurious homoplasious connections in a network in that we subject our data set to the following five stages of analysis. First, the pentanucleotide repeat locus DXYS156-Y was given twice the weight of other loci for both network algorithms on account of its apparent stability in major ethnic groups: over 90% of Europeans have 12 repeats at DXYS156-Y (Sajantila *et al.* 1996, Kayser *et al.* 1997, and this study), while about 85% of a sub-Saharan African sample (Sajantila *et al.* 1996) have 11 repeats, and 13 repeats so far have only been found in Asians (this study, excepting Ger53). Furthermore the data of Sajantila *et al.* (1996) includes typing of the unique YAP insertion event, which reduces the number of possible MP trees for their African data to just four trees, all of which agree in demanding four parallel mutations at DYS19, but only one parallelism at DXYS156-Y. Second, we have extracted only those Y-types from our data set that occur in at least two individuals. The most recent mutations, which are less useful for phylogenetic reconstruction have had little time to spread and are more likely to be represented by single individuals in a sample. In the presence of high homoplasy, there is a risk of connecting distinct clades via peripheral (recent) types when their ancestors are not present in the sample. This criterion also eliminates ancestral singletons which are rare due to age, but these are generally reconstructed as branching nodes in phylogenetic procedures. Third, we have subjected this smaller data set (195 individuals out of 298) to the reduced median network algorithm (Bandelt *et al.* 1995) in order to generate a solution space for which parallelisms are postulated according to character compatibility and frequency criteria to reduce the level of homoplasy. Fourth, we have subjected this output to the median-joining network algorithm with parameter ε set to 0 (Bandelt *et al.* 1998), which successively connects subnetworks by generating only few consensus (median) sequences. Fifth, the full data set is augmented by inferred ancestral types from the previous analysis and subjected to our combined network method (not shown). The data input was coded according to the one-step model for repeat changes, which gives a lower weight to one-repeat changes over multi-repeat changes, as inferred from pedigree studies (Brinkmann *et al.* Submitted).

3. RESULTS AND DISCUSSION

Y-Chromosomal Variation

For the Y-chromosomal analysis we have determined repeat numbers of the STR loci DYS19, DYS390, DYS391, DYS392, DYS393 and DXYS156-Y in Turks (n = 88), Germans (n = 80), Han Chinese (n = 72) and Japanese (n = 58). These 298 individuals were found to represent 161 distinct Y-chromosomal types (or 162, if a dinucleotide deletion in Ger70 is counted). The network of the Y-types represented in at least two individuals is displayed in Fig. 1. Of the 15-hypothesised ancestral nodes, eight actually exist as singletons in our sample. We have also confirmed that the full data set yields the same topology by subjecting the complete data set (including eight further southeast Europeans) to the same sort of analysis, but adding the 15 hypothesised ancestral types to the data set prior to running the two network algorithms. Only two of the 15-hypothesised nodes are superfluous in the total network (the empty nodes adjacent to Ger42 and Han40 in Fig. 1). The network in the vicinity of these two nodes is therefore less robust. Although appropriate robusticity measures for trees or networks of homoplasious population data (akin to bootstrapping for species data) have not yet been developed (Bandelt *et al.* 1995), we can appreciate the reliability of the overall topology by the evident ethnographical structure. The network contains numerous reticulations, and strikingly, these nearly always involve DYS390 or DYS19, indicating that these loci mutate particularly rapidly. A count of mutation events (Table 1) confirms that mutation rates between loci differ by up to an order of magnitude, suggesting that it is inadvisable to apply a uniform mutation rate to Y-STRs.

The node marked by an asterisk contains the majority state of the conservative DXYS156-Y in sub-Saharan Africans (Sajantila *et al.* 1996, Salem *et al.* 1996) and is

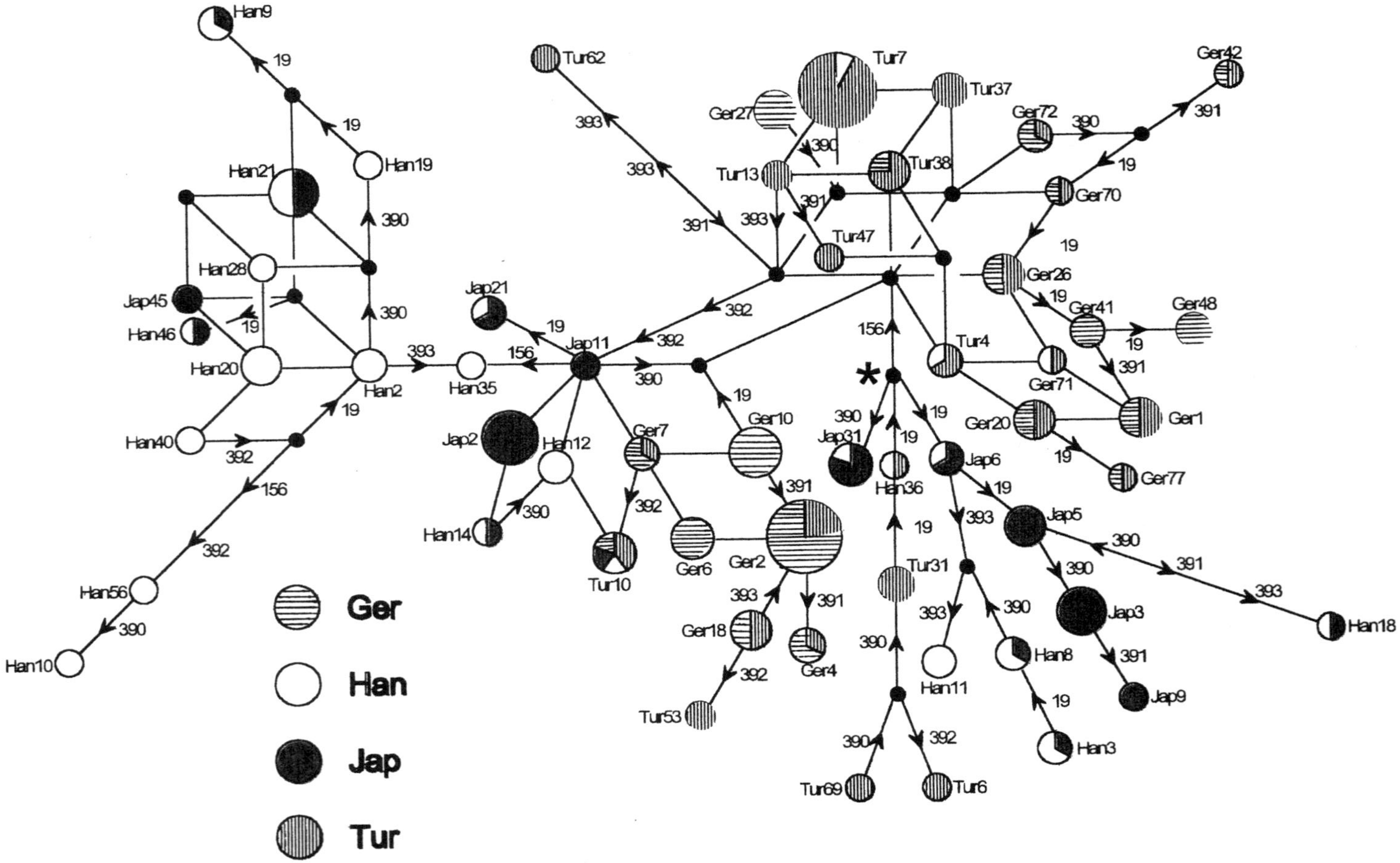

Figure 1. Phylogenetic network of Eurasian Y-STR types. Only types that occur at least twice in the data set (195 individuals out of the total data set of 298) are displayed. Types shared between populations are labelled by only one individual. Circle area is proportional to the number of individuals, and ethnic origins are depicted with graphics. The node marked by an asterisk represents the following sequence (amplicon length indicated in parentheses): DYS390: 24 repeats (215 bp), DYS19: 15 repeats (194 bp), DYS392: 11 repeats (248 bp), DYS393: 13 repeats (124 bp), DYS391: 10 repeats (279 bp), DXYS156-Y: 11 repeats (160 bp). Arrow point out the direction of a repeat gain.

Table 1. Estimated number of mutations at Y STR loci

	Minimum number of mutations	Maximum number of mutations
DYS19	15	28
DXYS156-Y	3	(3)
DYS390	13	28
DYS391	7	12
DYS392	7	15
DYS393	7	10

Note: The estimated number of mutations for each locus was determined from the phylogenetic network in Fig. 1 (58 Y-types). Minimum and maximum mutations were counted for a given locus by resolving reticulations respectively in favour of or against a given locus. The maximum number of mutations for DXYS156-Y is not comparable since this locus is weighted for the phylogeny (see subjects and Methods).

therefore probably identical to or in the immediate vicinity of one of the major Eurasian founding Y-chromosomal types. The cluster emanating from this node are, as indicated by the graphical coding, geographically specific for Caucasoids on the one hand and east Asians on the other. For example, the cluster defined by 11 repeats at DXYS156-Y is virtually absent in Europeans, but common in Asians (this study, Sajantila *et al.* 1996, Kayser *et al.* 1997, and Hammer, personal communication) and Africans (Sajantila *et al.* 1996, Kayser *et al.* 1997). Australians and highland Papuans also independently coalesce on this node (Forster *et al.* 1998). The vicinity of Jap11 also appears to include a major founding node for Asians, Europeans (Fig. 1) as well as Papuans and Australians (Forster *et al.* 1998). There is, at the present genetic resolution of six STRs, a branch which does not seem to fit the general ethnic pattern. The mainly Turkish branch linked to the node marked by an asterisk contains an Asian (Han36), but no Germans. This Turkish branch may represent (a) original founders of the Turkish Y pool, which did not penetrate into or did not survive in central Europe or (b) subsequent migration of males into our southern Turkish sample which did not affect central Europeans. With the given data, neither possibility can be dismissed. In scenario (b) two plausible historic explanations are worth consideration. The branch may conceivably be Asian, deriving from Turkic invaders. However a historic Asian migration to Turkey may be expected to have resulted in a branch with more Asian representatives. The presence of Han36 in this branch does not necessarily lend support since Han36 is only grouped into this branch due to the fast locus DYS19 (Table 1). The contact of southern Turkey to the Arabic world is also historically recorded (Rother 1971). A Near eastern origin of the branch, whether historic or prehistoric, receives support from the fact that 50/153 Egyptians have Y types with 11 repeats at DXYS156-Y and 14 or less repeats at DYS19 (Salem *et al.* 1996).

The ethnically mixed node represented by Tur10 in Fig. 1 is another apparent exception to the ethnic segregation of Y variation. However, this is simply due to insufficient resolution and typing of the 5' variable segment of DYS389 for all the data (Rolf, unpublished) resolves the reticulation between Jap11 and Tur10 by assigning 5 repeats to Germans and 4 repeats to all Asians descended from and including the node of Jap11. Six of the Turks (Turks 3, 10, 14, 39, 58 and 63) have this Asian motif and one further Turk (Tur39) has the Asian-specific 13 repeats at DXYS156-Y, but 5 repeats at DYS389. At least one further "Asian" Turk is the singleton Tur64 (not included in Fig. 1), which differs from Han8 by only one repeat at DYS391. The presence of the Turkish/Asian cluster descended from Jap11 in Fig. 1 may again be explained by ancestral diversity, but

Table 2. mtDNA groups in Europeans, Turks, and Asians

	Asian mtDNA groups								European mtDNA groups					African	unclass.
	n	A	B	C	D	E	F	M*	H,U,V	I,W,X	J	K	T	L1,L2	
W.Europe	891	0	0	0	0	0	0	1	596	44	105	61	75	7	2
Turkey	96	1	0	3	4	0	0	1	50	10	15	3	8	1	0
Mongolia	103	4	10	15	25	2	9	20	9	0	0	1	1	0	7
Tibet	54	6	3	2	9	4	9	21	0	0	0	0	0	0	0
Taiwan	66	5	13	3	7	2	15	16	1	0	0	0	0	0	5

Notes: European data from Richards *et al.* (1996) and Torroni *et al.* (1996); Turkish data from Richards *et al.* (1996), Calafell *et al.* (1996) and Comas *et al.* (1996); Mongolian data from Kolman *et al.* (1996); Tibetan data from Torroni *et al.* (1994); Taiwanese (Han) data from Horai *et al.* (1996). mtDNA group nomenclature according to Torroni *et al.* (1996) and Bandelt *et al.* (1998). mtDNA lineages M*, C, D, E together constitute the supergroup M.

the imperfect segregation of Turkish from Asian types may be more compatible with recent Asian immigration to Turkey. Arab immigration seems less likely in this case as 8 repeats at DXYS156-Y are not found in the Near Easterns of Salem *et al.* (1996). Hence, there is a minimum of 7/88 (8%) Turks who may indicate a male east Asian contribution to our southern Turkish sample.

mtDNA Variation

Three published data sets for Turkish mtDNA exist (Richards *et al.* 1996, Calafell *et al.* 1996, Comas *et al.* 1996) but have not previously been applied to the question of female Turkic introgression into the population of Asia Minor. The Eurasian mtDNA phylogeny has been extensively characterised in both the control region (using sequencing analysis) as well as by 14-enzyme RFLP analysis of the complete mtDNA genome, and Torroni and coworkers (Torroni *et al.* 1994, Torroni *et al.* 1996) have introduced a widely used shorthand for labeling the branches emanating from the African mtDNA founders in the Eurasian RFLP phylogeny. At the 14-enzyme RFLP resolution, the west European and east Asian branches are, unsurprisingly, almost completely disjoint phylogenetically from each other, and this is reflected in the mtDNA nomenclature (Table 2): only group X (represented mainly in Amerinds) as well as groups H, U, K, T and M (are shared between Asians and Europeans, while the other mtDNA groups are geographically specific. A proportion of 9% of the Turkish sample has East Asian mitochondrial types, and the Turkish mtDNA profile matches the Altaic-speaking Mongolians more closely than the other Asians (Table 2). Due to the circumstance that up to 18 individuals (groups H, U, V, K, T and unclassified) in the Mongolian sample are potentially indistinguishable from Caucasoid mtDNA, a figure of up to 11% of east Asian mtDNA input in the Turks may be more realistic.

4. CONCLUSION

This study has identified east Asian Y-chromosomal and mtDNA markers at a proportion of about 10% in the population of Turkey, confirming the conclusion of Cavalli-Sforza *et al.* (1994) that language replacement has occurred there with little gene replacement. Further studies using these markers can now investigate whether there are microgeographic patterns of Asian input in the Turkic-speaking populations. From a methodological point of view, it is evident that Y-STRs do allow the approximate reconstruction of DNA genealogies of a time depth of several tens of thousand years, provided some appropriate weighting is applied. However, STR genealogies will never be as free of homology as point mutation genealogies are (e.g. Hammer 1995, Jobling *et al.* 1997, Underhill *et al.* 1997) due to the high mutation rate of STRs (relative to the time depth of human evolution) coupled with locus-specific mutation rate heterogeneity, and a combination of both types of data is desirable. The advantage of the rapidly mutating STRs, apart from their forensic value, may however be the eventual quantification of absolute mutation rates in pedigree studies (Heyer *et al.* 1997, Kayser *et al.* 1997).

5. ACKNOWLEDGMENTS

We thank M. Hammer, M. Kayser, and C. J. Kolman for information on Mongolian data. This work was performed by A. Röhl in partial fulfillment of the requirements for the doctoral degree.

6. REFERENCES

Bandelt H-J, Forster P, Sykes BC, Richards MB (1995). Mitochondrial portraits of human populations using median networks. Genetics 141:743–753.

Bandelt H-J, Forter P, Röhl A (1999). Median-joining networks for inferring intraspecific phylogenies. Mol Biol Evol 16:38–48.

Calafell E, Underhill P, Tolun A, Angelicheva D, Kalaydjieva L (1996). From Asia to Europe: mitochondrial DNA sequence variability in Bulgarians and Turks. Ann Hum Genet 60:35–40.

Cavalli-Sforza LL, Menozzi P, Piazza A (1994). The History and Geography of Human Genes. Princeton University Press, New Jersey, p.245.

Chen H, Lowther W, Avramopoulos D, Antonarakis SE (1994). Homologous loci DXYS156X and DXYS156Y contain a polymorphic pentanucleotide repeat (TAAAA) and map to human X and Y chromosomes. Human Mutation 4:208–211.

Comas D, Calafell F, Mateu E, Perez-Lezaun A, Bertranpetit J (1996). Geographic variation in human mitochondrial control region sequence: the population history of Turkey and its relationship to the European populations. Mol Biol Evol 13:1067–1077.

Forster P, Harding R, Torroni A, Bandelt H-J (1996). Origin and evolution of Native American mtDNA variation: a reappraisal. Am J Hum Genet 59:935–945.

Forster P, Kayser M, Meyer E, Roewer L, Pfeiffer H, Benkmann H, Brinkmann B (1998). Phylogenetic resolution of complex mutational features at Y-STR DYS390 in aboriginal Australians and Papuans. Mol Biol Evol 15:1108–1114.

Hammer MF (1995). A recent common ancestry for Human Y chromosomes. Nature 378:376–378.

Hammer MF, Spurdle AB, Karafet MR, Wood ET, Novelletto A, Malaspina P *et al.* (1997) The geographic distribution of human Y chromosome variation. Genetics 145:787–805.

Heyer E, Puymirat J, Dieltjes P, Bakker E, de Knijff P (1997). Estimating Y chromosome specific microsatellite mutation frequencies using deep rooting pedigrees. Hum Mol Genet 6:799–803.

Horai S, Murayma K, Hayasaka K, Matsubayashi S, Hottori Y *et al.* (1996). mtDNA polymorphism in East Asian populations, with special reference to the peopling of Japan. Am J Hum Genet 59:579–590.

Jobling MA, Pandya A, Tyler-Smith C (1997). The Y chromosome in forensic analysis and paternity testing. Int J Legal Med 110:118–124.

Karafet T, de Knijff P, Wood E, Ragland J, Clark A, Hammer MF (1998). Different patterns of variation at the X- and Y-linked microsatellite loci DXYS156X and DXYS156Y in human populations. Hum Biol 70:979–992.

Kayser M, Caglia A, Corach D, Fretwell N, Gehrig N, Gehrig C, Graziosi G, Heidorn F *et al.* (1997). Evaluation of Y-chromosomal STRs: a multicenter study. Int J Legal Med 110:125–133 and 141–149.

De Knijff P, Kayser M, Caglia A, Corach D, Fretwell N, Gehrig C, Graziosi G *et al.* (1997). Chromosome Y microsatellites: population genetic and evolutionary aspects. Int J Legal Med 110:134–140.

Kolman CJ, Sambuughin N, Bermingham E (1996). Mitochondrial DNA analysis of Mongolian populations and implications for the origin of New World founders. Genetics 142:1321–1334.

Richards MB, Macaulay V, Sykes BC, Pettitt P, Forster P, Bandelt HJ (1997). Reply to Cavalli-Sforza and Minch. Am J Hum genet 61:251–253.

Richards MB, Corte-Real H, Forster P, Macaulay V, Wilkinson-Herbots H, Demaine A. Papiha S *et al.* (1996). Paleolithic and neolithic lineages in the European mitochondrial gene pool. Am J Hum Genet 59:185–203.

Rother L (1971). Die Städte der Cukurova: Adana-Mersin-Tarsus. Geographisches Institut der Universität Tübingen, Tübingen.

Ruhlen M (1987). A guide to the World's Languages. Stanford University Press, Stanford, California.

Sajantila A, Salem A-H, Savolainen P, Bauer K, Girig C. Pääbo S (1996). Paternal and maternal DNA lineages reveal a bottleneck in the founding of the Finnish population. Proc Natl Acad Sci USA 93:12035–12039.

Salem A-H, Badr FM, Gaballah MF, Pääbo S (1996). The genetics of traditional living: Y-chromosomal and mitochondrial lineages in the Sinai peninsula. Am J Hum Genet 59:741–743.

Torroni A, Miller JA, Moore LG, Zamudio S, Zhuang J, Droma T, Wallace DC (1994). Mitochondrial DNA analysis in Tibet: implications for the origin of the Tibetan population and its adaptation to high altitude. Am J Phys Anthropol 93:189–199.

Torroni A, Huoponen K, Francalacci P, Petrozzi M, Morelli L, Scozzari R, Obinu D *et al.* (1996). Classification of European mtDNAs from an analysis of three European populations. Genetics 144:1835–1850.

Underhill PA, Jin L, Lin AA, Quasim Mehdi S, Jenkins T, Vollrath D, Davis RW *et al.* (1997). Detection of numerous Y chromosome biallelic polymorphisms by denaturing high-performance liquid chromatography. Genome Res 7:996–1005.

Watson E, Forster P, Richards MB, Bandelt H-J (1997). Mitochondrial footprints of human expansions in Africa. Am J Hum Genet 61:691.

Zerjal T, Dashnyam, Pandya A, Kayser M, Roewer L, Santos FR, Schiefenhövel W *et al.* (1997). Genetic relationships of Asians and Northern Europeans by Y-chromosomal DNA analysis. Am J Hum Genet 60:1174–1183.

7

THE ORIGINS OF PAKISTANI POPULATIONS

Evidence from Y Chromosome Markers

S. Q. Mehdi[1], R. Qamar[1], Q. Ayub[1], S. Khaliq[1], A. Mansoor[1], M. Ismail[1], M. F. Hammer[2], P. A. Underhill[3], and L. L. Cavalli-Sforza[3]

[1]Biomedical and Genetic Engineering Division
Dr. A. Q. Khan Research Laboratory
P. O. Box 2891, Islamabad 44000
Pakistan.
[2]Laboratory of Molecular Systematics and Evolution
University of Arizona
Tucson, AZ 85721. USA
[3]Departments of Genetics
Stanford University School of Medicine
Stanford, CA 94305-5120
USA

ABSTRACT

In an attempt to investigate the origin of the present-day Pakistani populations, we have analysed a microsatellite locus (DYS19) and the *Alu* insertion polymorphism (YAP) in a sample of nine ethnic groups from the Northern (viz., Punjabi, Burusho, Pathan, Kalash, and Hazara) and the Southern (Brahui, Baloch, Makrani, and Sindhi) parts of the country. A north-south divide among the populations is evident. This separation is not only due to their geographic location but these two regional groups might have originated as a result of two distinct waves of migrations, one from North Africa and the other from Sub-Saharan region.

1. INTRODUCTION

Pakistan lies on the postulated coastal route from Africa to Australia and the earliest settlers probably came to this area some 60,000 years ago. Evidence of Neolithic settlements of modern humans have been found at Mehrgarh, dating back to the seventh millennium BC, in the southern Pakistani province of Balochistan (Jarrige 1991). This site predates the agrarian Harappan culture that flourished in the fertile Indus Valley around 2400–2000 BC. Although fossil evidence from these sites is scanty, some observers have found similarities with fossils found in the Caspian and Mediterranean regions (Sarkar 1981).

Invaders including the Aryans, Macedonians, Arabs, Mongols etc. have all contributed to the ethnic variety of Pakistan's population. In present-day Pakistan, the Balochis, Brahuis, Makranis and Sindhis constitute the major southern populations of Pakistan. The major northern ethnic groups include the Balti, Burushos, Hazaras, Kalash Kafirs, Kashmiris, Pathans and Punjabis.

A brief description of some major ethnic groups of Pakistan are as follows: The Balochis are believed to have originated in Aleppo (Syria) and migrated to Pakistan via Iran. The area they inhabit in southwestern Pakistan includes the Bolan River valley where the earliest evidence of Neolithic settlements have been unearthed (Jarrige 1991; Quddus 1990; Hughes-Buller 1991). The Brahuis occupy the Sarawan and Jhalawan regions of Kalat State in Balochistan and are said to be of Turko-Iranian origin (Quddus 1990; Hughes-Buller 1991). The Brahui language is of Dravidian origin (Grimes 1992). However, the Brahui tribes are separated by almost a thousand miles from their neighbouring Dravidian speakers in South India and Sri Lanka. The Makranis reside along the dry and arid Makran coast of Pakistan. The Greek historians referred to it as Gedrosia (Hughes-Buller 1991). Their language and ethnic affinity is with the Baloch tribes (Grimes 1992). Many have Negroid features and it is speculated that they took the coastal route out of Africa. The Sindhis from the southeastern province of Pakistan (Sindh) are of a mixed ethnic background (Ansari 1996; Burton 1851). This region was referred to as Sindomana by the Greeks and Sindhudesha by ancient Hindus. The advent of Islam into this region in 711 AD led to the settlement of Arabs, Turks and Persians (Burton 1851). Of the ancient populations, the Mehs (or Mohanas) and the Bhil have survived in areas surrounding the Indus River. In addition many Hindus and pure Sindhi individuals from clans such as the Abro, Channa and Sumro reside there.

The Balti are considered to be Tibetan in origin (A. H. Dani 1991) and their language is archaic Sino-Tibetan (Grimes 1992). They reside in the northeastern Himalayan region of Pakistan known as Baltistan. The Burusho are from Nagar and Hunza in the Hindukush mountains (Dani 1991; Biddulph 1977) and their language, Burushaski, is one of the remaining language isolates of the world (Grimes 1992). Although legends indicate that the Burusho are of Greek ancestry, and it has been suggested that they had accompanied Alexander the Great during his expedition to India, yet their origin is uncertain. The Kalash Kafir tribes have been isolated for centuries in the remote mountains of the Hindukush in the northwestern areas of Pakistan (Robertson 1974). This ethnic minority is now confined to the three valleys of Bumburet, Birir and Rambur near Chitral. Although little is known of their pagan origin, they are believed to be descendants of soldiers in Alexander's army. Some authors assign to them a Slav origin (Robertson 1974). The Pathans inhabit the North West Frontier Province (NWFP) of Pakistan. Legends claim that they are either a lost tribe of Israel or descendants of soldiers that came with Alexander the Great (Caroe 1992; Bellew 1979). The Hazaras have typical Mongoloid features and are thought to be of Tartar origin (Bellew 1979). In Balochistan, the Hazaras have settled mainly in Quetta and considerable numbers reside in Parachinar in the NWFP. The Punjab is the most populous province of Pakistan accounting for almost two thirds of the country's population. There is considerable ethnic admixture among the Punjabis because all the invaders who came to the Indian subcontinent passed through and mixed with the previous settlers. The major ethnic groups in the Punjab are the Jats, Rajputs, Arains, Gujars, Awans, and Meos (Ibbetson 1994).

The Parsis (Parsees) are Zoroastrians who settled in India (mainly Bombay) and Pakistan (Karachi) (Jussawalla *et al.* 1970). The Syeds (Sayyids or Sharif) are descendants of the Prophet Muhammad and therefore related to the Prophet Abraham (and thus the Iraqi Jews) (Holt *et al.* 1970).

The non-recombinant portion of the Y-chromosome has been useful in studies on human evolutionary history (Hammer 1995; Jobling and Tyler-Smith 1995; Underhill *et al.* 1996). In this study, nine of the above mentioned ethnic groups were examined for polymorphic sites on the Y-chromosome including the YAP element (DYS287), an A-G polymorphic transition -DYS271 and a tetranucleotide microsatellite DYS19. In addition 4 biallelic markers and the poly A tail associated with the YAP element were genotyped to further determine the origin of the Pakistani YAP positive individuals (Hammer *et al.* 1997).

2. MATERIALS AND METHODS

Blood samples were collected from unrelated Pakistani individuals and Epstein Barr virus (EBV) transformed lymphoblastoid cell lines were established. DNA was prepared from these cell lines by standard procedures (Sambrook *et al.* 1989). For the DYS19 study, a total of 406 individuals were examined. Of these 44 were Punjabi, 51 Burusho, 51 Pathan, 40 Kalash, 30 Hazara, 51 Brahui, 62 Balochi, 27 Makrani and 50 Sindhi. A total of 884 individuals were typed for the YAP insertion of which 97 were Baloch, 16 Baltis, 139 Brahui, 98 Burusho, 30 Hazara, 41 Kalash Kafirs, 15 Kashmiris, 27 Makrani Baloch, 32 Makrani Negroids, 86 Parsi, 112 Pathans, 46 Punjabi and 145 Sindhi. YAP positive individuals were further haplotyped in order to compare them with other world populations.

The reaction conditions and primers used for DYS19 (Y27H39) PCR amplification were those described previously (Roewer *et al.* 1992). Briefly the PCR conditions consisted of an initial denaturation at 94 °C for 2 minutes followed by 30 cycles of denaturation at 94 °C for 30 seconds, primer annealing at 55 °C for 30 seconds and primer extension at 72 °C for 1.5 minutes. After PCR amplification the amplimers were separated over-night at 10 W constant power, on a 4% polyacrylamide sequencing gel and visualized by silver staining (Bosari *et al.* 1995). The reaction conditions and primers used to detect the YAP (DYS287) were those reported earlier (Hammer 1995). After PCR amplification the samples were separated on an agarose gel and visualized following ethidium bromide staining.

3. RESULTS AND DISCUSSION

Five of the seven DYS19 alleles were found in the nine Pakistani populations. These included the alleles A (186 bp), B (190 bp), C (194 bp), D (198 bp) and E (202 bp). All nine ethnic groups had predominantly the DYS19 alleles B, C and D (Table 1). Alleles A and E were found at very low frequencies. However, unlike other populations, the Hazara exhibited a very low frequency of the C allele.

Among the northern Pakistani populations, the Punjabis and Pathans, allele C is the predominant DYSl9 allele. In the Burusho allele D was predominant whereas in the Kalash and Hazara allele B was the most common. All the four southern populations had B as the most common allele with a frequency of 0.45 in the Brahuis, 0.52 in the

Table 1. Frequency of DYS19 alleles in nine Pakistani ethnic groups

		Allelic Frequency					
Ethnic Group	n	A (186 bp)	B (190 bp)	C (194 bp)	D (198 bp)	E (202 bp)	H*j*
Northern Populations							
Punjabi	44	0.02	0.20	0.50	0.23	0.04	65.5
Burusho	51	0.02	0.20	0.31	0.37	0.10	71.7
Pathan	51	0.12	0.22	0.35	0.31	0.00	71.9
Kalash	40	0.00	0.38	0.28	0.22	0.12	71.4
Hazara	30	0.07	0.50	0.03	0.33	0.07	63.0
Southern Populations							
Brahui	51	0.06	0.45	0.39	0.08	0.02	63.5
Baloch	62	0.02	0.52	0.32	0.11	0.03	61.4
Makrani	27	0.07	0.37	0.33	0.22	0.00	70.1
Sindhi	50	0.10	0.40	0.18	0.26	0.06	72.6

Balochis, 0.37 in Makranis and 0.40 in the Sindhis. Allele C was the second most common in the Brahui (0.39), Balochi (0.32) and Makrani (0.33), whereas allele D was the second most frequent in the Sindhi (0.26). The Punjabis, Hazara, Brahui and the Baloch were the least heterozygous populations with heterozygosity values of <65.5%. The other five ethnic groups (Burusho, Pathan, Kalash, Makrani and Sindhi) had heterozygosity values more than 70.0% due to the even distribution of the DYS19 alleles in these populations. The Punjabis and the Pathans exhibited a high frequency of allele C, which is similar to other Asian populations examined (Hammer *et al.* 1997). The other two northern populations (Kalash and the Hazara) and the four southern populations (Baluch, Brahui, Makrani and the Sindhi) exhibited a high frequency of allele B. Similar high frequencies have been found in European, Middle Eastern and the North African populations.

The YAP insertion was found in the Parsis, in all the four southern populations and only in the Pathans from the north. All the other northern populations were YAP negatives (Fig. 1). The frequency of the YAP insertion was 9.4% in the Makrani Negroids, 8.2% in the Balochis, 7.4% in the Makrani Baloch tribes, 5.8% in the Parsi community, 2.2% in the Brahuis, 2.1% in the Sindhis, and 1.8% in the Pathans (Table 2).

Table 2. Frequency of YAP in various Pakistani ethnic groups

Ethnic Group	n	YAP +ive	YAP +ive (%)
Makrani Negroid	32	3	9.4
Baloch	97	8	8.2
Makrani Baloch	27	2	7.4
Parsi	86	5	5.8
Brahui	139	3	2.2
Sindhi	145	3	2.1
Pathan	112	2	1.8
Balti	16	0	0.0
Burusho	98	0	0.0
Hazara	30	0	0.0
Kalash	41	0	0.0
Kashmiri	15	0	0.0
Punjabi	46	0	0.0
TOTAL	884	26	2.9

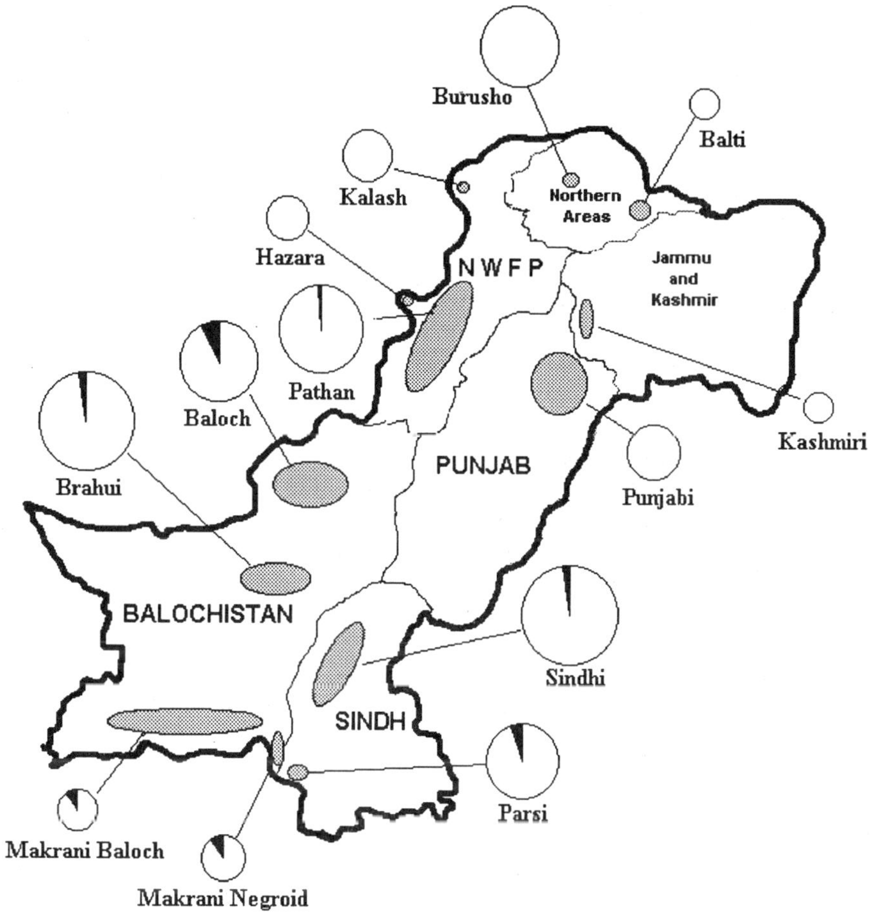

Figure 1. Distribution of YAP in Pakistan. The grey shaded area on the map represents the area of sampling. The area of the circles is proportional to the population sample size. The black area within each circle represents the percentage of YAP positive individuals.

Analysis of all these Pakistani ethnic groups reveal that the mean YAP positive frequency (2.9%) is low in comparison with the Arabs and higher compared to populations of Southeast Asia (Hammer *et al.* 1997). This low frequency in the Pakistani population follows a distinct worldwide clinal variation in YAP positive frequencies. The highest frequencies have been reported in Africa and there is a gradual decline in frequency from West to East Asia.

Further genotyping of the Pakistani YAP positive individuals was carried out by determining the prevalence of five sites associated with the YAP insertion, including PN1, PN2, SRY_{4064}, DYS271 and the poly A tail (Hammer MF 1995; Hammer MF *et al.* 1997). These five polymorphisms define YAP positive haplotypes 3, 4 and 5. Of the 8 positive Baloch samples, six had the YAP 4 haplotype while two individuals were of haplotype 5. YAP haplotype 5 was found only in the Baluch, Brahui and Makranis. The haplotype 4 was the only YAP positive haplotype observed in the Pathans and the Sindhis and was also present in the Baluch and the Makranis. YAP haplotype 3 was not found in any individual.

Of the six samples with haplotype 4, five had the DYS19 allele B, whereas the sixth had the DYS19 allele A. All the three Brahui samples were YAP haplotype 5 in association with the DYSl9 allele C. One of the Makrani sample had the YAP haplotype 4 and the DYSl9 allele A, while the other had the YAP haplotype 5 with DYSl9 allele C. All three Sindhi samples and both Pathan samples had the YAP haplotype 4 with the DYSl9 allele A. Previous studies have shown that the haplotype 4 was present predominantly in North Africa and spread from there to the Levant, Europe and West Asia. YAP haplotype 5 is present predominantly in Sub-Saharan Africa and spread from there to West Asia via a southern route through the Horn of Africa. Thus, the present day Pathans and the Sindhis are probably descendants of migrants from North Africa and Baloch, Brahui and the Makrani Baloch tribes seem to be of Sub-Saharan African origin. Analyzing the YAP haplotype data in association with the DYSl9 data further supports this. Previous surveys have shown that YAP haplotype 4 in association with the DYSl9 allele A was found at a high frequency in North Africa, at moderate frequencies in southern Europe and at low frequencies in northern Europe and Sub-Saharan Africa. The YAP haplotype 5 in association with the DYSl9 allele C was found at a high frequency in the Sub-Saharan Africa and at low frequency in North Africa. Before the present study the YAP haplotype 5 in association with the DYSl9 allele C was only observed in Africans, African Americans and one sample from Iran (Hammer *et al.* 1997). These results lend support to the hypothesis that the peopling of Pakistan has been influenced by separate migrations from West Asia, the Levant and Sub-Saharan Africa.

In addition to the Y chromosome polymorphisms, microsatellite markers provide useful information for reconstructing human phylogenetic trees reflecting human evolutionary history and migration patterns. Current opinion favours an African origin for modern humans (Cavalli-Sforza *et al.* 1994; Bowcock *et al.* 1994). A set of 35 microsatellite loci was typed across various world populations by multiplex PCR employing ABI Prism Linkage mapping sets. The results shown in Fig. 2 suggest that genetic relationships on a local geographical level may be dictated primarily by proximity rather than linguistic classification (Sun 1994; Mehdi *et al.* 1997). However, in the particular instance of the Basque and Burushaski, the phylogenetic tree provides evidence that these two

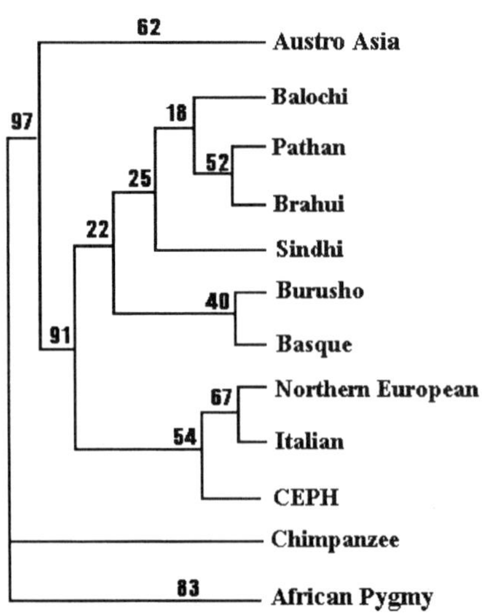

Figure 2. Neighbour joining tree based on D_{AS} genetic distance for 35 microsatellites. Bootstrap values, shown at each fork, indicate the degree of support for each branching.

language isolate groups may contain some historical connection despite the distance between them.

In conclusions, this preliminary study provides evidence that the North-South divide among the populations of Pakistan is not only due to their geographical inhabitation but these two regional populations might have originated as a result of two distinct waves of migrations, one from North Africa and other from Sub-Saharan region. There is further evidence that the invaders from the West and Eurasia might have brought groups with isolated languages and these groups due to their strict endogamy and isolation still show affinity to some of the isolated language groups of Europe. Further study with Y chromosomal SNPs haplotypes will improve our understanding about the origin of these highly diverse groups of the populations of Pakistan.

4. REFERENCES

Ansari, SSA, 1996, *The Musalman Races Found in Sindh, Baluchistan and Afghanistan*, Indus Publications, Karachi, Pakistan.

Bellew, HW, 1979, *The Races of Afghanistan*, Sang-e-Meel Publications, Lahore, Pakistan.

Biddulph, J, 1977, *Tribes of the Hindoo Koosh*, Indus Publications, Karachi. Pakistan.

Bosari, S, Marchetti, A, Battitta, F, Graziani, D, Borsani, G, Loda, M, Bevilacqua, G, Goggi, G, 1995, Detection of p53 mutations by single-strand conformation polymorphisms (SSCP) gel electrophoresis. A comparitive study of radioactive and non-radioactive silver-stained SSCP analysis, *Diagn. Mol. Pathol.* **4**:249–255.

Bowcock, AM, Ruiz-Linares, A, Tomfohrde, J, Minch, E, Kidd, JR, Cavalli-Sforza, LL, 1994, High resolution of human evolutionary trees with polymorphic microsatellites, *Nature* **368**:455–457.

Burton, RF, 1851, *Sindh and the Races that Inhabit the Valley of the Indus*, WH Allen & Co. Ltd., London, UK.

Caroe, O, 1992, *The Pathans*, Oxford University Press, Karachi, Pakistan.

Cavalli, LL, Menozzi, P, Piazza, A, 1994, The History and Geograpgy of Human genes, Princeton University Press, New Jersey, USA.

Dani, AH, 1991, *History of Northern Areas of Pakistan*, National Institute of Historical and Cultural Research, Islamabad, Pakistan.

Grimes, BF, 1992, *Ethnologue: Languages of the World*, Summer Institute of Linguistics, Inc., Dallas, Texas, USA.

Hammer, MF, 1995, A recent common ancestry for human Y chromosomes, *Nature* **378**:376–378.

Hammer, MF, Spurdle, AB, Karafet, T, Bonner, MR, Wood, ET, Novelletto, A, Malaspina, P, Mitchell, RJ, Horai, S, Jenkins, T, Zegura, SL, 1997, The geographic distribution of human Y chromosome variation, *Genetics* **145**:787–805.

Holt, PM, Lambton, AKS, Lewis, B, 1970, *The Cambridge History of Islam: Volume 2B*, Cambridge University Press, London, UK.

Hughes-Buller, R, 1991, *Imperial Gazetteer of India: Provincial Series: Baluchistan*, Sang-e-Meel Publications, Lahore, Pakistan.

Ibbetson, D, 1994, *Panjab Castes*, Sang-e-Meel Publications, Lahore, Pakistan.

Jarrige, J-F, 1991, Mehrgarh: its place in the development of ancient cultures in Pakisan, in: *Forgotten Cities on the Indus: Early Civilization in Pakistan from 8th to 2nd Millennium BC* (M, Jansen M, Mulloy M, Urban G (eds) Verlag Philipp von Zabern, Mainz, Germany, p 34–51.

Jobling, MA, Tyler-Smith, C, 1995, Fathers and sons: the Y chromosome and human evolution, *Trend. Genet.* **11**:449–456.

Jussawalla, DJ, Deshpande, VA, Haenszel, W, Natekar, MV, 1970, Differences observed in the site incidence of cancer, between the Parsi community and the total population of greater Bombay:a critical appraisal, *Br. J. Cancer* **24**:56–66.

Mehdi, SQ, Qamar, R, Khaliq, S, Mohyuddin, A, Mansoor, A, Ismail, M, Hameed, A, Ayub, Q, Sun, B, Underhill, P, Cavalli-Sforza, LL, Jin, L, 1997, Clinical and evolutionary genetics of Pakistani populations, in: *ICGEB 10th Anniversary Symposium. 1987–1997: Molecular Biology and Biotechnology for Development*, Trieste, Italy.

Quddus, SA, 1990, *The Tribal Baluchistan*, Ferozsons (Pvt.) Ltd., Lahore, Pakistan.

Robertson, GS, 1974, *The Kafirs of the Hindu-Kush*, Oxford University Press, Karachi, Pakistan.

Roewer, L, Arnemann, J, Spurr, NK, Grzeschik, KH, Epplen, JT, 1992, Simple repeat sequences on the human Y chromosome are equally polymorphic as their autosomal counterparts. *Hum. Genet.* **89**:389–394.

Sambrook, J, Fritsch, EF, Maniatis, T, 1989, *Molecular Cloning: A Laboratory Manual:1*, Cold Spring Harbor Laboratory Press, New York, USA.

Sarkar, SS, 1981, *Ancient Races of Baluchistan, Panjab and Sind*, Indus Publications, Karachi, Pakistan.

Sun, B, 1997, *The Use of Polymorphic Microsatellites to Determine Genetic Relationships Between Language Isolates and Neighboring Populations*, M. Sc. Thesis, Biology Department, Stanford University, Stanford, California, USA.

Underhill, PA, Jin, L, Zemans, R, Oefner, PJ, Cavalli-Sforza, LL, 1996, A pre-Columbian Y chromosome-specific transition and its implications for human history, *Proc. Natl. Acad. Sci. USA* **93**:196–200.

8

THE USE OF Y-CHROMOSOMAL DNA VARIATION TO INVESTIGATE POPULATION HISTORY

Recent Male Spread in Asia and Europe

Tatiana Zerjal[1], Arpita Pandya[1], Fabrício R. Santos[1], Raju Adhikari[1], Eduardo Tarazona[1], Manfred Kayser[2], Oleg Evgrafov[3], Lalji Singh[4], Kumarasamy Thangaraj[4], Giovanni Destro-Bisol[5], Mark G. Thomas[6], Raheel Qamar[7], S. Qasim Mehdi[7], Zoë H. Rosser[8], Matthew E. Hurles[8], Mark A. Jobling[8], and Chris Tyler-Smith[1]

[1]CRC Chromosome Molecular Biology Group
Department of Biochemistry
University of Oxford
South Parks Road, Oxford OX1 3QU, UK
[2]Max Planck Institute for Evolutionary Anthropology
Department of Evolutionary Genetics
Inselstrasse 22, D-04103 Leipzig, Germany
[3]Research Centre for Medical Genetics
Russian Academy of Medical Sciences
1 Moskvorechie, Moscow 115478, Russia
[4]Centre for Cellular and Molecular Biology
Uppal Road, Hyderabad 500 007, India
[5]Dipartimento di Biologia Animale e dell'Uomo
Universita "La Sapienza", Roma, Italy
[6]Center for Genetic Anthropology
The Departments of Anthropology and Biology
Darwin Building
University College London
Gower Street, London WC1E 6BT, UK
[7]Biomedical & Genetic Engineering Laboratories
P.O. Box 2891, 25 Mauve Area, G-9/1
Islamabad 44000, Pakistan
[8]Department of Genetics
University of Leicester
Adrian Building
University Road, Leicester LE1 7RH, UK

Genomic Diversity: Applications in Human Population Genetics
edited by Papiha, Deka, and Chakraborty, Kluwer Academic / Plenum Publishers, New York, 1999.

ABSTRACT

Y-chromosomal DNA lineages can be used to trace the origins of males in modern populations. A combination of biallelic markers has been used to identify "haplogroup 3" Y chromosomes, which are widespread and common in many European and Asian populations. Microsatellite analysis shows that the diversity of haplogroup 3 chromosomes is low, suggesting a recent spread.

1. INTRODUCTION

Since most of the Y chromosome does not recombine, the mutations that accumulate in Y-specific DNA as polymorphisms can be used to deduce the phylogenetic relationships of surviving chromosomes. If these mutations are unique events, the phylogeny can be represented simply as a tree; if they are recurrent, the phylogeny can still be represented, but as a more complex network. If there was complete mixing between populations at each generation, chromosomes lying in different sections of the tree or network would be distributed randomly among populations and geographical areas. Departures from random distributions are therefore due, in part, to the limited geographical mobility of the individuals carrying the Y chromosomes and can consequently provide information about past population movements and history. Since the Y chromosome is carried by men, this information is of course solely about male population history, which may differ from female.

Rare mutations producing base substitution or some insertion/deletion polymorphisms (biallelic markers) provide a convenient way of identifying Y chromosomes that have all descended from a single common ancestor, the individual in whom the mutation occurred. Although such mutations may recur, recurrence followed by increase in frequency to detectable levles will be rare. The geographical distribution of a set of chromosomes of this kind can, by itself, provide some information about population history, but more information is obtained if haplotypes combining the biallelic marker with additional markers are used. Microsatellites are particularly useful because their rapid mutation rates mean that variation will soon start to accumulate within the set of chromosomes, and will increase with time. If the microsatellite mutation rate is known (for example, from measurements in families), an estimate of the time that has elapsed since the most recent common ancestor of the chromosomes examined can be obtained, and with more uncertainty, the time that has elapsed since the mutation producing the biallelic marker.

Such considerations have led to a series of studies in which Y chromosome phylogeny is linked to the geographical distribution of Y variation. Several insights have emerged. The deepest split in the tree leads to one branch, now rare, that is restricted to Africa, and a second branch that is common in Africa and universal outside, suggesting that contemporary Y chromosomes arose in Africa (Hammer *et al.* 1998; Pandya 1998). Most African Y chromosomes (70% of the total in one survey; Hammer *et al.* 1998) carry the YAP Alu insertion, but, surprisingly, the most ancestral chromosomes on this branch have not been found in Africa but occur in Asian countries such as Japan and Tibet, leading to the suggestion that African Y chromosomes have been largely replaced by Asian ones, perhaps between about 20,000 and 30,000 years ago (Hammer *et al.* 1998). Most native South American and many North American Y chromosomes fall on a branch of the tree identified either by a combination of alphoid heteroduplex and

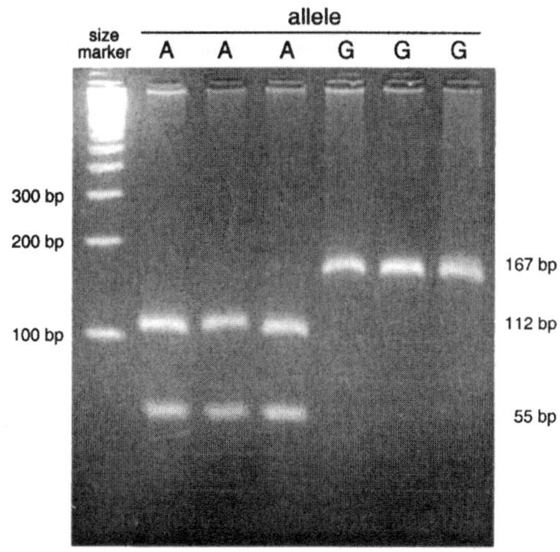

Figure 1. Assay for the *SRY-1532* polymorphism. DNA was amplified as described previously using the primers SRY1 and SRY2 (Kayser 1998; Santos *et al.* 1999), digested with *Dra*III, and fractionated by electrophoresis in a 3% NuSieve, 1% agarose gel containing ethidium bromide in 1 × TBE buffer. Tracks labelled 'A' contain the A allele, while those labelled 'G' contain the G allele.

microsatellite markers (Pena *et al.* 1995) or by a C → T transition at the *DYS199* locus (Underhill *et al.* 1996). This lineage has been traced back to a possible origin in central Siberia (Santos *et al.* 1999). In Asia and Europe, a lineage identified by the Tat T → C transition which appears to have originated recently in Asia is now widespread among northern populations (Zerjal *et al.* 1997), while another C → T transition 2627 bp upstream of the *SRY* gene identifies a smaller set of chromosomes that have begun to spread from Iberia into adjacent European countries, probably within the last 2,000 to 3,000 years (Hurles *et al.* submitted). Nevertheless, these lineages together account for only a small proportion of Asian and European Y chromosomes. In this chapter, we focus on an additional lineage that is common in many Asian and European populations and demonstrate that, despite its wide range and abundance, it shows only a low level of variability, suggesting that its extensive spread has been recent.

2. THE *SRY-1532* A/G/A TRANSITION

The A/G base substitution 1532 bp upstream of the *SRY* gene (Whitfield *et al.* 1995; Kwok *et al.* 1996) was discussed in the accompanying chapter (Tyler-Smith 1999) as an example of a recurrent base substitution. The base present in a DNA sample can conveniently be determined by PCR amplification followed by digestion with *Dra*III and agarose gel electrophoresis (Fig. 1). Samples containing the A allele produce two bands of 112 bp and 55 bp, while samples containing the G allele produce a single band of 167 bp. Both the A (ancestral; Whitfield *et al.* 1995) and the G alleles are found in association with both the ancestral and derived states of the markers M9 (Underhill *et al.* 1997) and 92R7 (Mathias *et al.* 1994). This situation can be explained more easily by recurrent mutation at the *SRY-1532* locus than by recurrent mutation at both the M9 and 92R7 loci or recombination between Y chromosomes. Chromosomes in which the *SRY-1532* A allele is associated with the ancestral states of M9 and 92R7 are designated "haplogroup 7", while chromosomes in which the A is associated with their derived states are "haplogroup 3" (Table 1, Fig. 2). Haplogroup 7 chromosomes were found among the San in Africa, but not elsewhere in the populations examined (Table 2), while haplogroup 3

Table 1. Haplogroup definitions

Haplogroup	SRY-1532	M9	92R7
7	A	C	C
3	A	G	T
other	G	C or G	C or T

chromosomes were widespread (see below). Thus, in practice, haplogroup 3 chromosomes can readily be identified, despite the recurrence of the mutation at *SRY-1532*.

3. WORLDWIDE DISTRIBUTION OF HAPLOGROUP 3 Y CHROMOSOMES

A sample of 1317 males from around the world was tested for the presence of haplogroup 3 Y chromosomes; some previously published results (Hurles *et al.* 1998; Santos *et al.* 1999) are included. 229 haplogroup 3 chromosomes were found, making up about 17% of the total. They were not found in the samples from Africa, Oceania or America, (with the exception of one chromosome that can be accounted for by recent admixture; Hurles *et al.* 1998) but were widespread in Asia and Europe (Table 2; Fig. 3). 19 out of the 22 Asian + European populations examined contained them, the only exceptions being the Basques in Europe and the Soliga (India) and Chinese in Asia. The Basques and Soliga are small isolated populations, and few individuals were tested from each, so the absence of haplogroup 3 chromosomes could be due to the origin of these populations from individuals lacking haplogroup 3 chromosomes, their subsequent loss by drift,

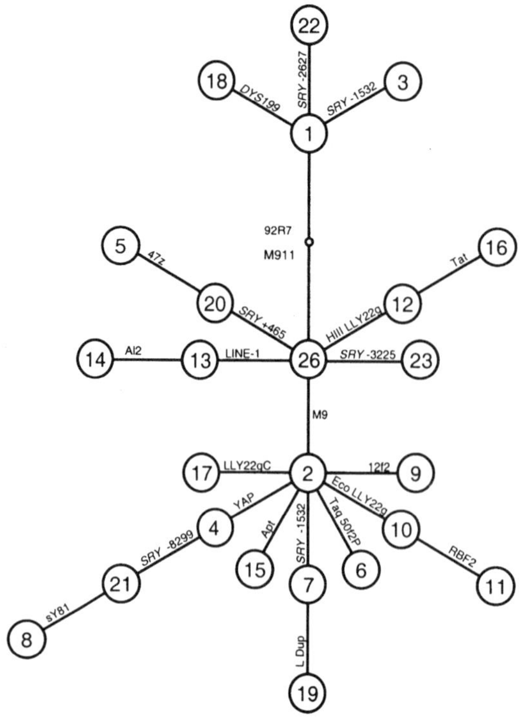

Figure 2. Network of Y-chromosomal haplogroups. Haplogroups are represented by numbers inside the circles; the polymorphisms that distinguish haplogroups are shown on the lines.

Table 2. Worldwide distribution of haplogroup 3 and haplogroup 7 Y chromosomes

Population	A (hg7)	G	A (hg3)	n	
Africa					
San	4	3	0	7	
Kenya	0	14	0	14	
Bamileke	0	40	0	40	
Pygmy	0	25	0	25	
Algeria	0	27	0	27	
Europe					
Iceland	0	22	6	28	
Norway	0	37	16	53	
Latvia	0	20	15	35	
UK	0	26	1	27	
Basque	0	26	0	26	
Germany	0	33	14	47	
Greece	0	33	3	36	
Turkey	0	20	4	24	
Russia	0	14	11	25	
Mari	0	34	14	48	
Asia					
Burusho (Pakistan)	0	68	26	94	
Sindhi (Pakistan)	0	35	38	73	
Parsi (Pakistan)	0	83	7	90	
India (Jaunpur)	0	121	31	152	
Soliga (India)	0	19	0	19	
Sri Lanka	0	18	6	24	
Nepal	0	24	15	39	
Mongolia	0	59	6	65	
Altai	0	11	12	23	Santos *et al.* 1999
Evenki	0	22	1	23	Santos *et al.* 1999
Ket	0	8	2	10	Santos *et al.* 1999
China	0	43	0	43	
Oceania					
Trobriand Islands	0	70	0	70	
Papua New Guinea	0	72	0	72	Hurles *et al.* 1998
Cook Islands	0	32	1	33	Hurles *et al.* 1998
Australia	0	3	0	3	
America					
Quechua	0	15	0	15	
Other	0	7	0	7	
Totals	4	1,084	229	1,317	

or the testing of too few individuals. It would be rash to conclude that haplogroup 3 chromosomes are absent from China after examining 43 males, but the frequency here is likely to be low.

Among the populations where haplogroup 3 chromosomes were detected, the frequency ranges from about 4% in the UK to 52% in the Sindhi from Pakistan and the Altai from Siberia. In 13 of the 22 populations, haplogroup 3 chromosomes make up one-fifth of the population or more. Thus they account for a major proportion of Asian and European Y chromosomes. Within Asia, there is considerable local variation in frequency from population to population, without a simple overall geographical pattern. This is particularly noticeable in the Pakistani populations, where the sample

Figure 3. Worldwide distribution of haplogroup 3 Y chromosomes. The population samples are represented by circles with an area proportional to sample size. Haplogroup 3 chromosomes are shown in black; all other haplogroups are in white.

sizes are large enough to provide reliable frequency estimates. Among the Parsi, the frequency was 8% (7/90), while, as noted above, it was 52% (38/73) in the Sindhi. Within Europe, there is a more even gradient in frequency from north-east (high) to south-west (low).

4. DIVERSITY OF HAPLOGROUP 3 Y CHROMOSOMES

The variation associated with 49 of the haplogroup 3 chromosomes was determined using a set of five Y-chromosomal microsatellites (Kayser *et al.* 1997): four tetranucleotide repeats (*DYS19, DYS390, DYS391* and *DYS393*), and one trinucleotide repeat (*DYS392*). *DYS392* and *DYS393* show little variation: 48/49 chromosomes have 11 and 13 repeats, respectively. *DYS19* and *DYS390* show clear unimodal distributions with maxima at 16 and 25 repeats, respectively. Only *DYS391* shows a roughly equal division between two repeat lengths, 10 and 11 (Fig. 4). These findings are consistent with descent from a recent ancestral chromosome with the haplotype (*DYS19*, *DYS390*, *DYS391*, *DYS392*, *DYS393*) 16, 25, 10, 11, 13 or 16, 25, 11, 11, 13. These haplotypes are the most common and second most common, respectively, among those analysed, again consistent with a recent origin from one of the two haplotypes. A median-joining network (Bandelt *et al.* 1999) was drawn from the 49 haplotypes. The high-frequency putative ancestral haplotypes lie towards the centre of the network, and each is linked to multiple additional haplotypes, while the lower-frequency haplotypes lie towards the outside of the network. These are additional characteristics expected of a young haplogroup in which a significant proportion of the chromosomes have not yet had time to accumulate mutations at any of the loci examined.

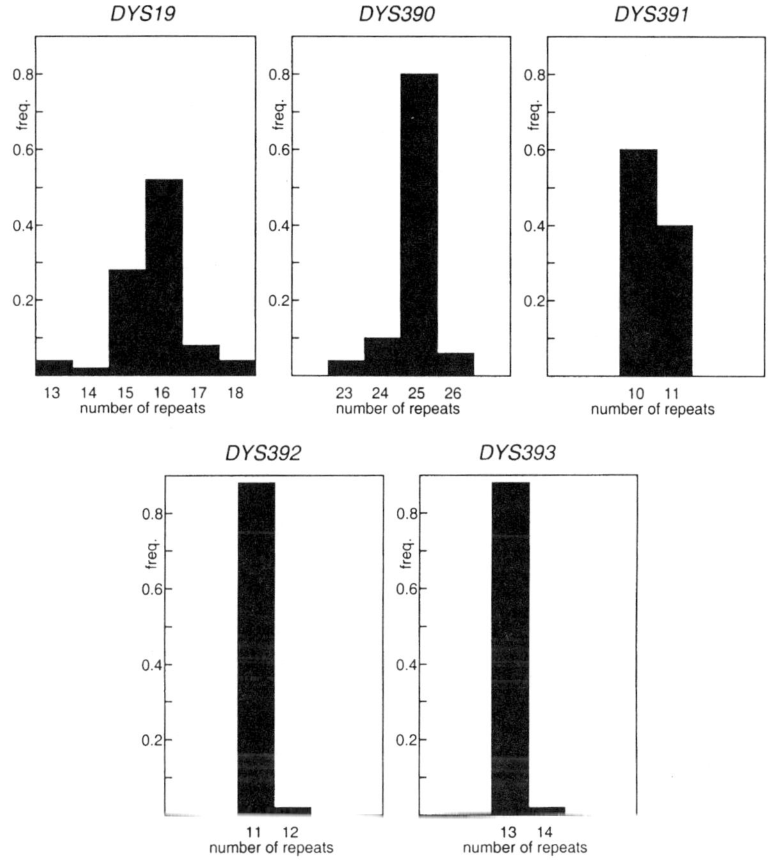

Figure 4. Frequency of microsatellite repeat size classes among 49 haplogroup 3 Y chromosomes. Microsatellite repeat numbers were determined using a single multiplex amplification for each individual (Thomas *et al.* in preparation).

The mean number of steps per chromosome per locus along the network from the putative ancestral haplotype 16, 25, 10, 11, 13 is 0.27. If the mutation rate measured by Heyer *et al.* (1997) at 0.0021 mutations per locus per generation is used (95% confidence intervals: 0.0006 to 0.0047), 0.27 steps per chromosome per locus can be translated into a time of 126 generations (95% confidence intervals 54–440). If the generation time for males is 30 years, these figures would correspond to 3 800 years (1600–13,000 years). There are several alternative ways of estimating the time that such diversity would take to accumulate. These have recently been compared for haplogroup 22 chromosomes (Hurles *et al.* submitted). Different methods gave similar results, and the largest source of error was the uncertainty in the mutation rate. Thus, while there may be additional sources of error that bias all methods, the available methods give consistent results.

5. DISCUSSION

The combination of an A allele at *SRY-1532* with the derived states for M9 and/or 92R7 identifies a unique group of Y chromosomes, haplogroup 3. These chromosomes are carried by a substantial proportion of the males in many Asian and European populations, but have low levels of diversity. These characteristics suggest a recent spread over a wide geographical area. Is it possible to identify an origin for the chromosomes and link their spread to known palaeoanthropological and archaeological data, or historical events?

Where did the chromosomes originate? The Y-chromosomal phylogeny (Fig. 2) can provide some information about this. Haplogroup 3 chromosomes are derived from haplogroup 1 chromosomes, and must therefore have arisen in a location occupied by haplogroup 1 chromosomes in the past. Haplogroup 1 chromosomes are now widespread in Asia and Europe, occurring in most of the same populations as haplogroup 3 chromosomes (results not shown). If they were present in similar areas in the past, this would suggest an origin in Asia or Europe, but does not indicate a specific location. More information can be obtained by examining haplogroup 1 microsatellite haplotypes. These are distinct in Asia and Europe (Jobling *et al.* 1997; Hurles *et al.* submitted), with European haplogroup 1 chromosomes showing low levels of diversity. In a network containing haplogroup 1 and haplogroup 3 chromosomes from both Asia and Europe, the haplogroup 3 chromosomes were linked to the Asian haplogroup 1 chromosomes, and not directly to the European haplogroup 1 chromosomes (results not shown). This suggests that haplogroup 3 arose in Asia and, therefore, that it subsequently spread from Asia to Europe, rather than vice-versa. Within the haplogroup 3 network (Fig. 5), Asian and European chromosomes are both found throughout the network and there is no evidence for localisation of haplotypes to particular geographical areas. Even the putative ancestral haplotypes are equally distributed between Asia and Europe, and microsatellite diversity is similar in the two areas. The network therefore does not suggest a specific geographical location within Asia for the origin of haplogroup 3 chromosomes.

The languages spoken by populations carrying haplogroup 3 chromosomes are diverse (Table 2), and include members of the Indo-European, Uralic, Altaic, and Dravidian families, as well as isolates such as Burushaski and Ket. Such linguistic diversity could arise from recent gene flow between populations, an origin for the haplogroup that predates the linguistic differentiation, or some combination of these.

Overall, the restricted geographical distribution of haplogroup 3 chromosomes, including their absence from Africa, implies an origin in Asia or Europe after the entry of modern humans < 60,000 years ago. The low microsatellite diversity suggests an origin <13,000 years ago, a finding consistent with the coalescent analysis of Hammer *et al.* (1998) which estimated an origin for a similar set of chromosomes ("haplotype 1D") at about 10,000 years ago. Such dates are considerably later than the initial colonisation of most parts of Asia and Europe. By 10,000 years ago, the last ice age had ended and farming was beginning. The detailed archaeological record available from Europe and the simple geographical pattern of haplogroup 3 chromosomes make it easier to look for associations in Europe than in Asia. In Europe, the earliest Neolithic sites are found in the south-east and a progressive spread towards the north-west is seen. This pattern contrasts with that seen among haplogroup 3 chromosomes in Europe, which indicates a spread from the east, but from a more northerly region (Fig. 3). While an association with the Neolithic in Asia cannot entirely be ruled out because of our lack of knowledge, the entry of haplogroup 3 chromosomes into Europe is likely to post-date the Neolithic. The distribution of haplogroup 3 chromosomes in Europe resembles, to some extent, the synthetic map of Europe obtained by Cavalli-Sforza *et al.* (1994) from 95 autosomal loci using the third principal component. It is more distinct from the distribution of Tat C allele (haplogroup 16) chromosomes (Zerjal *et al.* 1997), which have a more

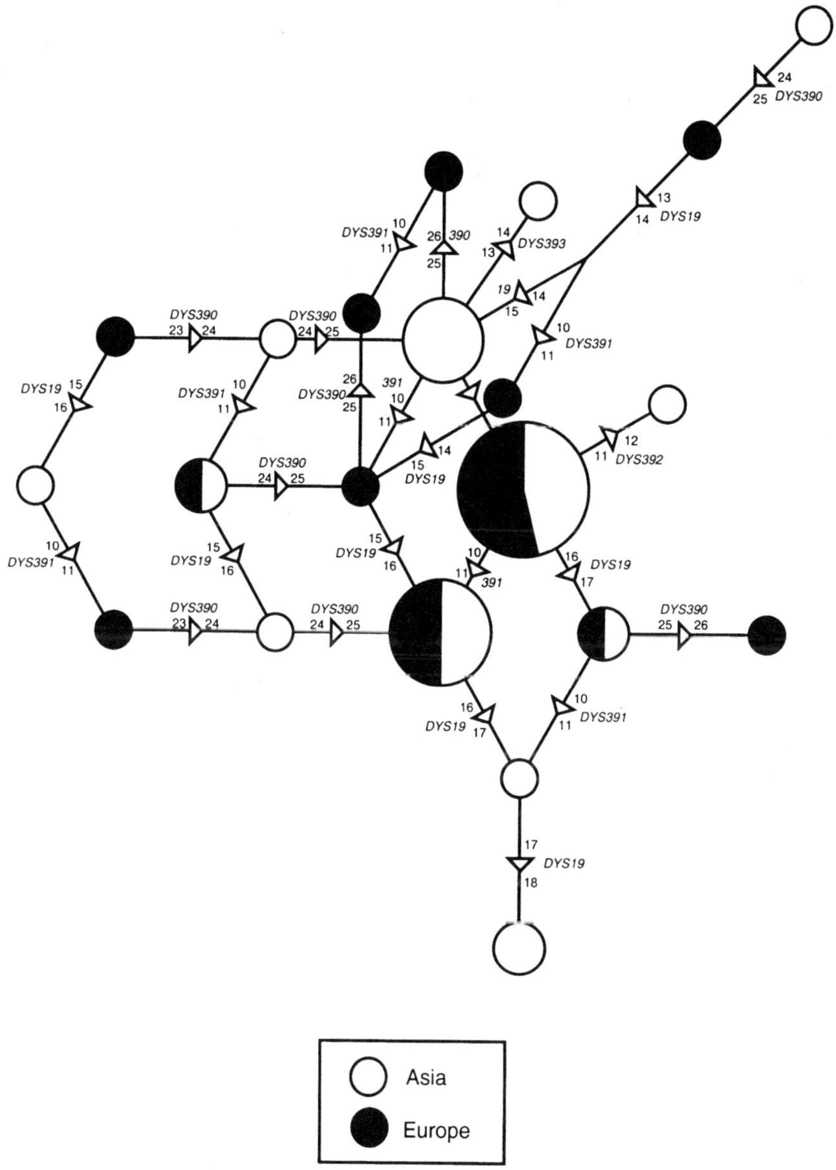

Figure 5. Median-joining network of haplogroup 3 Y chromosome haplotypes. Microsatellite haplotypes are represented by circles with an area proportional to their frequency in the sample of 49 chromosomes. Microsatellite changes are indicated on the lines joining the circles. Haplogroup 3 chromosomes from Asia are shown in white, those from Europe in black.

northern distribution and in Europe are found mainly in Uralic-speaking populations. Haplogroup 16 chromosomes may therefore have entered Europe with Uralic-speakers, although the date for this is not well-established. Haplogroup 3 chromosomes show a less sharp gradient and extend further into Europe, implying an earlier entry. One candidate event could be the Kurgan expansion postulated by Gimbutas (1970). According to this scenario, pastoral nomads who had domesticated the horse could have spread from a region to the north of the Caspian Sea bringing Indo-European languages into Europe and the Indian subcontinent. However, this is only one of several hypotheses to account for the spread of the Indo-European languages (Cavalli-Sforza *et al.* 1994, Chapter 5) and the subject continues to be debated. Y-chromosomal DNA analysis and other branches of molecular anthropology can now contribute to this debate.

6. ACKNOWLEDGMENTS

We thank the DNA donors for making this work possible, Wulf Schiefenhövel for providing the Trobriand samples and Lutz Roewer for providing the German samples, Hans Bandelt, Peter Forster and Arne Röhl for providing their Network 1.8 program, and Qasim Ayub, Lars Beckman and Andrea Novelletto for comments on the manuscript. TZ was supported by a Wellcome Trust studentship, RA by a Wellcome Trust vacation scholarship, AP by the BBSRC, FRS by The Leverhulme Trust, RQ and SQM by The Wellcome Trust, ZHR by the BBSRC, MEH by the MRC, MAJ by The Wellcome Trust and CTS by the CRC.

7. REFERENCES

Bandelt, H-J, Forster, P, Rohl, A (1999) Median-joining networks for inferring intraspecific phylogenies. *Mol. Biol. Evol.* **16**, 37–48.

Cavalli-Sforza, LL, Menozzi, P, Poazza, A (1994) *The history and geography of human genes.* Princeton University Press, Princeton.

Gimbutas, M (1970) Proto-Indo-European culture: the Kurgan culture during the fifth, fourth and third millennia BC. In *Indo-European and Indo-Europeans.* G Cardona, HM Hoenigswald, A Senn, eds., pp 155–195. University of Pennsylvania Press, Philadelphia.

Hammer, MF, Karafet, T, Rasanayagam, A, Wood, ET, Altheide, TK, Jenkins, T, Griffiths, RC, Templeton, AR, Zegura, SL (1998) Out of Africa and back again: nested cladisitc analysis of human Y chromosome variation. *Mol. Biol. Evol.* **15**, 427–441.

Heyer, E, Puymirat, J, Dieltjes, P, Bakker, E, de Knijff, P (1997). Estimation of Y chromosome specific microsatellite mutation frequencies using deep rooting pedigrees. *Hum. Mol. Genet.* **6**, 799–803.

Hurles, ME, Irven, C, Nicholson, J, Taylor, PG, Santos, FR, Loughlin, J, Jobling, MA, Sykes, BC (1998) European Y-chromosomal lineages in Polynesians: a contrast to the population structure revealed by mtDNA. *Am. J. Hum. Genet.* **63**, 1793–1806.

Jobling, MA, Pandya, A, Tyler-Smith, C (1997) The Y chromosome in forensic analysis and paternity testing. *Int. J. Legal Med.* **110**, 118–124.

Kayser, M, Caglia, A, Corach, D, Fretwell, N, Gehrig, C, Graziosi, G, Heidorn, F, Herrmann, S, Herzog, B, Hidding, M, Honda, K, Jobling, M, Krawczak, M, Leim, K, Meuser, S, Meyer, E, Oesterreich, W, Pandya, A, Parson, W, Penacino, G, Perez-Lezaun, A, Piccinini, A, Prinz, M, Schmitt, C, Schneider, PM, Szibor, R, Teifel-Greding, J, Weichhold, G, de Knijff, P, Roewer, L (1997) Evaluation of Y-chromosomal STRs: a multicenter study. *Int. J. Legal Med.* **110**, 125–133.

Kayser, M (1998) *Y-chromosomal DNA polymorphism: characterisation, population genetics and application to molecular anthropology and forensic individualisation.* Ph.D. thesis, Humboldt-University Berlin.

Kwok, C, Tyler-Smith, C, Mendonca, BB, Hughes, I, Berkovitz, GD, Goodfellow, PN, Hawkins, JR (1996) Mutation analysis of the 2 kb 5′ to *SRY* in XY females and XY intersex subjects. *J. Med. Genet.* **33**, 465–468.

Mathias, N, Bayes, M, Tyler-Smith, C (1994) Highly informative compound haplotypes for the human Y chromosome. *Hum. Mol. Genet.* **3**, 115–123.

Pandya, A (1998) *Human Y chromosomal DNA variation.* DPhil thesis, Oxford University.

Pena, SDJ, Santos, FR, Bianchi, NO, Bravi, CM, Carnese, FR, Rothhanner, F, Gerelsaikhan, T, Munkhtuya, B, Oyunsuren, T (1995) A major founder Y-chromosome haplotype in Amerindians. *Nature Genet.* **11**, 15–16.

Santos, FR, Pandya, A, Tyler-Smith, C, Pena, SDJ, Schanfield, M, Leonard, WR, Osipova, L, Crawford, MH, Mitchell, RJ (1999) The Central Siberian origin for Native American Y-chromosomes. *Am. J. Hum. Genet.* **64**, 619–628.

Tyler-Smith, C (1999) Y-chromosomal DNA markers *(this volume)*.

Underhill, PA, Jin, L, Zemans, R, Oefner, PJ, Cavalli-Sforza, LL (1996) A pre-Colombian Y chromosome-specific transition and its implications for human evolutionary history. *Proc. Natl. Acad. Sci. USA* **93**, 196–200.

Underhill, PA, Jin, L, Lin, AA, Mehdi, SQ, Jenkins, T, Vollrath, D, Davis, RW, Cavalli-Sforza, LL, Oefner, PJ (1997) Detection of numerous Y chromosome biallelic polymorphisms by denaturing high-performance liquid chromatography. *Genome Res.* **7**, 996–1005.

Whitfield, LS, Sulston, JE, Goodfellow, PN (1995) Sequence variation of the human Y chromosomes. *Nature* **378**, 379–380.

Zerjal, T, Dashnyam, B, Pandya, A, Kayser, M, Roewer, L, Santos, FR, Fretwell, N, Jobling, MA, Harihara, S, Shimizu, K, Semjidmaa, D, Sajantila, A, Salo, P, Crawford, MH, Ginter, EK, Evgrafov, OV, Tyler-Smith, C (1997) Genetic relationships of Asians and Northern Europeans, revealed by Y-chromosomal DNA analysis. *Am. J. Hum. Genet.* **60**, 1174–1183.

9

A MITOCHONDRIAL DNA DATABASE

Applications to Problems of Nomenclature and Population Genetics

P. Francalacci[1], R. Montiel[2], and A. Malgosa[2]

[1]Dip. di Zoologia e Antropologia Biologica
Università di Sassari
[2]Dep. de Biologia Animal, Biologia Vegetal i Ecologia
Universitat Autónoma de Barcelona

ABSTRACT

The analysis of the mitochondrial DNA variability has been of great interest in the study of human population history. There are some compilations of sequences and polymorphic sites in different databases. Here we present a database of the variability of the control region sequence (segment I), formatted in a Microsoft Excel file. The organisation and use of the database are described. The database can be used for easily grouping published sequences, based on the correspondence already established with RFLP based haplogroups, showing further subdivision in monophyletic clusters (subhaplogroups) and pointing out the mutation hierarchy. Guidelines for the subhaplogroup nomenclature are exposed. Finally, a possible use of the database for timing migrational events in two Mediterranean islands is discussed.

1. INTRODUCTION

Since the end of the 1980s, variability of the mitochondrial DNA (mtDNA) has been widely studied in human populations for evolutionary purposes because of its special features such as the unilinear maternal transmission, the absence of recombination, and the high mutational rate. The mitochondrial genome can be divided in two main domains: a control region (also called d-loop) of about 1,000 bp, subdivided in two hypervariable segments (HVS-I and HVS-II of about 400 bp each) spaced by a more conservative segment, and a longer coding region of about 15,000 bp, that contains 37 genes strictly packed. mtDNA evolves very fast, up to 5–10 fold higher than the nuclear

Genomic Diversity: Applications in Human Population Genetics
edited by Papiha, Deka, and Chakraborty, Kluwer Academic / Plenum Publishers, New York, 1999.

DNA (Brown *et al.* 1979) and the control region has a mutation rate about 10 times higher than the coding region (Greenberg *et al.* 1983). The substitution rate, esteemed to 0.19–3.2% per million years in the coding region (Giles *et al.* 1980; Wallace *et al.* 1987; Wallace and Torroni 1992) raises to 8.4% per million years in the control region (Greenberg *et al.* 1983; Horai and Hayasaka 1990; Vigilant *et al.* 1989). The variability is not evenly distributed within the control region. The number of polymorphic sites is about the double in the HVS-I compared to HVS-II, and within the two segments the mutation rate is site-dependent, with some extremely variable positions and some more constant sites (Wakeley 1993; Piercy *et al.* 1993).

Early studies investigated the mitochondrial polymorphisms by RFLP analysis (Cann *et al.* 1987; Torroni *et al.* 1990) on the whole molecule, although most of the polymorphisms were detected within the coding portion, being the majority of the genome. More recently, the analysis has focused on the control region, especially on HVS-I, whose variability is higher than HVS-II (Vigilant *et al.* 1989; Vigilant 1990).

The restriction analysis, involving mainly the more conservative coding region, suggests that some neutral mutations that presumably occurred only once in the evolutionary history, allowing the grouping of different lineages in clusters defined by restriction sites, called haplogroups. The haplogroups follow the alphabetic nomenclature initiated by Torroni *et al.* (1992) and are based on high-definition restriction analysis. This analysis involves the use of a set of 14 endonucleases encompassing about 15–20% of the whole mitochondrial genome (Fig. 1). The haplogroups are continent specific, and can be used to track back the migrational patterns of human population at a continent level.

The information obtained from the non-coding region can be higher than that from the coding region, and the sequence analysis, a more precise detection technique, i.e. sequencing, allows the analysis to go deeper in the microevolutionary level, up to the individual differentiation. For this reason, the number of studies on the control region variability (especially on HVS-I) has recently greatly increased, coming from different fields, such as anthropology, genetics and forensic medicine.

On the other hand, this high variability makes it possible both retromutation and independent mutations on the same site, which complicates the phylogenetic analysis of the data. In fact, a nucleotide change can be frequent in human populations due to two very different reasons: either it represents an ancient mutational event and it is present in all the individuals derived from that particular ancestor, or it is a site with an extremely high mutational rate, thus representing independent mutational events in different lineages, with no phylogenetic value. The analyses which do not take into account this effect, giving the same weight to all the nucleotide changes detected, such as pairwise analysis and maximum parsimony analysis based on d-loop sequences, give poor resolution in the study of populations at the microevolutionary level. Sequences detected in different populations were often clustered together, with no phylogenetic relationship (Cann *et al.* 1989; Piercy *et al.* 1993).

The observation that mtDNA has no recombination and both the control and the coding region are obviously inherited together, lead recently some authors to combine both RFLP analysis and d-loop sequencing on the same sample (Graven *et al.* 1995, for African populations; Torroni *et al.* 1996, for European populations; Torroni *et al.* 1993a,b, Kolman *et al.* 1996 and Castrì 1998, for Asian populations, including native Americans). The combined approach shed further light on the evolution of the mitochondrial genome, joining the advantages of both methods: the information of the sequencing is not lost, but it is arranged according to its phylogenetic value through the association with monophyletic restriction sites.

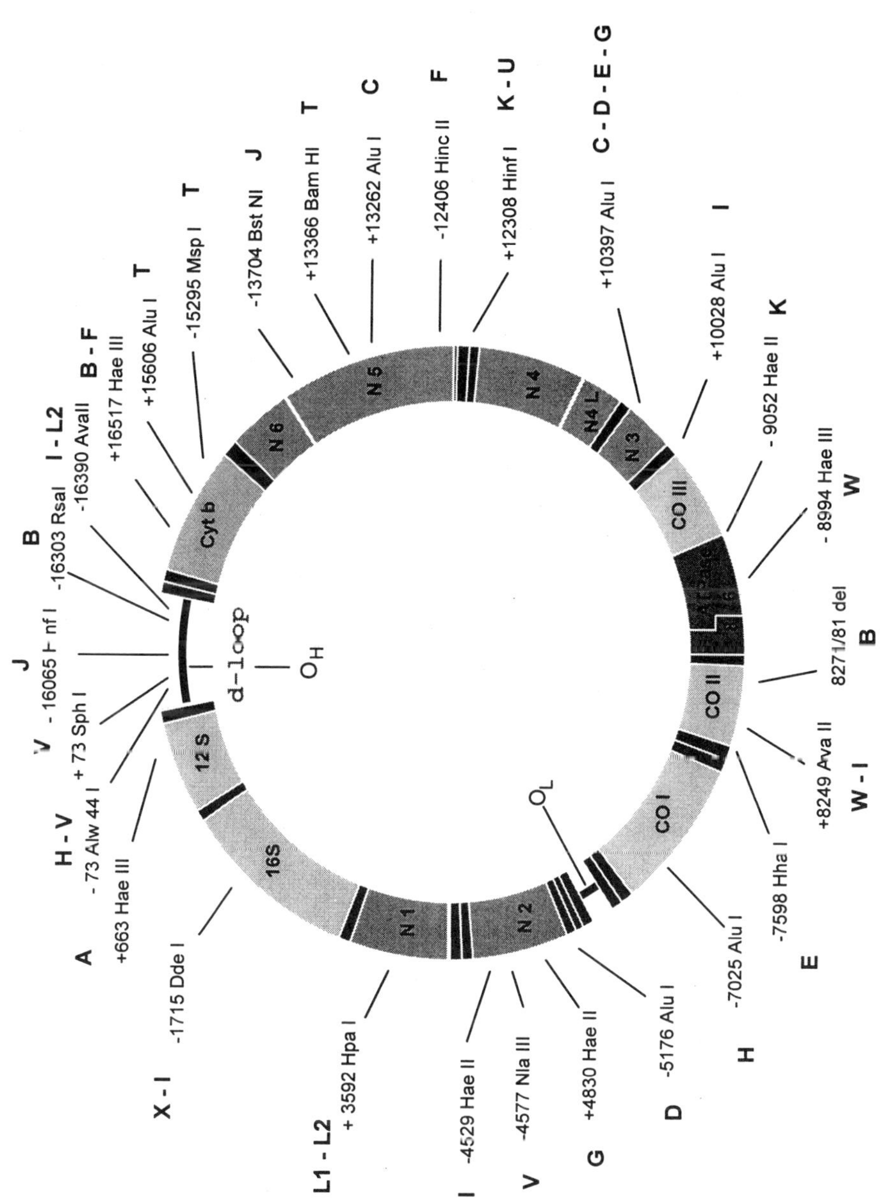

Figure 1. Localization of haplogroup specific restriction sites in the human mitochondrial genome.

Table 1 shows the established relationship between control region sequence and restriction polymorphisms in the mitochondrial genome. The univocal association of some specific control region changes with RFLP defined haplogroups allows the tentative but reasonable classification of previously published sequences even in absence of restriction analysis.

2. MITOCHONDRIAL SEQUENCE DATABASES

The amount of data on mitochondrial variability available in literature, with special reference to d-loop sequences, has increased exponentially over the last few years (Handt *et al.* 1998). Some authors compiled mtDNA variant databases, freely available to the scientific community via World Wide Web. However, the possible use of these databases is strictly dependent on the format in which the data are presented.

MITOMAP is a database compiled at the Centre for Molecular Medicine, Emory University, Atlanta, Georgia (Kogelnik *et al.* 1996, 1997, 1998), available at the address http://infinity.gen.emory.edu/mitomap.html. It reports detailed information on the mitochondrial genome, including information of evolutionary interest, such as control and coding regions sequence polymorphisms, common continent specific variants, polymorphic restriction sites (high resolution screening). A complete bibliography allows the recovery of the original published data, but the sequences are not reported and, therefore, a direct use of the data is not possible using the available software for phylogenetic analysis.

A collection of mtDNA d-loop sequences compiled by the research group of the University of Munich (Handt *et al.* 1998) is available at http://www.zi.biologie.uni-muenchen.de/~meyers/mtdna.html. This database contains the alignment of HVS-I and HVS-II sequences, numbered according to the Cambridge Reference Sequence (CRS) (Anderson *et al.* 1981) from different populations from all the continents. The data are divided into an information file (containing the information of the individual sample; geographical origin, linguistic affiliation, the original labelling and reference) and two sequence files (one for HVS-I and one for HVS-II). The sequences aligned here can be used in program packages for maximum parsimony, pairwise difference analysis and median network analysis. The sequence files report only unique lineages (which differ in at least one position) and, in the case that two or more different individuals share the same sequence, an identification code refers to the specific sequence. However, the organisation of the database, with the whole sequence in a continuous text string, and the lack of immediate correspondence between the lineage and the corresponding individual data make the application of this database difficult for retrieving the frequency of a specific mutation in human populations and for the observation of the mutation distribution within the RFLP haplogroups.

For this reason, we describe here a database directly derived from the data contained in Handt *et al.* (1998) database, but with a different arrangement of the sequence information and nucleotide data. The quality and completeness of the data rely entirely on those of the source database.

3. ORGANIZATION OF THE DATABASE

The database reports 3294 human sequences of the HVS-I from the position 16001 to 16400. Each individual is associated with the corresponding sequence, and two iden-

Table 1. Correlation between coding region restriction sites and d-loop variants in human mtDNAs

Haplogr.	Distrib.	16069	16126	16129	16148	16172	16187	16188	16189	16223	16224	16227	16230	16264	16270	16278	16290	16292	16294	16298	16304	16311	16319	16320	16327	16362	16390	16391	72	73	199	250	295	Coding Region RFLP sites
CRS		C	T	G	C	T	C	C	T	C	T	A	A	C	C	C	C	C	C	T	T	T	G	C	C	T	G	G	A	T	T	T	C	
Ancestor		.	.	.	T	.	T	.	C	T	.	.	.	.	.	T	.	.	.	.	.	C	.	T	.	.	.	.	G	.	.	.	.	
L1a	Africa	.	.	.	T	C	T	G	C	T	.	.	G	.	.	.	.	.	.	.	.	C	.	T	.	.	.	.	G	.	.	.	.	+3592 HpaI
L1b	Africa	.	C	.	.	.	T	.	C	T	.	.	.	T	T	T	.	.	.	.	.	C	.	.	.	.	.	.	G	.	.	.	.	+3592 HpaI
L2	Africa	.	.	.	.	.	.	.	.	T	.	.	.	.	.	T	.	.	.	.	.	.	.	.	.	.	A	.	G	.	.	.	.	+3592 HpaI
L3b	Africa	.	.	.	.	.	.	.	.	T	.	.	.	.	.	T	.	.	.	.	.	.	.	.	.	.	.	.	G	.	.	.	.	
L3a	Africa	.	.	.	.	.	.	.	.	T	.	.	.	.	.	.	.	.	.	.	.	.	.	.	.	.	.	.	G	.	.	.	.	
X	Eur.–Amer.	.	.	.	.	.	.	.	.	T	.	.	.	.	.	T	.	.	.	.	.	.	.	.	.	.	.	.	G	.	.	.	.	–1715DdeI
W	Europe	.	.	.	.	.	.	.	.	T	.	.	.	.	.	.	.	T	.	.	.	.	.	.	.	.	.	.	G	.	.	.	.	+8249 AvaII; –8994 HaeIII
I	Europe	.	.	A	.	.	.	.	.	T	.	.	.	.	.	.	.	.	.	.	.	.	.	.	.	.	.	A	G	.	C	C	.	+1715 DdeI; +10028 AluI; –4529 HaeII; +8249 AvaII
A	Asia–Amer.	.	.	.	.	.	.	.	.	T	.	.	.	.	.	.	T	.	.	.	.	.	A	.	.	C	.	.	G	.	.	.	.	+663 HaeIII
C	Asia–Amer.	.	.	.	.	.	.	.	.	T	.	.	.	.	.	.	.	.	.	C	.	.	.	.	T	.	.	.	G	.	.	.	.	+10397 AluI; –13259/+13262 HincII/AluI
D	Asia–Amer.	.	.	.	.	.	.	.	.	T	.	.	.	.	.	.	.	.	.	.	.	.	.	.	.	C	.	.	G	.	.	.	.	+10397 AluI; –5176 AluI
E	Asia	.	.	.	.	.	.	.	.	T	.	G	.	.	.	T	.	.	.	.	.	.	.	.	.	C	.	.	G	.	.	.	.	+10397 AluI; –7598HhaI
G	Asia	.	.	.	.	.	.	.	.	T	.	.	.	.	.	.	.	.	.	.	.	.	.	.	.	C	.	.	G	.	.	.	.	+10397 AluI; +4830/+4831 HaeII/HhaI
B	Asia–Amer.	.	.	.	.	.	.	.	C	.	.	.	.	.	.	.	.	.	.	.	.	.	.	.	.	.	.	.	G	.	.	.	.	8271-8281 9bp del; +16517 HaeIII
F	Asia	.	.	.	.	.	.	.	C	.	.	.	.	.	.	.	.	.	.	.	C	.	.	.	.	.	.	.	G	.	.	.	.	–12406 HpaI/HincII; +16517 HaeIII
J	Europe	T	C	.	.	.	.	.	.	.	.	.	.	.	.	.	.	.	.	.	.	.	.	.	.	.	.	.	G	.	.	.	T	–13704 BstNI
T	Europe	.	C	.	.	.	.	.	.	.	.	.	.	.	.	.	.	.	T	.	.	.	.	.	.	.	.	.	G	.	.	.	.	+13366 BamHI; +15606 AluI; –15925 MspI
U	Europe	.	.	.	.	.	.	.	.	.	.	.	.	.	.	.	.	.	.	.	.	.	.	.	.	.	.	.	G	.	.	.	.	+12308 HinfI
K	Europe	.	.	.	.	.	.	.	.	.	C	.	.	.	.	.	.	.	.	.	.	C	.	.	.	.	.	.	G	.	.	.	.	–9052 HaeII/HhaI; +12308 HinfI
V	Europe	.	.	.	.	.	.	.	.	.	.	.	.	.	.	.	.	.	.	C	.	.	.	.	.	.	.	.	.	C	.	.	.	–4577 NlaIII
H	Europe	.	.	.	.	.	.	.	.	.	.	.	.	.	.	.	.	.	.	.	.	.	.	.	.	.	.	.	.	.	.	.	.	–7025 AluI

tical sequences can be found in the database. The data are reported in a MS Excel spreadsheet. The first column reports the sequence code given in the source database (Handt *et al.* 1998), followed by the continent (two digits) and the population (three digits) assignation. The second column reports the full name of the population, the geographical area and the language appartenance. The third reports the labelling as in the original paper, and the bibliographic reference. The columns from 4 to 252 report the polymorphic nucleotide positions. The nucleotide is indicated when it is different from the CRS, otherwise it is represented by a dash. Since the number of polymorphic sites detected in all the HVS-I sequences exceeded the maximum number of columns allowed by the program (256 columns), the polymorphic sites at the extremities of the sequence were unified into two columns, one with the polymorphic sites from 16401 to 16430 and one from 16381 to 16400. The length mutations (insertions and deletions) have been excluded from the database.

The aim of this database is to give an immediate view of the variability of mtDNA in a format comparable to the one used for publication, to allow the easy grouping of lineages sharing specific point mutations, and to determine the frequency of a nucleotide change in humans. For this reason, we decided to use a database format easy to manage and sufficiently flexible in order to export the data in different format for further analysis, selecting a commercial program widely distributed and easily accessible although not explicitly written for this application. The limitation to 256 columns is of little importance, especially considering that the most informative nucleotides are usually not located at the edges of the HVS-I.

4. AVAILABILITY OF THE DATABASE

The complete database, and subsets of smaller databases divided per continents, are available in compressed format (*.zip) as winzip archives. The files can be requested to antropos@ssmain.uniss.it or to montiel@habilis.uab.es or can be downloaded over the internet at the following addresses: http://www.uniss.it/~sciweb/dizab/database.html, http://www.uab.es/~iban0/database.html.

New HVS-I sequences can be submitted for the inclusion in the database at the above mentioned e-mail addresses.

5. HOW TO USE THE DATABASE

The standard Excel commands can be used to manage the sequences present in the database. To retrieve sequences showing a specific mutation is advisable to follow the path: sort > sort by > header row > and subsequently select the desired position. To sort the sequences with up to three positions, mark “then by”. In case, when further subdivisions are necessary, we advise to copy the already selected sequences in a new file and to repeat the sort commands.

6. APPLICATION OF THE DATABASE

Since the main purpose of the database is to present only the polymorphic sites of the HVS-I, its applications are different from those of the previously published databases.

The organisation of one site per column (except for the first and last columns) allows a better visualisation and reorganisation of the data. In addition, it permits a better way to search individuals either by their sequence, population or language appartenance. In a general sense, the database is useful for finding out the occurrence of specific substitutions in specific populations, although the database could have certain number of particular applications. Among the various usages of the database here described, we point out two possible applications:

1) Haplogroup classification and nomenclature;
2) Individuation of population specific mutations.

6.1. Haplogroup Classification and Nomenclature

The database can be used for easily grouping published sequences, according to the correspondence already established with RFLP based haplogroups (Table 1).

In some cases, two different haplogroups cannot be distinguished on the base of the HVS-I polymorphisms. This happens with the following haplogroups: D-G; L3a-M; L3b-X; U-H. However, since most of the haplogroups are continent specific, the geographic origin of the samples allows the correct haplogroup assignation, with the exception of the D-G haplogroups (both Asian) and U-H haplogroups (both Europeans). In this case, in absence of restriction and/or HVS-II data, the sequences cannot be separated.

The haplogroups nomenclature has been established over the time, as soon as new sequences and RFLP patterns were available, and before a general review of the phylogeny of the human mtDNA. Haplogroups were defined in the case that the restriction patterns clustered an appreciable number of lineages, and were named with an alphabetic letter (Torroni *et al.* 1992, 1994, 1996). For this reason, the haplogroup definition (intended as a cluster of related sequences) is arbitrary. Some haplogroups fall in a higher taxonomic rank in relation to others. Further research showed, for example, that the Amerindian haplogroups C and D are subclusters of the larger Asian haplogroup M, the haplogroup U includes the haplogroup K, and the two European haplogroups J and T have a common ancestor, that could be named pre-JT. A general revision with a strict phylogenetic approach of the haplogroup names would probably be confusing, given the already established and well known alphabetic nomenclature. However, at least the nomenclature at a lower level than haplogroup should avoid possible ambiguities.

The haplogroups subdivision of the sequences allows the differentiation of the control region variants that are relatively ancient and stable from those that are recent or have undergone to repeated mutations. The hierarchy of the mutations within haplogroups permits the subdivision in subgroups whose origin is more recent, and whose distribution may be different in respect to the haplogroup as a whole. The study of the subhaplogroups can shed light on local migration patterns and can help to date the occupation of a geographical region.

To avoid the nomenclature problems at the subhaplogroup level, a cladistic approach has been proposed (Richards *et al.* 1998), according to the following scheme: capital letter $\subset$ number $\subset$ small letter $\subset$ number, such as A $\subset$ A1 $\subset$ A1a $\subset$ A1a1. The subclusters are specified by gradually more recent mutations in respect to those characterising the haplogroup.

However, the proposed nomenclature presents the problem that some subhaplogroups are specified by multiple mutations, while it is very likely that the origin of the mutations was subsequent in the time, and it is theoretically possible to individualise a further hierarchy in the clusters, raising problems of nomenclature. As an example, Table 2 (derived from the sequence reported in the database) shows the nomenclature for the subgroups of haplogroup J1, characterised by the transitions 16069T, 16126C and 16261T. Richards *et al.* (1988) proposed a further subdivision, named J1a (specified by two other mutations 16145A and 16231C), and J1b (defined by the mutations 16145A and 16222T). However, it seems very likely that the transition 16145A is ancestral to the others, since it is common to both clusters, and there is at least one individual harbouring this mutation only (without both 16231C and 16222T), that can be considered the ancestral sequence. In this case, the subhaplogroups should be more opportunely renamed as in Table 2. In general, a subdivision should be specified by a single mutation, while in the case that the mutations specifying a cluster are more than one, the nomenclature should go ahead for the corresponding number of steps, to leave the possibility to name the ancestral lineages once the new available data could clearly point out the hierarchy in the mutations.

The application of this rule might appear redundant in case that all the ancestral lineages have been lost by drift, as probably could have happened in the case of the subhaplogroup T1, defined by the pattern 16163G-16186T-16189C, since no intermediate lineages have never been observed. However renaming it T1a1 (following the above mentioned rule) would still save the cladistic approach and would keep open the possibility of naming as T1 or T1a the ancestral lineages in case that they would be actually retrieved. In fact, this rule will be useful when the absence of intermediate lineages is due only to lack of available data.

The clusters can be individualised using the sorting option of the program managing the database, and the application of the hierarchical rule allows the definition of the subhaplogroups, some of them already defined (Richards *et al.* 1998, Macaulay *et al.* 1999), and other newly proposed here (Table 3). A Reduced Median Network graph shows the phylogenetic relationships among haplogroups (Fig. 2).

For the subhaplogroup definition, the following guidelines should be followed:

a) To define a subcluster by a single mutation;
b) To select clusters showing internal variability;
c) To avoid, whenever possible, the choice of "fast" sites.

The rule a), as previously exposed, allows avoiding nomenclature problems in case of further phylogenetical information as soon as new sequences become available, rule b) avoids the selection of cluster made from individuals strictly related among them, and the rule c) limits the possibility of polyphyletic clusters due to recurrent mutations. The transitions 16129-16189-16311-16362 are considered "fast" (high mutation rate) (Hasegawa *et al.* 1993; Richards *et al.* 1998, Wakeley 1993) and should be taken into account for subhaplogroup characterisation only when other features point out the monophyletic origin of the cluster.

6.2. Individuation of Population Specific Mutations

As an example of the application of the use of the database, we report here a study of the peopling of the Mediterranean islands of Corsica and Sardinia. We analysed 242 unrelated healthy subjects: 56 were Corsicans collected from Corte and its neighbourhood, 96 were Sardinians, 51 collected from Gallura, a Northern area linguistically and historically related with Corsica, and 45 samples from Barbagia, the hilly central area of

Table 2. Nomenclature of the haplogroup J desumed from sequences included in the database

		16063	16069	16093	16114	16126	16145	16147	16150	16163	16172	16176	16183	16189	16192	16193	16201	16222	16231	16256	16260	16261	16265	16274	16278	16287	16291	16296	16300	16304	16309	16311	16319	16327	16355	16362	16363	16366	16368
0528 EU CRS	G. Britain	T	C	T	C	T	G	C	C	A	T	C	A	T	C	C	C	C	T	C	C	C	A	G	C	C	C	C	A	T	A	T	G	C	C	T	C	C	T
[pre-JT]		.	.	.	.	[C]	.	.	.	.	.	.	.	.	.	.	.	.	.	.	.	.	.	.	.	.	.	.	.	.	.	.	.	.	.	.	.	.	.
[J]		.	[T]	.	.	[C]	.	.	.	.	.	.	.	.	.	.	.	.	.	.	.	.	.	.	.	.	.	.	.	.	.	.	.	.	.	.	.	.	.
[J1]		.	[T]	.	.	[C]	.	.	.	.	.	.	.	.	.	.	.	.	.	.	.	[T]	.	.	.	.	.	.	.	.	.	.	.	.	.	.	.	.	.
0658 EU SwN	Switzerland	.	T	.	.	C	.	.	.	.	.	.	.	.	.	.	.	.	.	.	.	T	.	.	.	.	.	.	.	.	.	.	.	.	.	.	.	.	.
0682 EU Eur	Europe	.	T	C	.	C	.	.	.	.	.	.	.	.	.	.	.	.	.	.	.	T	.	A	.	.	.	.	.	.	.	.	.	.	.	.	.	.	.
1084 EU Can	Canarian	.	T	.	.	C	.	.	.	G	.	.	.	.	.	.	.	.	.	.	.	T	.	.	.	.	.	.	.	.	.	.	.	.	T	.	.	.	.
2251 EU SpN	N. Spain	?	T	.	.	C	.	.	.	.	.	.	.	.	.	.	.	.	.	.	.	T	.	.	.	.	.	.	.	.	.	.	.	.	.	.	.	.	.
1642 AS Ana	Anatolia	.	T	.	.	C	.	.	.	.	.	.	.	.	.	.	.	.	.	.	.	T	.	.	.	.	.	.	.	.	.	.	.	.	.	.	.	.	.
[J1a]		.	[T]	.	.	[C]	[A]	.	.	.	.	.	.	.	.	.	.	.	.	.	.	[T]	.	.	.	.	.	.	.	.	.	.	.	.	.	.	.	.	.
1641 AS Ana	Anatolia	.	T	.	.	C	A	.	.	.	.	.	.	.	.	.	.	.	.	.	.	T	.	.	.	T	.	.	.	.	.	.	.	.	.	.	.	.	.
[J1a1]		.	[T]	.	.	[C]	[A]	.	.	.	.	.	.	.	.	.	.	.	[C]	.	.	[T]	.	.	.	.	.	.	.	.	.	.	.	.	.	.	.	.	.
0538 EU Bri	G. Britain	.	T	.	.	C	A	.	.	.	.	.	.	.	.	.	.	.	C	.	.	T	.	.	.	.	.	.	.	.	.	.	.	.	.	.	.	.	.
0553 EU Bri	G. Britain	.	T	.	.	C	A	.	.	.	.	.	.	.	.	.	.	.	C	.	.	T	.	.	.	.	.	.	.	.	.	.	.	.	.	.	.	.	.
0794 EU Dan	Danmark	.	T	.	.	C	A	.	.	.	.	.	.	.	.	.	.	.	C	.	.	T	.	.	.	.	.	.	.	.	.	.	.	.	.	.	.	.	.
0812 EU Ger	Friesland	.	T	.	.	C	A	.	.	.	.	.	.	.	.	.	.	.	C	.	.	T	.	.	.	.	.	.	.	.	.	.	.	.	.	.	.	.	.
0498 EU Spa	Spain	.	T	.	.	C	A	.	.	.	.	.	.	.	.	.	.	.	C	.	.	T	.	.	.	.	.	.	.	.	.	.	.	.	.	.	.	.	.
0498 AF Ber	Berber	.	T	.	.	C	A	.	.	.	.	.	.	.	.	.	.	.	C	.	.	T	.	.	.	.	.	.	.	.	.	.	.	.	.	.	.	.	.
[J1a1.1]		.	[T]	.	.	[C]	[A]	.	.	.	.	.	.	[C]	.	.	.	.	[C]	.	.	[T]	.	.	.	.	.	.	.	.	.	.	.	.	.	.	.	.	.
0995 EU Bas	Basque	.	T	.	.	C	A	.	.	.	.	.	.	C	.	.	.	.	C	.	.	T	.	.	.	.	.	.	.	.	.	.	.	.	.	.	.	.	.
0622 EU SwO	Switzerland	.	T	.	.	C	A	.	.	.	.	.	C	C	.	.	.	.	C	.	.	T	.	.	.	.	.	.	.	.	.	.	.	.	.	.	.	.	.
0622 EU SwO	Switzerland	.	T	.	.	C	A	.	.	.	.	.	C	C	.	.	.	.	C	.	.	T	.	.	.	.	.	.	.	.	.	.	.	.	.	.	.	.	.
0622 EU SwO	Switzerland	.	T	.	.	C	A	.	.	.	.	.	C	C	.	.	.	.	C	.	.	T	.	.	.	.	.	.	.	.	.	.	.	.	.	.	.	.	.
[J1a2]		.	[T]	.	.	[C]	[A]	.	.	.	.	.	.	.	.	.	.	[T]	.	.	.	[T]	.	.	.	.	.	.	.	.	.	.	.	.	.	.	.	.	.
1173 EU And	Andalusia	?	T	.	.	C	A	.	.	.	.	.	.	.	.	.	.	T	.	.	.	T	.	.	.	.	.	.	.	.	.	.	.	.	.	.	.	.	.
1173 EU And	Andalusia	?	T	.	.	C	A	.	.	.	.	.	.	.	.	.	.	T	.	.	.	T	.	.	.	.	.	.	.	.	.	.	.	.	.	.	.	.	.
1174 EU Tus	Tuscany	?	T	.	.	C	A	.	.	.	.	.	.	.	.	.	.	T	.	.	.	T	.	A	.	.	.	.	.	.	.	.	.	.	.	.	.	.	.
[J1a2.1]		.	[T]	.	.	[C]	[A]	.	.	.	[C]	.	.	.	.	.	.	[T]	.	.	.	[T]	.	.	.	.	.	.	.	.	.	.	.	.	.	.	.	.	.
0608 EU Bri	G. Britain	.	T	.	.	C	A	.	.	.	C	.	.	.	.	.	.	T	.	.	.	T	.	.	.	.	.	.	.	.	.	.	.	.	.	.	.	.	.
0943 EU Ice	Iceland	.	T	.	.	C	A	.	.	.	C	.	.	.	.	.	.	T	.	.	.	T	.	.	.	.	.	.	.	.	.	.	.	.	.	.	.	.	.
0937 EU Ice	Iceland	.	T	.	.	C	A	.	.	.	C	.	.	.	T	.	.	T	.	.	.	T	.	.	.	.	.	.	.	.	.	.	.	.	.	.	.	.	.
0937 EU Ice	Iceland	.	T	.	.	C	A	.	.	.	C	.	.	.	T	.	.	T	.	.	.	T	.	.	.	.	.	.	.	.	.	.	.	.	.	C	.	.	.
1640 AS Ana	Anatolia	.	T	.	.	C	A	.	.	.	C	.	.	.	.	.	.	.	.	.	.	T	.	.	.	.	.	.	.	.	.	.	.	.	.	C	.	.	.
[J2]		.	[T]	.	.	[C]	.	.	.	.	.	.	.	.	.	[T]	.	.	.	.	.	.	.	.	.	.	.	.	.	.	.	.	.	.	.	.	.	.	.
[J2a]		.	[T]	.	.	[C]	.	.	.	.	.	.	.	.	.	[T]	.	.	.	.	.	.	.	.	[T]	.	.	.	.	.	.	.	.	.	.	.	.	.	.

Table 3. Haplogroup and subhaplogroup specific nucleotide changes in HVS-1

	16051	16069	16093	16111	16124	16126	16129	16140	16145	16148	16153	16163	16168	16172	16186	16187	16188	16189	16193	16216	16217	16219	16222	16223	16224	16227	16230
CRS.	A	C	T	C	T	T	G	T	G	C	G	A	C	T	C	C	C	T	C	A	T	A	C	C	T	A	A
Anc.	.	.	.	.	.	.	.	.	.	T	.	.	.	.	.	T	.	C	.	.	.	.	.	T	.	.	.
L1a	.	.	.	.	.	.	.	.	.	T	.	.	.	C	.	T	G	C	.	.	.	.	.	T	.	.	G
L1a1	.	.	.	.	.	.	A	.	.	T	.	.	.	C	.	T	G	C	.	.	.	.	.	T	.	.	G
L1a1.1	.	.	.	.	.	.	A	.	.	T	.	.	T	C	.	T	G	C	.	.	.	.	.	T	.	.	G
L1b	.	.	.	.	.	C	.	.	.	.	.	.	.	.	.	T	.	C	.	.	.	.	.	T	.	.	.
L1b1	.	.	.	.	.	C	.	.	.	.	.	.	.	.	.	T	.	C	.	.	.	.	.	T	.	.	.
L2	.	.	.	.	.	.	.	.	.	.	.	.	.	.	.	.	.	.	.	.	.	.	.	T	.	.	.
L3b	.	.	.	.	.	.	.	.	.	.	.	.	.	.	.	.	.	.	.	.	.	.	.	T	.	.	.
L3b1	.	.	.	.	.	.	.	.	.	.	.	.	.	.	.	.	.	.	.	.	.	.	.	T	.	.	.
L3b1.1	.	.	.	.	.	.	.	.	.	.	.	.	.	.	.	.	.	.	.	.	.	.	.	T	.	.	.
L3b2	.	.	.	.	.	.	.	.	.	.	.	.	.	.	.	.	.	.	.	.	.	.	.	T	.	.	.
L3a	.	.	.	.	.	.	.	.	.	.	.	.	.	.	.	.	.	.	.	.	.	.	.	T	.	.	.
L3a1	.	.	.	.	C	.	.	.	.	.	.	.	.	.	.	.	.	.	.	.	.	.	.	T	.	.	.
L3a2	.	.	.	.	.	.	.	.	.	.	.	.	.	.	.	.	.	.	.	.	.	.	.	T	.	.	.
L3a3	.	.	.	.	.	.	.	.	.	.	.	.	.	.	.	.	.	.	.	.	.	.	.	T	.	.	.
X	.	.	.	.	.	.	.	.	.	.	.	.	.	.	.	.	.	.	.	.	.	.	.	T	.	.	.
W	.	.	.	.	.	.	.	.	.	.	.	.	.	.	.	.	.	.	.	.	.	.	.	T	.	.	.
I	.	.	.	.	.	.	A	.	.	.	.	.	.	.	.	.	.	.	.	.	.	.	.	T	.	.	.
A	.	.	.	.	.	.	.	.	.	.	.	.	.	.	.	.	.	.	.	.	.	.	.	T	.	.	.
A1	.	.	.	T	.	.	.	.	.	.	.	.	.	.	.	.	.	.	.	.	.	.	.	T	.	.	.
A1a	.	.	.	T	.	.	.	.	.	.	.	.	.	.	.	.	.	.	.	.	.	.	.	T	.	.	.
A1b	.	.	.	T	.	.	.	.	.	.	.	.	.	.	.	.	.	.	.	.	.	.	.	T	.	.	.
A1c	.	.	.	T	.	.	.	.	.	.	.	.	.	.	.	T	.	.	.	.	.	.	.	T	.	.	.
C	.	.	.	.	.	.	.	.	.	.	.	.	.	.	.	.	.	.	.	.	.	.	.	T	.	.	.
C1	.	.	.	.	.	.	.	.	.	.	.	.	.	.	.	.	.	.	.	.	.	.	.	T	.	.	.
D	.	.	.	.	.	.	.	.	.	.	.	.	.	.	.	.	.	.	.	.	.	.	.	T	.	.	.
D1	.	.	.	.	.	.	.	.	.	.	.	.	.	.	.	.	.	.	.	.	.	.	.	T	.	.	.
E	.	.	.	.	.	.	.	.	.	.	.	.	.	.	.	.	.	.	.	.	.	.	.	T	.	G	.
G	.	.	.	.	.	.	.	.	.	.	.	.	.	.	.	.	.	.	.	.	.	.	.	T	.	.	.
B	.	.	.	.	.	.	.	.	.	.	.	.	.	.	.	.	.	C	.	.	.	.	.	.	.	.	.
B1	.	.	.	.	.	.	.	.	.	.	.	.	.	.	.	.	.	C	.	.	C	.	.	.	.	.	.
B1a	.	.	.	.	.	.	.	.	.	.	.	.	.	.	.	.	.	C	.	.	C	.	.	.	.	.	.
B2	.	.	.	.	.	.	.	.	.	.	.	.	.	.	.	.	.	C	.	C	.	.	.	.	.	.	.
B2a	.	.	.	.	.	.	.	.	.	.	.	.	.	.	.	.	.	C	.	c	.	.	.	.	.	.	.
B2a1	.	.	.	.	.	.	.	.	.	.	.	.	.	.	.	.	.	C	.	C	.	.	.	.	.	.	.
B3	.	.	.	.	.	.	.	C	.	.	.	.	.	.	.	.	.	C	.	.	.	.	.	.	.	.	.
B3a	.	.	.	.	.	.	.	C	.	.	.	.	.	.	.	.	.	C	.	.	.	.	.	.	.	.	.
B3b	.	.	.	.	.	.	.	C	.	.	.	.	.	.	.	.	.	C	.	.	.	.	.	.	.	.	.
F	.	.	.	.	.	.	.	.	.	.	.	.	.	.	.	.	.	C	.	.	.	.	.	.	.	.	.
F1	.	.	.	.	.	.	.	.	.	.	.	.	.	C	.	.	.	C	.	.	.	.	.	.	.	.	.
JT	.	.	.	.	.	C	.	.	.	.	.	.	.	.	.	.	.	.	.	.	.	.	.	.	.	.	.
J	.	T	.	.	.	C	.	.	.	.	.	.	.	.	.	.	.	.	.	.	.	.	.	.	.	.	.
J1	.	T	.	.	.	C	.	.	.	.	.	.	.	.	.	.	.	.	.	.	.	.	.	.	.	.	.
J1a	.	T	.	.	.	C	.	.	A	.	.	.	.	.	.	.	.	.	.	.	.	.	.	.	.	.	.
J1a1	.	T	.	.	.	C	.	.	A	.	.	.	.	.	.	.	.	.	.	.	.	.	.	.	.	.	.
J1a1.1	.	T	.	.	.	C	.	.	A	.	.	.	.	.	.	.	.	C	.	.	.	.	.	.	.	.	.
J1a2	.	T	.	.	.	C	.	.	A	.	.	.	.	.	.	.	.	.	.	.	.	.	T	.	.	.	.
J1a2.1	.	T	.	.	.	C	.	.	A	.	.	.	.	C	.	.	.	.	.	.	.	.	T	.	.	.	.
J2	.	T	.	.	.	C	.	.	.	.	.	.	.	.	.	.	.	.	T	.	.	.	.	.	.	.	.
J2a	.	T	.	.	.	C	.	.	.	.	.	.	.	.	.	.	.	.	T	.	.	.	.	.	.	.	.
T	.	.	.	.	.	C	.	.	.	.	.	.	.	.	.	.	.	.	.	.	.	.	.	.	.	.	.
T1a1	.	.	.	.	.	C	.	.	.	.	.	G	.	.	T	.	.	C	.	.	.	.	.	.	.	.	.
T2	.	.	.	.	.	C	.	.	.	.	.	.	.	.	.	.	.	.	.	.	.	.	.	.	.	.	.
T2a	.	.	.	.	.	C	.	.	.	.	.	.	.	.	.	.	.	.	.	.	.	.	.	.	.	.	.
T2a1	.	.	.	.	.	C	.	.	.	.	.	.	.	.	.	.	.	.	.	.	.	.	.	.	.	.	.
U	.	.	.	.	.	.	.	.	.	.	.	.	.	.	.	.	.	.	.	.	.	.	.	.	.	.	.
U1	.	.	.	.	.	.	.	.	.	.	.	.	.	.	.	.	.	C	.	.	.	.	.	.	.	.	.
U2	G	.	.	.	.	.	C	.	.	.	.	.	.	.	.	.	.	.	.	.	.	.	.	.	.	.	.
U3	.	.	.	.	.	.	.	.	.	.	.	.	.	.	.	.	.	.	.	.	.	.	.	.	.	.	.
U4	.	.	.	.	.	.	.	.	.	.	.	.	.	.	.	.	.	.	.	.	.	.	.	.	.	.	.
U5	.	.	.	.	.	.	.	.	.	.	.	.	.	.	.	.	.	.	.	.	.	.	.	.	.	.	.
U6	.	.	.	.	.	.	.	.	.	.	.	.	.	C	.	.	.	.	.	.	.	G	.	.	.	.	.
K	.	.	.	.	.	.	.	.	.	.	.	.	.	.	.	.	.	.	.	.	.	.	.	.	C	.	.
K1	.	.	C	.	.	.	.	.	.	.	.	.	.	.	.	.	.	.	.	.	.	.	.	.	C	.	.
V	.	.	.	.	.	.	.	.	.	.	.	.	.	.	.	.	.	.	.	.	.	.	.	.	.	.	.
V1	.	.	.	.	.	.	.	.	.	.	A	.	.	.	.	.	.	.	.	.	.	.	.	.	.	.	.
H	.	.	.	.	.	.	.	.	.	.	.	.	.	.	.	.	.	.	.	.	.	.	.	.	.	.	.

16231	16243	16247	16249	16261	16264	16266	16270	16278	16290	16292	16293	16294	16296	16298	16304	16309	16311	16319	16320	16325	16327	16343	16355	16356	16360	16362	16390	16391
T	T	A	T	C	C	C	C	C	C	C	A	C	C	T	T	A	T	G	C	T	C	A	C	T	C	T	G	G
.	.	.	.	.	.	.	.	T	.	.	.	.	.	.	.	.	C	.	T	.	.	.	.	.	.	.	.	.
.	.	.	.	.	.	.	.	.	.	.	.	.	.	.	.	.	C	.	T	.	.	.	.	.	.	.	.	.
.	.	.	.	.	.	.	.	.	.	.	.	.	.	.	.	.	C	.	T	.	.	.	.	.	.	.	.	.
.	.	.	.	.	.	.	.	.	.	.	.	.	.	.	.	.	C	.	T	.	.	.	.	.	.	.	.	.
.	.	.	.	.	T	.	T	T	.	.	.	.	.	.	.	.	C	.	.	.	.	.	.	.	.	.	.	.
.	.	.	.	.	T	.	T	T	.	.	G	.	.	.	.	.	C	.	.	.	.	.	.	.	.	.	.	.
.	.	.	.	.	.	.	.	T	.	.	.	.	.	.	.	.	.	.	.	.	.	.	.	.	.	.	A	.
.	.	.	.	.	.	.	.	T	.	.	.	.	.	.	.	.	.	.	.	.	.	.	.	.	.	.	.	.
.	.	.	.	.	.	.	.	T	.	.	.	T	.	.	.	.	.	.	.	.	.	.	.	.	.	.	.	.
.	.	.	.	.	.	.	.	T	.	.	.	T	.	.	.	G	.	.	.	.	.	.	.	.	.	.	.	.
.	.	.	.	.	.	.	.	T	.	.	.	.	.	.	.	.	.	.	.	.	.	.	.	.	.	C	.	.
.	.	.	.	.	.	.	.	.	.	.	.	.	.	.	.	.	.	.	.	.	.	.	.	.	.	.	.	.
.	.	.	.	.	.	.	.	.	.	.	.	.	.	.	.	.	.	.	.	.	.	.	.	.	.	.	.	.
.	.	.	.	.	.	.	.	.	.	.	.	.	.	.	.	.	.	.	T	.	.	.	.	.	.	.	.	.
.	.	.	.	.	.	.	.	.	.	.	.	.	.	.	.	.	T	.	.	.	.	.	.	.	.	.	.	.
.	.	.	.	.	.	.	.	T	.	.	.	.	.	.	.	.	.	.	.	.	.	.	.	.	.	.	.	.
.	.	.	.	.	.	.	.	.	.	T	.	.	.	.	.	.	.	.	.	.	.	.	.	.	.	.	.	.
.	.	.	.	.	.	.	.	.	.	.	.	.	.	.	.	.	.	.	.	.	.	.	.	.	.	.	.	A
.	.	.	.	.	.	.	.	.	T	.	.	.	.	.	.	.	.	A	.	.	.	.	.	.	.	C	.	.
.	.	.	.	.	.	.	.	.	T	.	.	.	.	.	.	.	.	A	.	.	.	.	.	.	.	C	.	.
.	.	.	.	.	.	.	.	.	T	.	.	.	.	.	.	.	.	A	.	.	.	.	T	.	.	C	.	.
.	.	.	.	.	.	.	.	.	T	.	.	.	.	.	.	.	.	A	.	.	.	.	.	.	T	C	.	.
.	.	.	.	.	.	.	.	.	T	.	.	.	.	.	.	.	.	A	.	.	.	.	.	.	.	C	.	.
.	.	.	.	.	.	.	.	.	.	.	.	.	.	C	.	.	.	.	.	.	T	.	.	.	.	.	.	.
.	.	.	.	.	.	.	.	.	.	.	.	.	.	C	.	.	.	.	.	C	T	.	.	.	.	.	.	.
.	.	.	.	.	.	.	.	.	.	.	.	.	.	.	.	.	.	.	.	.	.	.	.	.	.	C	.	.
.	.	.	.	.	.	.	.	.	.	.	.	.	.	.	.	.	.	.	.	C	.	.	.	.	.	C	.	.
.	.	.	.	.	.	.	.	T	.	.	.	.	.	.	.	.	.	.	.	.	.	.	.	.	.	C	.	.
.	.	.	.	.	.	.	.	.	.	.	.	.	.	.	.	.	.	.	.	.	.	.	.	.	.	C	.	.
.	.	.	.	.	.	.	.	.	.	.	.	.	.	.	.	.	.	.	.	.	.	.	.	.	.	.	.	.
.	.	.	.	.	.	.	.	.	.	.	.	.	.	.	.	.	.	.	.	.	.	.	.	.	.	.	.	.
.	.	.	.	T	.	.	.	.	.	.	.	.	.	.	.	.	.	.	.	.	.	.	.	.	.	.	.	.
.	.	.	.	.	.	.	.	.	.	.	.	.	.	.	.	.	.	.	.	.	.	.	.	.	.	.	.	.
.	.	.	.	T	.	.	.	.	.	.	.	.	.	.	.	.	.	.	.	.	.	.	.	.	.	.	.	.
.	.	G	.	T	.	.	.	.	.	.	.	.	.	.	.	.	.	.	.	.	.	.	.	.	.	.	.	.
.	.	.	.	.	.	.	.	.	.	.	.	.	.	.	.	.	.	.	.	.	.	.	.	.	.	.	.	.
.	C	.	.	.	.	.	.	.	.	.	.	.	.	.	.	.	.	.	.	.	.	.	.	.	.	.	.	.
.	.	.	.	.	.	A	.	.	.	.	.	.	.	.	.	.	.	.	.	.	.	.	.	.	.	.	.	.
.	.	.	.	.	.	.	.	.	.	.	.	.	.	.	C	.	.	.	.	.	.	.	.	.	.	.	.	.
.	.	.	.	.	.	.	.	.	.	.	.	.	.	.	C	.	.	.	.	.	.	.	.	.	.	.	.	.
.	.	.	.	.	.	.	.	.	.	.	.	.	.	.	.	.	.	.	.	.	.	.	.	.	.	.	.	.
.	.	.	.	.	.	.	.	.	.	.	.	.	.	.	.	.	.	.	.	.	.	.	.	.	.	.	.	.
.	.	.	.	T	.	.	.	.	.	.	.	.	.	.	.	.	.	.	.	.	.	.	.	.	.	.	.	.
.	.	.	.	T	.	.	.	.	.	.	.	.	.	.	.	.	.	.	.	.	.	.	.	.	.	.	.	.
C	.	.	.	T	.	.	.	.	.	.	.	.	.	.	.	.	.	.	.	.	.	.	.	.	.	.	.	.
C	.	.	.	T	.	.	.	.	.	.	.	.	.	.	.	.	.	.	.	.	.	.	.	.	.	.	.	.
.	.	.	.	T	.	.	.	.	.	.	.	.	.	.	.	.	.	.	.	.	.	.	.	.	.	.	.	.
.	.	.	.	T	.	.	.	.	.	.	.	.	.	.	.	.	.	.	.	.	.	.	.	.	.	.	.	.
.	.	.	.	.	.	.	.	.	.	.	.	.	.	.	.	.	.	.	.	.	.	.	.	.	.	.	.	.
.	.	.	.	.	.	.	.	C	.	.	.	.	.	.	.	.	.	.	.	.	.	.	.	.	.	.	.	.
.	.	.	.	.	.	.	.	.	.	.	T	.	.	.	.	.	.	.	.	.	.	.	.	.	.	.	.	.
.	.	.	.	.	.	.	.	.	.	.	T	.	.	.	.	.	.	.	.	.	.	.	.	.	.	.	.	.
.	.	.	.	.	.	.	.	.	.	.	T	T	.	.	.	.	.	.	.	.	.	.	.	.	.	.	.	.
.	.	.	.	.	.	.	.	.	.	.	T	T	.	.	C	.	.	.	.	.	.	.	.	.	.	.	.	.
.	.	.	.	.	.	.	.	.	.	.	T	.	.	.	C	.	.	.	.	.	.	.	.	.	.	.	.	.
.	.	.	.	.	.	.	.	.	.	.	.	.	.	.	.	.	.	.	.	.	.	.	.	.	.	.	.	.
.	.	.	C	.	.	.	.	.	.	.	.	.	.	.	.	.	.	.	.	.	.	.	.	.	.	.	.	.
.	.	.	.	.	.	.	.	.	.	.	.	.	.	.	.	.	.	.	.	.	.	.	.	.	.	.	.	.
.	.	.	.	.	.	.	.	.	.	.	.	.	.	.	.	.	.	.	.	.	.	G	.	.	.	.	.	.
.	.	.	.	.	.	.	.	.	.	.	.	.	.	.	.	.	.	.	.	.	.	.	.	C	.	.	.	.
.	.	.	.	.	.	.	T	.	.	.	.	.	.	.	.	.	.	.	.	.	.	.	.	.	.	.	.	.
.	.	.	.	.	.	.	.	.	.	.	.	.	.	.	.	.	.	.	.	.	.	.	.	.	.	.	.	.
.	.	.	.	.	.	.	.	.	.	.	.	.	.	.	.	.	C	.	.	.	.	.	.	.	.	.	.	.
.	.	.	.	.	.	.	.	.	.	.	.	.	.	.	.	.	C	.	.	.	.	.	.	.	.	.	.	.
.	.	.	.	.	.	.	.	.	.	.	.	.	.	C	.	.	.	.	.	.	.	.	.	.	.	.	.	.
.	.	.	.	.	.	.	.	.	.	.	.	.	.	C	.	.	.	.	.	.	.	.	.	.	.	.	.	.
.	.	.	.	.	.	.	.	.	.	.	.	.	.	.	.	.	.	.	.	.	.	.	.	.	.	.	.	.

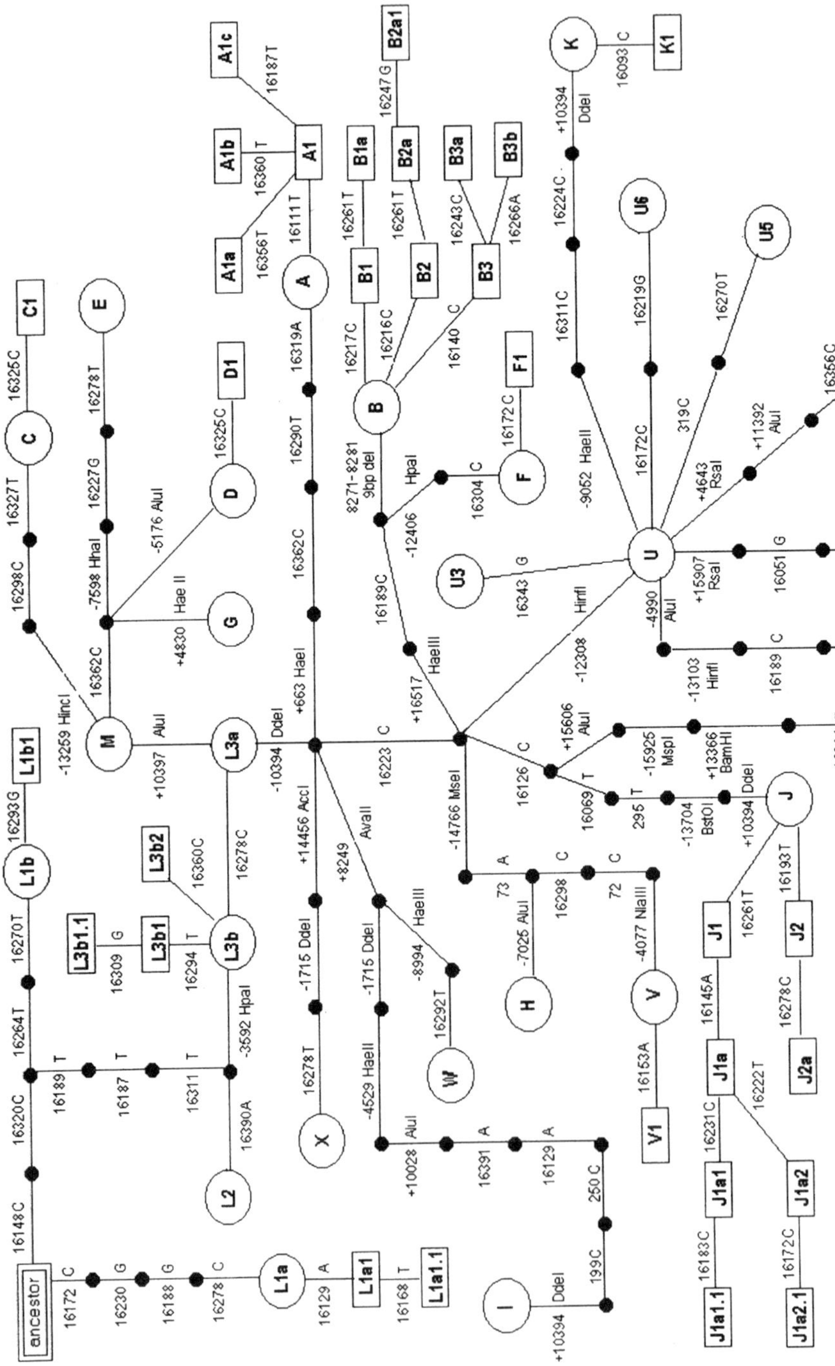

Figure 2. Phylogeny of the haplogroups and subhaplogroups in human populations. Legend: ○ = haplogroup; □ = subhaplogroup; □ = node.

the island, culturally and genetically very isolated (Cavalli-Sforza *et al.* 1994), and 90 from central Catalonia, an area of the Iberian peninsula known for having contributed to the peopling of the western Mediterranean islands. All maternal grandmothers of the examined individuals were born in the same area. The data were compared with 48 individuals from Tuscany (geographically and historically close to the two islands), that were previously analysed (Torroni *et al.* 1996) with the same methodology.

The samples were submitted to low-resolution restriction analysis, by amplifying and cleaving the mtDNA fragments harbouring haplogroup specific mutations. The haplogroup frequencies in the five populations are reported in Table 4. All but 12 mtDNAs (95.0%) can be grouped in eight of the nine Caucasian haplogroups, labelled H, I, J, K, T, U, V, W and X, confirming that these haplogroups encompass virtually all the mitochondrial lineages present in Europe. The lacking haplogroup W shows a low frequency in European populations, and its absence could be due to the sample size.

No differentiation among Tuscan, Corsican and Gallurese (Northern Sardinia) populations could be noted, neither with regard to the frequency of the more common haplogroups, nor to the haplogroup distribution pattern. This observation confirms the remarkable homogeneity in Europe for mtDNA, already noted with sequencing analysis of the d-loop (Francalacci *et al.* 1996). In spite of the homogeneity of mtDNA variability, a geographical pattern can be observed in Barbagia (Central Sardinia), where the haplogroup H shows a very high frequency, similar to the frequency of H (estimated by d-loop sequences) reported in various populations of N-W Europe, such as Galicians (Salas *et al.* 1998) Basques (Bertranpetit *et al.* 1995; Corte-Real *et al.* 1996) and Britons (Piercy *et al.* 1993). In addition, haplogroup V, originated in south-western Europe about 12,000–16,000 years ago from haplogroup H and later spread to Central-Northern Europe and North Africa (Torroni *et al.* 1998), has been observed both in the Catalonian and in the Barbagian sample, but it is absent in Gallura, Corsica and Tuscany. The East to West cline on the distribution of haplogroup H is a result of an ancient population growth, probably related to the Palaeolithic occupation and the lineages related to this haplogroup are more frequent in Western European regions, and in the isolated areas of Sardinia. A connection between Atlantic European populations and the first occupants of Sardinia is also shown by the presence of haplogroup V in Barbagia. On the other hand, people from Tuscany contributed significantly in historical period at the peopling of Corsica and, secondarily, Gallura, as reflected by the mtDNA haplogroup distribution, that has been enriched by the lineages coming from Middle East,

Table 4. Frequency distribution of haplogroups in five Southern European populations

Haplogroup	Tuscany (n = 48)	Corsica (n = 56)	Gallura (n = 51)	Barbagia (n = 45)	Catalonia (n = 90)
H	41.7	42.9	41.2	64.4	45.7
I	4.2	0.0	3.9	0.0	0.0
J	14.6	10.7	9.8	2.2	13.3
K	6.3	12.5	5.9	2.2	3.3
T	10.4	8.9	9.8	8.9	13.3
U	10.4	10.7	27.5	4.5	13.3
V	0.0	0.0	0.0	8.9	3.3
W	2.1	0.0	0.0	0.0	0.0
X	8.3	8.9	0.0	0.0	2.2
others	2.1	5.4	2.0	8.9	5.6

Table 5. Sequence diversity of HVS-1 in Gallura (Northern Sardinia)

		16029	16030	16042	16051	16069	16093	16126	16129	16134	16147	16172	16189	16191	16192	16193	16208	16209	16214	16219	16223	16224	16227	16239	16249	16256	16257	16261	16265	16270	16278	16292	16294	16296	16298	16304	16311	16343	16356
Ref		T	A	G	A	C	T	T	G	C	C	T	T	C	C	C	G	T	C	A	C	T	A	C	T	C	C	C	A	C	C	C	C	C	T	T	T	A	T
[I]		.	.	.	.	.	.	.	[A]	.	.	.	.	.	.	.	.	.	.	.	[T]	.	.	.	.	.	.	.	.	.	.	.	.	.	.	.	.	.	.
	G11	.	.	.	.	.	.	.	A	.	T	.	.	.	.	.	.	.	.	.	T	.	.	.	.	.	.	.	.	.	.	.	.	.	.	.	.	.	.
[J]		.	.	.	.	[T]	.	[C]	.	.	.	.	.	.	.	.	.	.	.	.	.	.	.	.	.	.	.	.	.	.	.	.	.	.	.	.	.	.	.
	G5	.	.	.	.	T	.	C	.	.	.	.	.	.	.	.	.	.	.	.	.	.	.	.	.	.	.	.	.	.	.	.	.	.	.	.	.	.	.
[J2a]		.	.	.	.	[T]	.	[C]	.	.	.	.	.	.	.	[T]	.	.	.	.	.	.	.	.	.	.	.	.	.	.	[T]	.	.	.	.	.	.	.	.
	G21	.	.	.	.	T	C	C	.	.	.	.	.	.	.	T	.	.	.	.	.	.	.	.	.	.	.	.	.	.	T	.	.	.	.	.	.	.	.
	G9	.	.	.	.	T	C	C	.	.	.	.	.	.	.	T	.	.	.	.	.	.	.	.	.	.	.	.	.	.	T	.	.	.	.	.	.	.	.
	G2	.	.	.	.	T	C	C	.	.	.	.	.	.	.	T	A	.	.	.	.	.	.	.	.	.	.	.	.	.	T	.	.	.	.	.	.	.	.
[T2a]		.	.	.	.	.	.	[C]	.	.	.	.	.	.	.	.	.	.	.	.	.	.	.	.	.	.	.	.	.	.	.	.	[T]	[T]	.	[C]	.	.	.
	G15	.	.	.	.	.	.	C	.	.	.	.	.	.	.	.	.	.	.	.	.	.	.	.	.	.	.	.	.	.	.	.	T	T	.	C	.	.	.
[T2a1]		.	.	.	.	.	.	[C]	.	.	.	.	.	.	.	.	.	.	.	.	.	.	.	.	.	.	.	.	.	.	.	.	[T]	[illegible]	.	[C]	.	.	.
	G17	.	.	.	.	.	.	C	.	.	.	.	.	T	.	.	.	.	T	.	.	.	.	.	.	.	.	.	.	.	.	.	T	.	.	C	.	.	.
	G8	.	.	.	.	.	.	C	.	.	.	.	.	.	.	.	.	.	.	.	.	.	.	.	.	.	.	.	.	.	.	.	T	.	.	C	.	.	.
[U]		.	.	.	.	.	.	.	.	.	.	.	.	.	.	.	.	.	.	.	.	.	.	.	.	.	.	.	.	.	.	.	.	.	.	.	.	.	.
	G28	C	T	.	.	.	.	.	.	.	.	.	.	.	.	.	.	.	.	.	.	.	.	.	.	.	.	.	.	.	.	.	.	.	.	.	.	.	.
[U1]		.	.	.	.	.	.	.	.	.	.	.	C	.	.	.	.	.	.	.	.	.	.	.	C	.	.	.	.	.	.	.	.	.	.	.	.	.	.
	G22	.	.	.	.	.	.	.	A	.	.	.	.	.	.	T	.	.	.	.	.	.	.	.	[C]	.	.	.	.	.	.	.	.	.	.	.	.	.	.
[U5]		.	.	.	.	.	.	.	.	.	.	.	.	.	.	.	.	.	.	.	.	.	.	.	.	.	.	.	.	[T]	.	.	.	.	.	.	.	.	.
	G6	.	.	.	.	.	.	.	.	.	.	.	C	.	T	.	.	.	.	.	.	.	.	.	.	.	.	.	.	T	.	.	.	.	.	.	.	.	.
	G16	.	.	.	.	.	.	.	.	.	.	.	C	.	T	.	.	.	.	.	.	.	.	T	.	.	.	.	.	T	.	.	.	.	.	C	.	.	.
	G27	.	.	.	.	.	.	.	.	.	.	.	.	.	T	.	.	.	.	.	.	.	G	.	.	.	.	.	.	T	.	.	.	.	.	.	.	.	.
	G23	.	.	.	.	.	.	.	.	.	.	.	.	.	T	.	.	.	.	.	.	.	.	.	.	.	.	.	.	T	.	.	.	.	.	.	.	.	.
	G25	.	.	.	.	.	.	.	.	.	.	.	.	.	.	.	.	.	.	.	.	.	.	.	.	T	.	.	.	T	.	.	.	.	.	.	.	.	.
[K]		.	.	.	.	.	.	.	.	.	.	.	.	.	.	.	.	.	.	.	.	[C]	.	.	.	.	.	.	.	.	.	.	.	.	.	.	C	.	.
	G3	.	.	.	.	.	.	.	.	.	.	.	.	.	.	.	.	.	.	.	.	C	.	.	.	.	.	.	.	.	.	.	.	.	.	.	[C]	.	.
[H]		.	.	.	.	.	.	.	.	.	.	.	.	.	.	.	.	.	.	.	.	.	.	.	.	.	.	.	.	.	.	.	.	.	.	.	.	.	.
	G4	.	.	.	.	.	.	.	.	.	.	.	.	.	.	.	.	.	.	.	.	.	.	.	.	.	.	.	.	.	.	.	.	.	.	.	.	.	.
	G7	.	.	A	.	.	.	.	.	T	.	.	.	.	.	.	.	.	.	.	.	.	.	.	.	.	.	.	.	.	.	.	.	.	.	.	.	.	.
	G10	.	.	.	.	.	.	.	.	.	.	.	.	.	.	.	.	.	.	.	.	.	.	.	.	.	.	.	.	.	.	.	.	.	.	.	.	.	.
	G12	.	.	.	.	.	C	.	.	.	.	.	.	.	.	.	.	.	.	.	.	.	.	.	.	.	.	.	G	.	.	.	.	.	.	.	.	.	.
	G13	.	.	.	.	.	.	.	.	.	.	.	.	.	.	.	.	.	.	.	.	.	.	.	.	.	T	.	.	.	.	.	.	.	.	.	.	.	.
	G14	.	.	.	.	.	.	.	.	.	.	.	.	.	.	.	.	.	.	.	.	.	.	.	.	.	.	.	.	.	.	.	.	.	.	.	.	.	.
	G20	.	.	.	.	.	.	C	.	.	.	.	.	.	.	.	.	.	.	.	.	.	.	.	.	T	.	.	.	.	.	.	.	.	.	.	.	.	.
	G26	.	.	.	.	.	C	.	.	.	.	.	.	.	.	.	.	.	.	.	.	.	.	.	.	.	.	.	G	.	.	.	.	.	.	.	.	.	.
	G29	.	.	.	.	.	.	.	.	.	.	.	.	.	.	.	.	.	.	.	.	.	.	.	.	.	.	.	.	.	.	.	.	.	.	.	.	.	.
	G30	.	.	.	.	.	.	.	.	.	.	.	.	.	.	.	.	.	.	.	.	.	.	.	.	.	.	.	.	.	.	.	.	.	.	.	.	.	.
[Other]	G1	.	.	.	.	.	.	.	.	.	.	.	.	.	.	.	.	C	.	.	.	.	.	.	.	.	.	T	.	.	.	.	.	.	.	.	.	.	.

following the Neolithic expansion (Comas *et al.* 1996; Francalacci *et al.* 1996; Richards *et al.* 1996).

The peopling of the two Mediterranean islands seems to have undergone through at least two separate waves, a first one of upper Palaeolithic origin, mostly represented in the mitochondrial variability in central Sardinia, and a second one more recent, coming from the Italian peninsula, with a more significant impact in Corsica and Northern Sardinia. It would be interesting to attempt a dating of the two peopling events. Obviously, the newcomers reflected a mitochondrial variability older than the actual timing of colonisation of the two islands. However, a mutation arising subsequently to the occupation (thus absent in the source population) can give origin to a cluster of lineages (or subhaplogroup) whose internal variability can be used for dating the minimum time of the arrival in the area, since the population should have been in place at least for sufficient period to develop the observed variability within the cluster. To assess this possibility, we sequenced 27 individuals from the Gallurese sample (Table 5).

No specific pattern could be observed within the sequences belonging to haplogroup H, I and K. Most of the lineages classified into the haplogroup U by RFLP analysis, belong to the subhaplogroup U5 (characterised by the transitions 16270T and 16192T) and two of them have also the transition 16189C. A control in the database indicates that this mutation is not characteristic of Northern Sardinia, since it has been observed also in other European samples. Similar consideration can be given to the samples belonging to haplogroup T. The sequences show that they can be assigned to the subhaplogroups T2a and T2a1, whose distribution is widespread in Europe. More interestingly, 3 out of 4 of the samples belonging to haplogroup J present the transition 16104C that individuates a separate cluster. A survey in the database shows that this mutation is not present elsewhere. As a matter of fact, this cluster appears to be a local evolution of the subhaplogroup J2a (and can be named J2a1), and the subsequent mutation 16207A shown by the individual G2 must have occurred after the arrival in Northern Sardinia. The sample size in this case is too low for a reasonable estimate of the cluster expansion date, but this approach seems to be valid when further sequencing information becomes available.

7. ACKNOWLEDGMENTS

We are indebted to Toomas Kisivild, Loredana Castrì and Martin Richards for useful discussion during the preparation of the manuscript. The experimental work was supported by MURST 60%, ex.40%, CNR 97.00712.PF36 (L. Terrenato) grants and Italo-Spanish Integrated Action Project.

8. REFERENCES

Anderson S, Bankier AT, Barrel BG, De Brujin MH, Coulson AR, Sanger F, Schreier PH, Smith AJH, Staden R, Joung G (1981) Sequence and organization of the human mitochondrial genome. Nature 290:457–465.

Bertranpetit J, Sala J, Calafell F, Underhill PA, Moral P, Comas D (1995) Human mitochondrial DNA variation and the origin of Basques. Ann Hum Genet 59:63–81.

Brown WM, George MM, Wilson AC (1979) Rapid evolution of animal Mitochondrial DNA. Proc Natl Acad Sci USA 76:1967–1971.

Cann RL, Stoneking M, Wilson AC (1987) Mitochondrial DNA and human evolution. Nature 325:31–36.

Castrì L (1998) PhD thesis, University of Bologna.

Cavalli-Sforza LL, Menozzi P, Piazza A (1994) The history and geography of human genes. Princeton University Press, Princeton.

Comas D, Calafell F, Mateu E, Pérez-Lezaun A, Bertranpetit J (1996) Geographic variation in human mitochondrial DNA control region sequence: the population history of Turkey and its relationship to the European populations. Mol Biol Evol 13:1067–1077.

Côrte-Real HBSM, Macaulay VA, Richards MB, Hariti G, Issad MS. Cambon-Thomsen A, Papiha S, Bertranpetit J, Sykes BD (1996) Genetic diversity in the Iberian peninsula determined from mitochondrial sequence analysis. Ann Hum Genet 60:331–350.

Francalacci P, Bertranpetit J, Calafell F, Underhill PA (1996) Sequence diversity of the control region of mitochondrial DNA in Tuscany and its implications for the peopling of Europe. American Journal of Physical Anthropology 100:443–460.

Giles RE, Blanc H, Cann HM, Wallace DC (1980). Maternal inheritance of human mitochondrial DNA Proc Natl Acad Sci USA 77:6715–6719.

Greenberg BD, Newbold JE, Sugino A (1983) Intraspecific nucleotide sequence variability surrounding the origin of replication in human mitochondrial DNA. Gene 21:33–49.

Graven L, Passarino G, Semino O, Boursot P, Santachiara-Benerecetti S, Langaney A, Excoffier L (1995) Evolutionary correlation between control region sequence and restriction polymorphisms in the mitochondrial genome of a large Senegalese Mandenka population. Mol Biol Evol 12:334–335.

Handt O, Meyer S, von Haessler A (1998) Compilation of human mtDNA control region sequences. Nucleic Acids Res 26:126–129.

Hasegawa M, Di Rienzo A, Kocher TD, Wilson AC (1993) Toward a more accurate time scale for the human mitochondrial DNA tree. J Mol Evol 37:347–354.

Horai S, Hayasaka K (1990) Intraspecific nucleotide sequence differences in the major noncoding region of human mitochondrial DNA. Am J Hum Genet 462:828–842.

Kogelnik AM, Lott MT, Brown MD, Navathe SB, Wallace DC (1996) MITOMAP: a human mitochondrial genome database. Nucleic Acids Res 24:177–179.

Kogelnik AM, Lott MT, Brown MD, Navathe SB, Wallace DC (1997) MITOMAP: an update on the status of the human mitochondrial genome database. Nucleic Acids Res 25:196–199.

Kogelnik AM, Lott MT, Brown MD, Navathe SB, Wallace DC (1998) MITOMAP: a human mitochondrial genome database—1998 update. Nucleic Acids Res 26:112–115.

Kolman CJ, Sambuughin N, Bermingham E (1996) Mitochondrial DNA analysis of Mongolian populations and implications for the origin of New World founders. Genetics 142:1321–1334.

Piercy R, Sullivan KM, Benson N, Gill P (1993) The application of mitochondrial DNA typing to the study of white Caucasian genetic identification. Int J Leg Med 106:85–90.

Richards M, Côrte-Real H, Forster P, Macaulay V, Wilkinson-Herbots H, Demaine A, Papiha S, Hedges R, Bandelt H-J, Sykes B (1996) Paleolithic and Neolithic lineages in European mitochondrial gene pool. Am J Hum Genet 59:185–203.

Richards MB, Macaulay VA, Bandelt H-J, Sykes BC (1998) Plylogeography of mitochondrial DNA in western Europe. Ann Hum Genet 62:241–260.

Salas A, Comas D, Lareu MV, Bertranpetit J, Carracedo A (1998) mtDNA analysis of the Galician population: a genetic edge of European variation. Eur J Hum Genet 6:365–375.

Torroni A, Scurr TG, Cabell MF, Brown MD, Neel JV, Larsen M, Smith DG, Vullo CM, Wallace DC (1993a) Asian affinities and continental radiation of the four founding Native American mtDNAs. Am J Hum Genet 53:563–590.

Torroni A, Sukernik RI, Schurr TG, Starikovskaya YB, Cabell MF, Crawford MH, Comuzzie AG, Wallace DC (1993b) mtDNA variation aboriginal Siberians reveals distinct genetic affinities with Native Americans. Am J Hum Genet 53:591–608.

Torroni A, Huoponen K, Francalacci P, Petrozzi M, Morelli L, Scozzari R, Obinu D, Savontaus ML, Wallace DC (1996) Classification of European mtDNAs from an analysis of three European populations. Genetics 144:1835–1850.

Torroni A, Bandelt H-J, D'Urbano L, Lahermo P, Moral P, Sellitto D, Rengo C, Forster P, Savontaus M-L, Bonné-tamir B, Scozzari R (1998) mtDNA analysis reveals a major Late Paleolithic population expansion from Southwester to Northeastern Europe. Am J Hum Genet 62:1137–1152.

Vigilant L, Pennington R, Harpending H, Kocher TD, Wilson AC (1989) Mitochondrial DNA sequences in single hairs from Southern African population. Proc Nat Acad Sci USA 86:9350–9354.

Vigilant L (1990) Control region sequences from African populations and the evolution of human mitochondrial DNA. PhD. Thesis, University of California at Berkeley.

Wakeley J (1993) Substitution rate variation among sites in hypervariable region I of human mitochondrial DNA. Science 253:1503–1507.

Wallace DC, Torroni A (1992) American e Indian prehistory as written in the mitochondrial DNA: a review. Hum Biol 64:403–416.

Wallace DC, Ye J, Neckelmann SN, Singh G, Webster KA, Greenberg BD (1987) Sequence analysis of cDNAs for the human and bovine ATP synthase b subunit: mitochondrial DNA genes sustain seventeen times more mutations. Curr Genet 12:81–90.

10

THE TRANS-CAUCASUS AND THE EXPANSION OF THE CAUCASOID-SPECIFIC HUMAN MITOCHONDRIAL DNA

Ene Metspalu, Toomas Kivisild, Katrin Kaldma, Jüri Parik, Maere Reidla, Kristiina Tambets, and Richard Villems

Department of Evolutionary Biology
Tartu University and Estonian Biocentre
Tartu, Estonia

ABSTRACT

The topology of the network of western Eurasian mitochondrial DNA (mtDNA) lineage clusters in the context of their expansion and spread in this geographic area is analysed. Special attention is devoted to the inner nods of the reconstructed median network tree, ancestral to mtDNA lineage clusters H and V, to the Caucasus and Trans-Caucasus area populations and to the problem of timing of the expansion of the Caucasoid-specific mtDNA lineage clusters in western Europe *versus* in the Trans-Caucasus. It appears on several examples that typical for Western Europe mtDNA lineage clusters exhibit significantly earlier expansion in the Trans-Caucasus area. Furthermore, the lineage cluster, radiating from the pre-HV node, is significantly more frequent and divergent in the Trans-Caucasus populations than it is in Europe. Meanwhile, a comparison of the Central Asian and the Trans-Caucasus area populations shows that the former have a significant share of eastern Asian-specific mtDNA lineages, which are almost absent in the latter. Finally, a picture starts to emerge, revealing an ancient Indian—Trans-Caucasian—European continuum of a significant proportion of human maternal lineage clusters, dating back to the period between the Upper and Lower Pleniglacials.

1. INTRODUCTION

When listing specific merits of the investigation of the phylogeny of human mtDNA lineages, properties like maternal inheritance, homoplasy and lack of recombination are usually stressed. Furthermore, as far as the mitochondrial DNA D-loop is concerned, neutral molecular evolution is usually assumed. From time to time, all the four mentioned special features of mtDNA are questioned, but it seems that the deviations from these rules are in general rare. However, there is one more, equally important but less empha-

Genomic Diversity: Applications in Human Population Genetics
edited by Papiha, Deka, and Chakraborty, Kluwer Academic / Plenum Publishers, New York, 1999.

sised aspect: unlike many other genetic markers, a network of combinatorial haplogroups has been worked out for mtDNA polymorphisms, allowing to differentiate between the major human races.

In world-wide mtDNA sequence data bases, Europeans are relatively "oversampled". This is, however, a very superficial generalisation: even in western and central Europe, there are wide undersampled areas, including France, Poland, southeastern Europe etc. In fact, many of the European and neighbouring populations are only marginally or not at all investigated. Being this gap only quantitative, it would not be of a significant hindrance. However, very limited data on Slavs, on Balkan area populations and specifically on the populations, living in the Caucasus and the Trans-Caucasus area, make this gap qualitative.

The classification of European-type mtDNA lineages (Torroni *et al.* 1994; Torroni *et al.* 1996) led to the defining of a limited number of major (H, J, K, T, U) and minor (I, V, W, X) Caucasoid-specific mtDNA haplogroups (lineage clusters), which describe at least 95% of mtDNA variations in Europe and in the Middle and Near East. The same haplogroups seem to be well represented also in northern Africa and in the Middle and Near East, but they are virtually absent in populations, living in sub-Saharan Africa and in eastern Eurasia, Australia and Polynesia (for review see: Wallace 1995). They are also absent in American Natives, except for haplogroup X, which has found its way to some northern American Amerind populations (Brown *et al.* 1998).

Perhaps the most intriguing question at present is the origin of these mtDNA clusters and the timing of their spread in Europe. The first massive study of the European mtDNA pool (Richards *et al.* 1996) postulated overwhelmingly Palaeolithic presence and expansion of the major extant mtDNA clusters in Europe: the observed variability suggested expansion times mostly from Middle to Upper Palaeolithic. Further similar mtDNA studies brought additional details into the picture but did not alter the basic conclusion (e.g. Torroni *et al.* 1998) This view is at least seemingly in conflict with the demic diffusion model of the Neolithic spread of new genes into Europe, carried by arriving from the Fertile Crescent farmers (summarised in Cavalli-Sforza *et al.* 1994). The debate is still ongoing and recent interpretation of the diversity of nuclear DNA markers in Europe (Chikhi *et al.* 1998) suggests a largely Neolithic ancestry of the European genes.

Here we summarise our own and published by others experimental data on some hitherto under-investigated populations and analyse the results in the context of the diversity and phylogeny of western Eurasian mtDNA lineages. We pay specific attention to the Trans-Caucasus area and show that a number of mtDNA lineage clusters, shared by those present also in Europe, started to expand in this area much earlier.

Some definitions and parallel terms used in this paper.

To avoid misunderstandings, we note that:

a) *haplogroup* and *lineage cluster* are used as synonyms;
b) *lineage* is used as an equivalent to *haplotype*;
c) as far as direct sequence analysis of the mtDNA is concerned (i.e. mostly the sequence of the first hypervariable region (HVR I)), we speak about a haplotype as about *an array of mutations*, accumulated in mtDNA genome, relative to its *ancestral state*;
d) all nucleotide positions (nps) are according to the Cambridge Reference Sequence (Anderson *et al.* 1981). In figures, an exact nature of mutation is specified only in case of transversions (e.g. 16,318 AG means an A to G transition at np 16,318);
e) for HVR I sequence, all mutation positions are shown less 16,000 relative to CRS; for RFLP positions, an exact np of loss or gain of a restriction enzyme cutting site is given. The latter does not necessarily coincide with a mutation,

which created a particular RFL polymorphic site; most of the precise locations of these particular mutations are now known (Richards *et al.* 1998).

2. THE GLOBAL NETWORK OF mtDNA LINEAGES

The accompanying paper (T. Kivisild *et al.*, this book) brings western Eurasian maternal lineages together with those found in India. Therefore, the present contribution mentions India only occasionally. Nevertheless, the reconstruction of human mtDNA lineages in the two papers is based on an identical ground: on parsimonious median networks (Bandelt *et al.* 1995). In contemporary variations of such networks their "skeleton" is largely based on informative restriction fragment length polymorphisms, whereas its "flesh" comes from the pattern of mutations, found in mtDNA hypervariable region I. One of such trees was recently published (Richards *et al.* 1998); here we refer to a slightly different variety (see Fig. 1 in Kivisild *et al.*, this book). We unified our haplogroup nomenclature with that in Richards *et al.* (1998); minor differences are commented below.

Bearing in mind this phylogenetic reconstruction, we first analyse the topology of the skeleton network from the point of view of the origin of the major Caucasoid-specific mtDNA lineage clusters. There are two such nodes: R* and Ö (see Fig. 1 in Kivisild *et al.*, this book). The first is a branching point for a large variety of not only Caucasoid-specific, but also Indian- and eastern Eurasian-specific mtDNA clusters. These lineage clusters encompass more than 90% of mtDNA haplotypes, found so far in Europe. The significance of this node is discussed in detail in Kivisild *et al.* (this book). Below we reconstruct and discuss its immediate descendant: node HV* (Richards *et al.* 1998). This node is designated as HV* to indicate that mtDNA haplogroups H and V descend from it. And because the latter two encompass about a half of the extant western European mtDNA lineages, this node must be considered as highly important in the understanding of the demographic history of the Caucasoid populations in general.

Secondly, in this paper we analyse a few mtDNA lineage clusters in greater details, in order to illustrate the approach we suggest for the coalescence time estimation. It brings us close to the recently reactivated debate on the ancestry of the European gene pool (e.g. Richards *et al.* 1996; Jackes *et al.* 1997; Chikhi *et al.* 1998)

3. THE ORIGIN OF HAPLOGROUPS H AND V

Caucasoids (*pro* Caucasians) is not a term used universally. Here we apply it in a traditional sense, covering not really well defined groups of populations, speaking largely Indo-European, Afro-Asiatic and Finno-Ugric languages. It is quite common to add here Indian populations as well. The mtDNA polymorphisms gathered so far is strongly biased: while the western and northern Europe, except France, is investigated in some depth, less is known about Slavic populations and even less about Caucasians living in the Caucasus area and in the Trans-Caucasus. Without them, however, the picture cannot be complete even as the first approximation.

We carried out such an analysis on Ossetes, Armenians and Georgians (summary n = 536, where n is a number of samples typed for haplogroup-specific RFLPs and HVR I sequenced). Here, we are particularly interested in selected aspects of these results. First, it turned out that all three populations possess a significant proportion of mtDNA lineages which belong neither to haplogroup H nor V, deriving nevertheless from a node, defined as HV* (Richards *et al.* 1998). The frequency of such lineages in all three populations is much higher than detected thus far anywhere else in Europe: around 8% (for comparison, see Table 2 in Kivisild *et al.*, this book). Median network tree of this mtDNA lineage cluster is shown in Fig. 1.

To this reconstruction, we included data from Macaulay *et al.* (1999), where a precise RFLP typing was performed, as well as a few other sequences from available data bases (Miller and Dawson 1998), what we added relying on HVR I sequence alone. We did it because the particular mutations: transversion at np 16,220 AC and transition at np 16,067 (see Fig. 1), are very rare or even absent in a world-wide mtDNA sequence databases, making such an identification reliable. A subcluster arising at np 16,311 (Fig. 1) is different: there is a large number of individuals in Europe possessing transition at this np and our analysis of a variety of Europeans shows that an overwhelming majority of such mtDNAs in a super-cluster HV* belong to haplogroup H. We also included our Indian data (Kivisild *et al.*, ms. submitted and Kivisild *et al.*, this book). Note that individuals from the Caucasus area and Anatolia (Armenians, Ossetes, Georgians and Turks) encompass about a half of the sample. However, any comprehensive comparison is at present premature since most of the data available from literature does not include full RFLP typing, like we carried out for Estonians, Armenians, Ossetes, Georgians, Czechs, Gypsies and Indians, which are included into the network, presented in Fig. 1.

Our analysis reveals that node HV*, an ancestral state to the most common for Europe haplogroup H (as well as for haplogroup V, see Fig. 1 in Kivisild *et al.*, this book), is diverse in the Trans-Caucasus area and also in India. Its expansion has probably started independently from a few founders. The node position itself is represented by a few Georgian, Armenian, Kashmir and Czech individuals, while lineages with mutations at np 16,067, 16,220 AC and 16,311 probably represent independent founders in addition to the node. These three additional founders were likely already present before the expansion phase(s), apparent from Fig. 1, took place. It suggests the occurrence of another, even earlier expansion phase. The presence of several sub-founders makes the coalescence age estimate for the total sample somewhat ambiguous; it is around 40,000 years before present (BP). The three representative sub-founders (the ancestral state and the derivatives after mutations at nps 16,067 and 16,311, all suggest values which are older than the Last Glacial Maximum LGM): around 25,000–35,000 yrs. BP. Note that Europeans (Estonian, Czech, Bulgarian, German, UK, North Irish), although rare in this picture, are distributed over all of the identified subclusters, while Indians from northwestern India (Kashmir, Uttar Pradesh and Lombadis from Punjab) derive from the central node only (Fig. 1).

Taken together, we suggest that the sum of these "pre-HV" lineages can be considered as a separate lineage cluster Ö, identified as lacking the *Alu*I site at np 7,025, possessing nucleotide A at np 00073 and can be further differentiated the ancestral to it node R* by loss of the *Tru*I site at np 14,467 and by possessing A at np 00073. Compared to haplogroup V, haplogroup Ö lineages have *Nla*III site at np 4,577. The fact that this lineage cluster, ancestral to about a half of all maternal lineages present in western Europe, is frequent and divergent in the Trans-Caucasus area as well as present in northwestern India, suggests that it started its expansion (and possibly even originated) from this geographic area. A few western European haplogroup Ö lineages thus far detected, are present in all sub-clusters of this haplogroup and may have been carried to western Europe any time since Europe was inhabited by modern humans.

In this context, it would be intriguing to reveal the presence of haplogroup Ö in yet another possible "fusion" area—in the Central Asia. Recently, about 200 mtDNA HVR I sequences of various populations of this area became available (Comas *et al.* 1998). Although this study lacked RFLP typing, we reconstructed a tentative median network phylogenetic tree of the published sequences (Fig. 2). Apart of some unavoidable ambi-

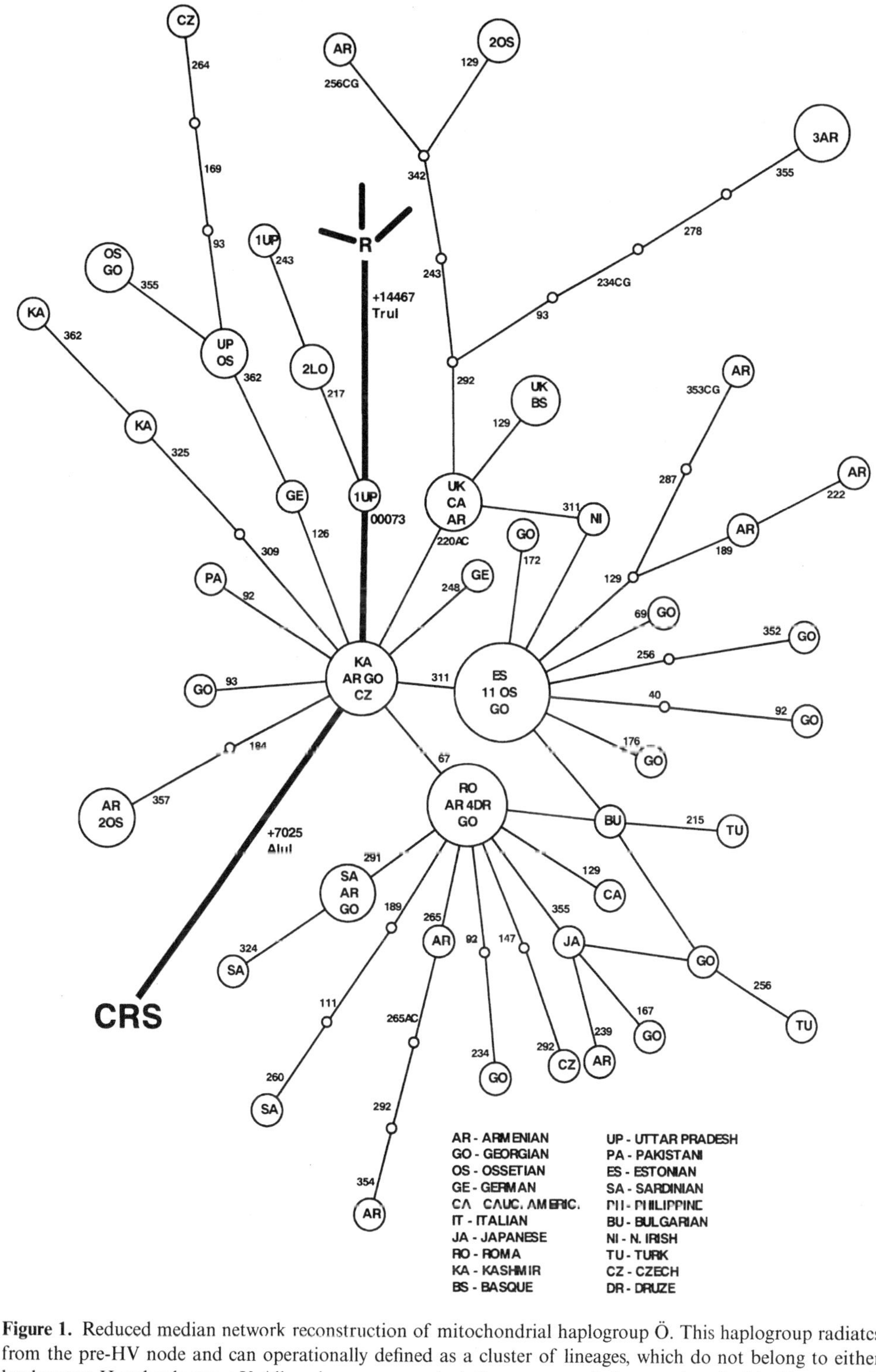

Figure 1. Reduced median network reconstruction of mitochondrial haplogroup Ö. This haplogroup radiates from the pre-HV node and can operationally defined as a cluster of lineages, which do not belong to either haplogroup H or haplogroup V. All coalescence age calculations were carried out as described in Forster *et al.* (1996).

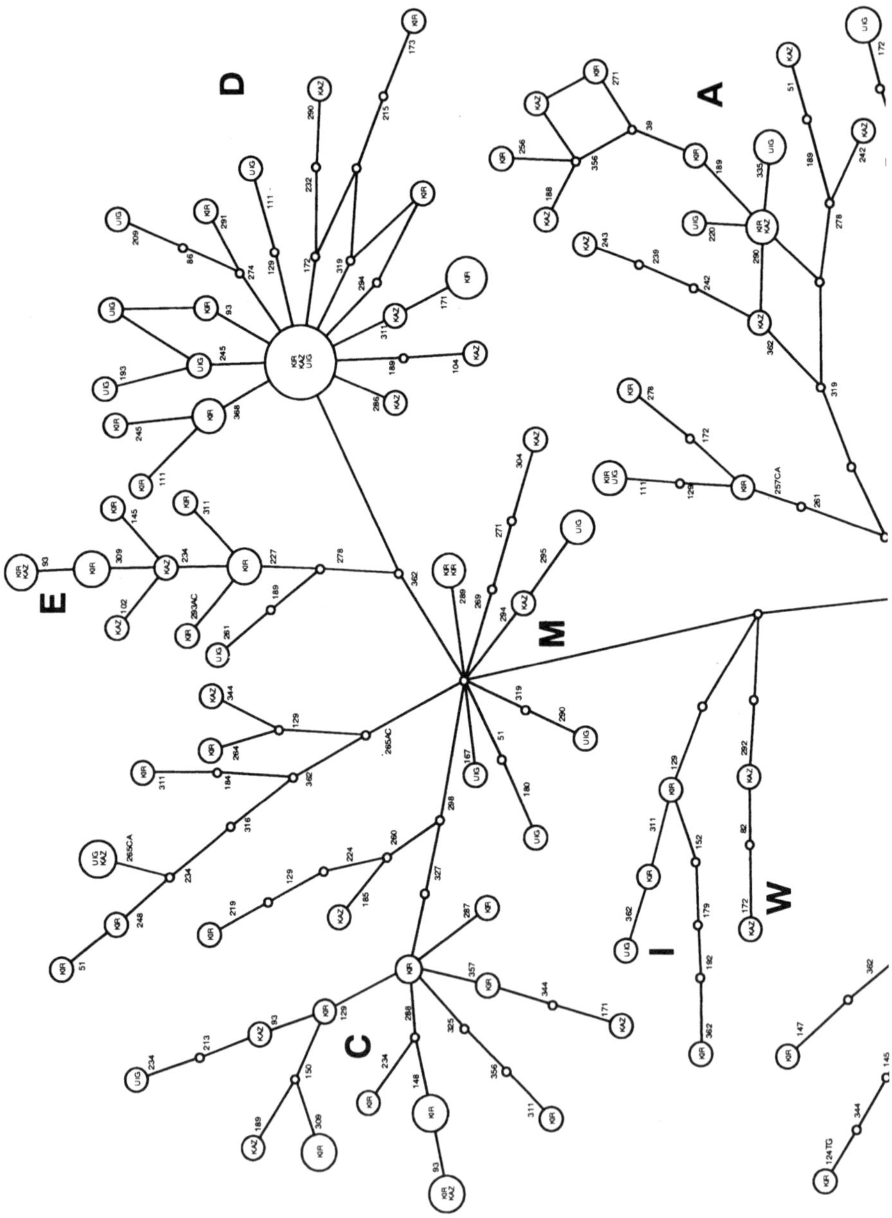
D
A
E
M
C
I
W

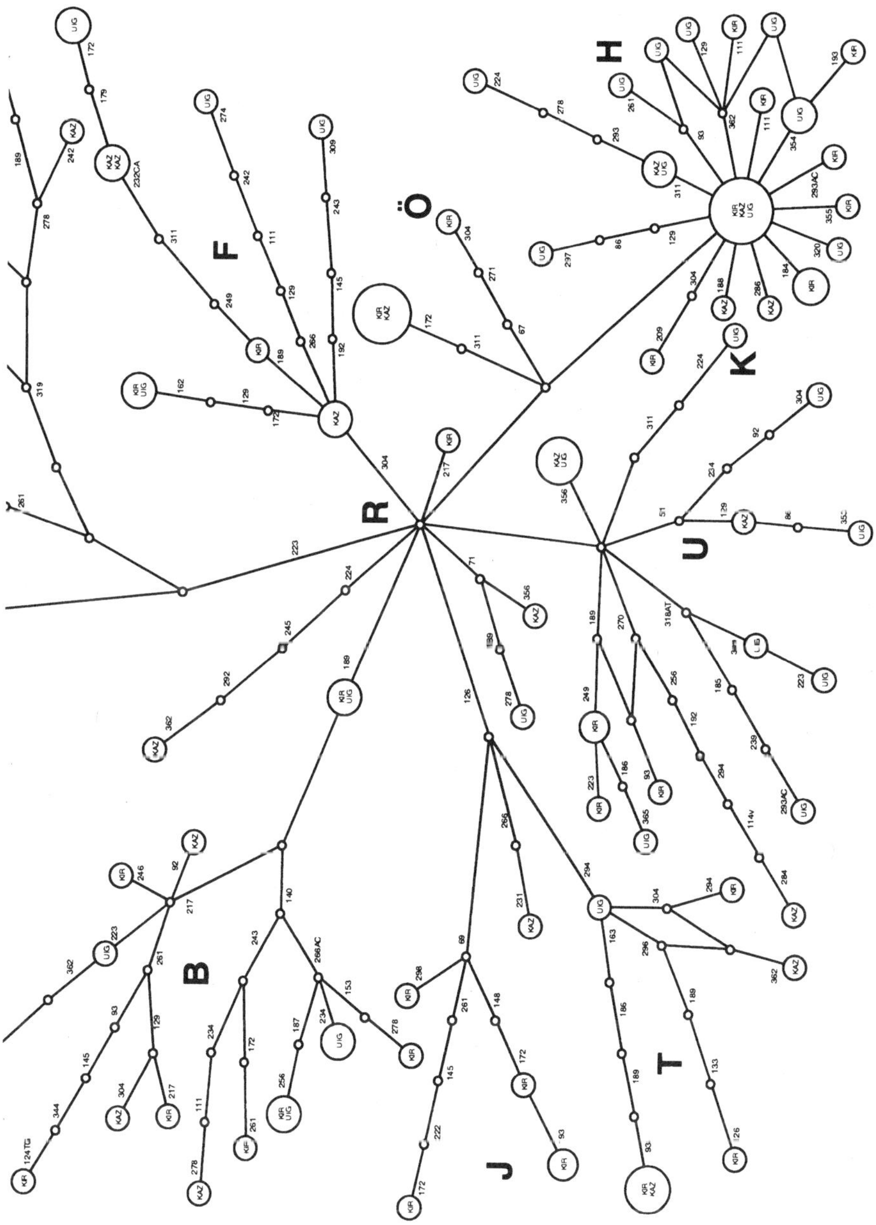

Figure 2. Tentative arranging of the Central Asian mtDNA HVR I sequences into reduced median network. Sequence information from Comas *et al.* (1998).

guities, this reconstruction fully agrees with the conclusion drawn by the authors: the populations studied possess mtDNA lineages, typical both for western and eastern Eurasia. In a set of three populations studied: Altaic language group Kazakhs, Uighurs and Kirgizis, there is one individual having transition at np 16,067 and therefore almost certainly belongs to haplogroup Ö. Some of those possessing transition at np 16,311 (motif 16,311 and 16,172, see Fig. 2) may belong to haplogroup Ö as well. The results do show that haplogroup Ö is present also in non Indo-Iranian language group Central Asians, albeit at lower frequency than in the Trans-Caucasian region. On the other hand, the frequency of haplogroup H in Central Asia is also lower: around 15% in these particular populations, compared to about 25% in average for Armenians, Georgians and Ossetes. We note that Ossetes, provided they are direct descendants of historic Alans, probably lived in Central Asia before moving westward about 2,000 BP (Frye and Litvinsky 1996).

Thus, in conclusion we can say that haplogroup Ö is present among Indo-European-, Afro-Asiatic-, Finno-Ugric-, Kartvelian-, North Caucasian- and Altaic-speaking populations. Among Indo-European linguistic super-family, this haplogroup can be found in Europe, in Indo-Iranian-speaking Ossetes as well as among nortwestern Indian populations of different localities. The fact that the three Caucasus area populations encompass these mtDNA lineages at nearly equal diversity and frequency (Fig. 1), while revealing several sharp frequency differences in some other mtDNA clusters (Metspalu *et al.*, ms in preparation), is somewhat puzzling. Ossetes are "newcomers" to the area—their arrival to the Caucasus is traditionally linked to the invasion of Hunns in late 4th century, pushing more northern Alans to south, behind Kuban and Terek rivers, from where several later invaders caused their further movement to the High Caucasus. The arrival of Indo-European-speaking Armenians to the political landscape of Trans-Caucasus some 2,600 years ago, although well documented historically, is nevertheless unclear from the point of view of their origin. It can be linked to the massive invasion of Thracians to Anatolia during the Late Iron Age, but possibly also to an invasion from the north—from the Pontus area. And, even if either of the two scenarios is valid, it is still unclear whether the linguistic change (Urartians, in contrast to Armenians, were linguistically not Indo-Europeans) caused any profound changes in the repertoire of maternal lineages of the local people. Finally, Kartvelians (Georgians) are considered to be autochthonous Caucasus-area population. All that lends credence to a possibility that the mtDNA gene pool in this geographic region obtained the founder of this haplogroup long ago, likely early in Upper Palaeolithic. After that, the population perhaps underwent several expansion events, at least one before the LGM. Note that we found haplogroup R individuals (defined as derivatives of the node R*, not belonging to haplogroups D, F, J, T, U and K, see Kivisild *et al.*, this book) also in all three Caucasus populations, albeit at a significantly lower frequencies than in India. Unfortunately, we lack data on Afghani and Iranian populations.

Taking now together all derivatives of mtDNA super-cluster HV* (i.e. H, V and Ö), one can see that haplogroup H is the most frequent in Europe, frequent in the Trans-Caucasus and present in northwestern India at a low frequency. Its sister-group, haplogroup V, is present in Europe at a low frequency, virtually absent in the Caucasus and Anatolia, not sampled so far in India and the corresponding lineages are not present in the published so far Central Asian data set. Meanwhile, haplogroup Ö (one can say—the rest of the super-cluster HV*) is uniformly (~8%) present in the Caucasus and found at a low but still well detectable frequency both in India and in Europe and seems to be also present in Central Asia. Further sampling of different populations in Iran, Afghanistan and Iraq, perhaps also in Turkey, southeastern Europe and in the Ukraine, will possibly add valuable details to the emerging picture of the historic spread of the descendants of HV* in the western Eurasia.

4. CASE STUDIES

Super-cluster HV* is only one of the derivatives of R* (see Fig. 1 in Kivisild *et al.*, this book). All other "classical" European-specific haplogroups that radiate from this node—T, J and U + K, are also present in the populations, living in the Trans-Caucasus (Table 2 in Kivisild *et al.*, this book). Thus, it is interesting to compare, where enough data available, the topology of individual haplogroups of various localities (Europe, the Caucasus area, India, Central Asia etc). Although this analysis does not necessarily reveal the absolute time depth of mtDNA lineage clusters, it does give a reliable comparative estimate of the relative time since their expansion in various geographic localities. Specifically, it allows to share light on the debate about the Palaeolithic *versus* Neolithic ancestry of the European gene pool. Below we analyse several mtDNA lineage clusters, derivatives of R*.

4.1. Sub-Haplogroup U3

This sub-cluster is characterised by transition at np 16,343 and is rare in Europe (see e.g. Fig. 6 in Richards *et al.* 1998). It seems to be absent also in the published (Comas *et al.* 1998) Central Asian sample (Fig. 2). In contrast to that, we found this haplogroup present in all three Trans-Caucasus populations studied by us. Furthermore, this sub-cluster was also found among Adygei population, living in the northern Caucasus (Macaulay *et al.* 1999). Because mutations at np 16,343 are relatively rare, and bearing in mind geographic proximity, we included also several putative sub cluster U3 sequence variants from Turkey (sequences from Calafell *et al.* 1996; Comas *et al.* 1996). The topology of the corresponding cluster (Fig. 3) reveals a star-like expansion event. The Adygei sub-sample (Macaulay *et al.* 1999) is limited to a single haplotype, whereas Armenians, Turks and Georgians are nearly equally divergent. Omitting Adygei sequences from calculations, we obtained $\rho = 1.42$ and time since the beginning of expansion about 30,000 yrs. BP (29,500 ± 5,000), i.e. well before the LGM.

4.2. Haplogroup J in the Trans-Caucasus Populations

Haplogroup J has complex topology in Europe with several sub-founders (Richards *et al.* 1998). This cluster is also well represented in Armenians, Georgians and Ossetes (Fig. 4). The basic topology of the European (see Fig. 7 in Richards *et al.* 1998) and of the Trans-Caucasus clusters overlap, but there the similarity ends. In Europe, the estimated expansion times for the node J* (which is also clearly the predominant sub-cluster of haplogroup J in Europe) is about 8,000 yrs. BP (Richards *et al.* 1998 and our unpublished results). For the same sub-cluster in the Caucasus—Anatolian area, the time since its expansion is about 27,000 yrs. BP. Sub-cluster J2, infrequent in Europe and present there predominantly in the Mediterranean area, is equal in frequency to J* in the Caucasus area.

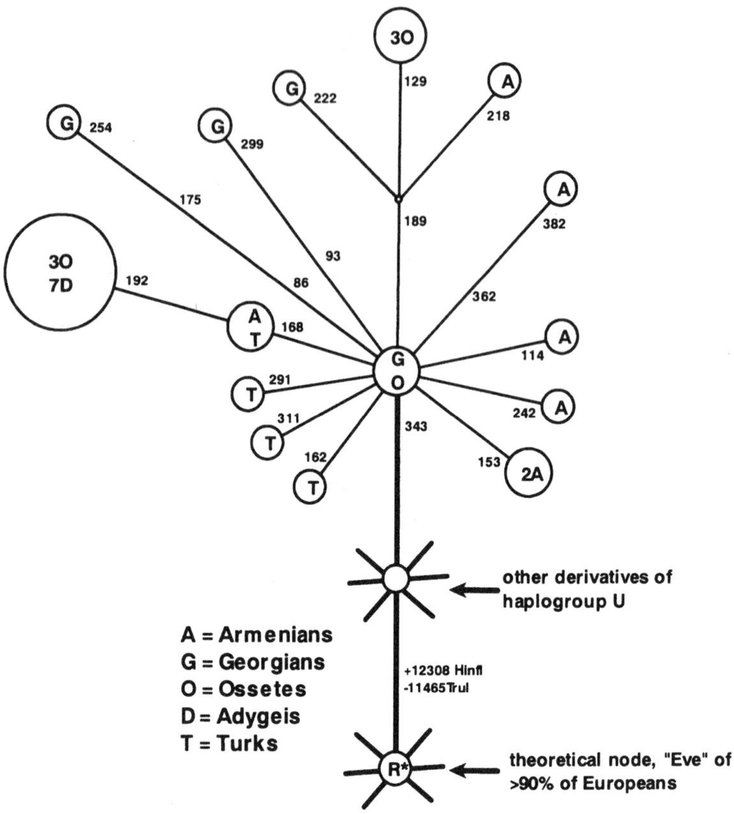

Figure 3. Reconstruction of human mtDNA haplogroup U sub-cluster U3 lineages network in the Caucasus area and in Turkey.

4.3. Sub-Haplogroup T1 in the Caucasus—Trans-Caucasus Populations

According to Richards *et al.* (1998), the European haplogroup T lineages started to expand at about 46,000 yrs. BP, making this haplogroup almost as ancient as some sub-clusters of haplogroup U. Still, this haplogroup has a well-resolved sub-founder, giving rise to a "Neolithic" (~9,000 yrs. BP) sub-cluster T1 (see Fig. 8 in Richards *et al.* 1998). In the Caucasus, the same estimate for Armenians and Georgians (we omitted Ossetians from the calculation because of their clustered spread) suggests expansion time around 27,000 yrs. BP. As for T* in Richards *et al.* (1998), we believe that this sub-cluster of lineages should also be sub-divided between several sub-founders, reducing significantly its apparent coalescence age. Adygei sequence is from Macaulay *et al.* (1999).

5. CONCLUSIONS

The three selected case studies for this analysis, together with the conclusions from the previous chapter about haplogroup Ö, allow to draw a few suggestions. First, we found that distinct lineage clusters, which seem to have expanded in Europe in Neolithic, have clear signs of much earlier, pre-LGM diversification in the Caucasus. Second,

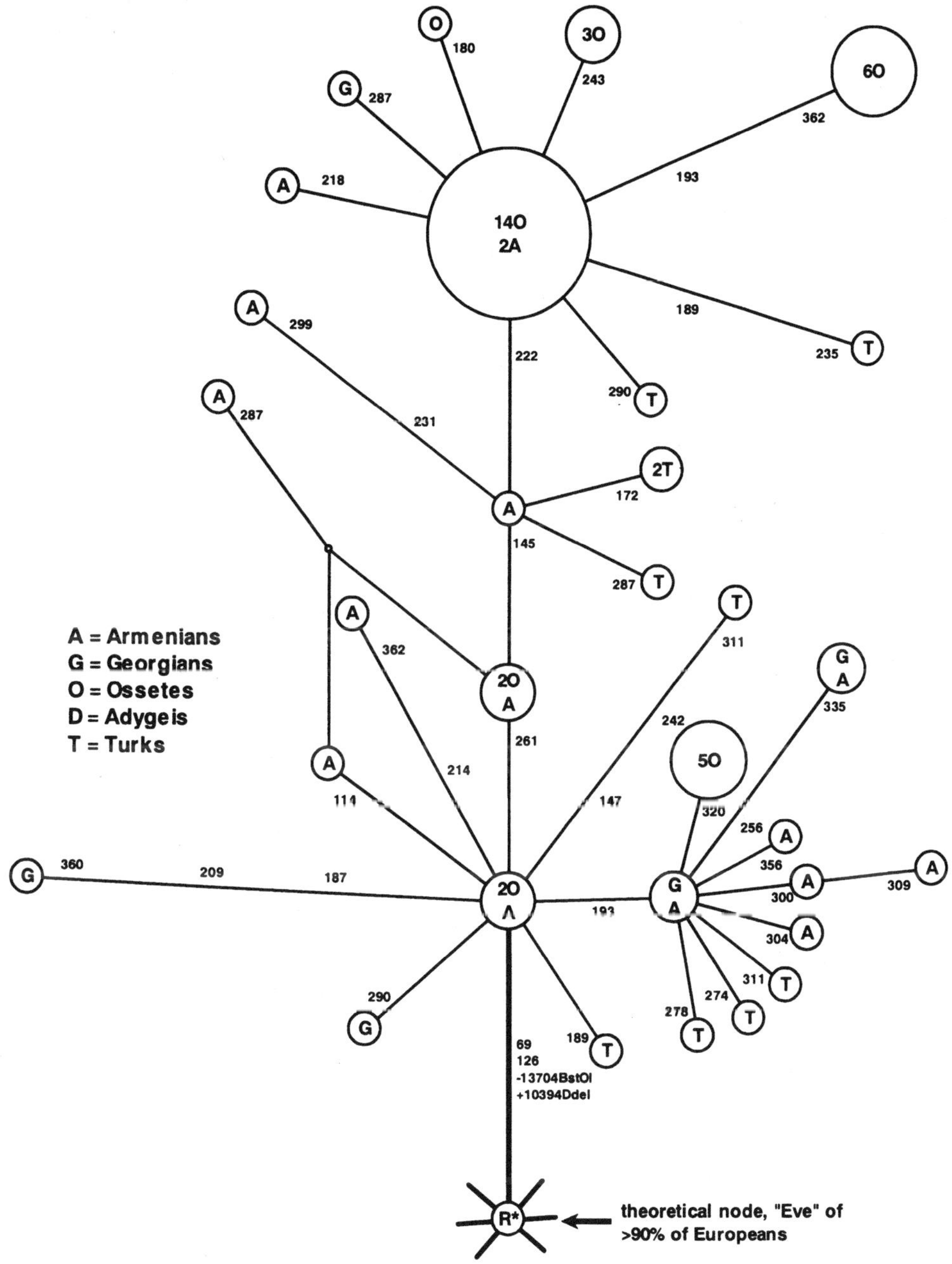

Figure 4. Reduced median network tree for mtDNA haplogroup J lineages found in Armenians, Georgians, Ossetes and Turks. The latter HVR I sequences are from Richards *et al.* (1996) Calafell *et al.* (1996), and Comas *et al.* (1996).

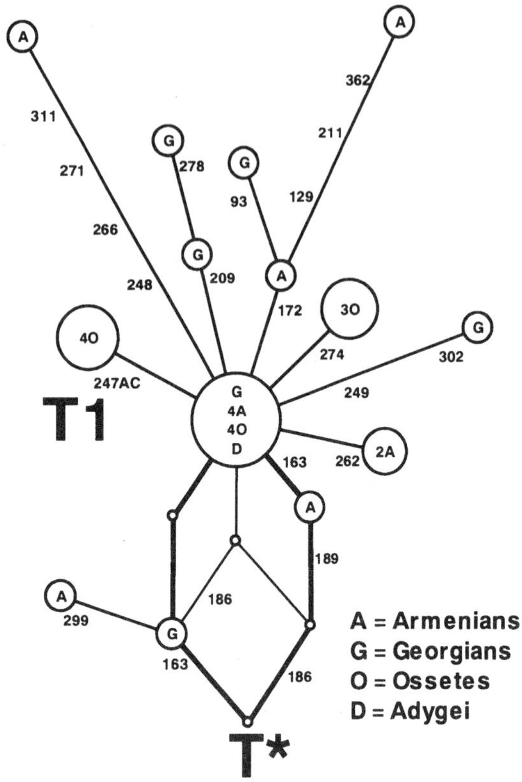

Figure 5. Median network tree for mtDNA haplogroup T sub-cluster T1 for Armenians, Georgians, Ossetes and Adygeis.

expansion time estimates for different mtDNA lineage clusters in the populations of the Caucasus area yield very close values. This seems to make sense: Palaeolithic population expansion waves were likely initiated by changes (climatic or else), which affected equally and simultaneously all present at that time founders. Indeed, it would much complicate any explanation, if all the varieties of mtDNA lineages have had different expansion times in the same area. Third, maternal lineages of the three different Caucasus area populations are in principle close to those found in western Europeans. Frequencies of certain haplogroups vary, those of many sub-clusters even more so, but none of the three populations studied possess at any significant quantities mtDNA lineages, which are either novel or belong to haplogroups, characteristic for eastern Asia or Africa.

Finally, it seems to us that the deconstruction of mtDNA lineage clusters into sub-founder-based units largely solves the problem of the initial controversy between the experimental data behind the "Neolithic demic diffusion" and "predominantly Palaeolithic mtDNA gene pool" in Europe. Not to add that irrespective of the area where any of those expansions took place, the expanding maternal lineages were phylogenetically close. It appears that neither India nor Central Asia—two localities with equally high incidence of haplogroup M, have contributed significantly into the western Eurasian mtDNA gene pool during many tens of thousands of years. Hence, there is little doubt that the Upper Palaeolithic mtDNA lineages in western Eurasia were qualitatively the same as they are at present. Therefore, their time depth is identical, irrespective of location, It does not, however, mean that there was no substantial gene flow from Anatolia to western Europe during early Neolithic. Much earlier expansion phases, observable in the same mtDNA lineage clusters and sub-clusters in the populations, living at present in the Trans-Caucasus area and in Turkey, compared to those in western Europe, support this suggestion.

6. REFERENCES

Bandelt H-J, Forster P, Sykes BC, Richards MB (1995) Mitochondrial portraits of human populations using median networks. Genetics 141:743–753.

Brown MD, Hosseini SH, Torroni A, Bandelt H-S, Allen JC, Schurr TG, Scozzari R, Cruciani F, Wallace DC (1998) MtDNA haplogroup X: An ancient link between Europe/West Asia and North America? Am J Hum Genet 63:1852–1861.

Calafell F, Underhill P, Tolun A, Anglicheva D, Kalaydjieva L (1996) From Asia to Europe: mitochondrial DNA sequence variability in Bulgarians and Turks. Ann Hum Genet 65:35–49.

Cavalli-Sforza LL, Menozzi P, Piazza A (1994) The history and geography of human genes. Princeton, NJ: Princeton University Press.

Chikhi L, Destro-Bisol G, Bertorelle G, Pascali V, Barbujani G (1998) Clines of nuclear DNA markers suggest a largely Neolithic ancestry of the European gene pool. Proc Natl Acad Sci USA 95:9053–9058.

Comas D, Calafell F, Mateu E, Perez-Leuzan A, Bertranpetit J (1996) Geographic variation in human mitochondrial DNA control region sequence: The population history of Turkey and its relationship to the European populations. Mol Biol Evol 13:1067–1077.

Comas D, Calafell F, Mateu E, Perez-Lezaun A, Bosch E, Martinez-Arias R, Clarimon J, Facchini F, Fiori G, Luiselli D, Pettener D, Bertranpatit J (1998) Trading genes along the Silk Road: Mitochondrial DNA sequences and the origin of Central Asian Populations. Am J Hum Genet 63:1827–1838.

Forster P, Harding R, Torroni A, Bandelt H-J (1996) Origin and evolution of Native American mtDNA variation: A reapprisal. Am J Hum Genet 59:935–945.

Frye RN, Litvinsky BA (1996) In: History of Humanity, Vol III, pp 461–464 Eds. J Hermann, E. Zürcher, UNESCO and Routlege, Clays Ltd, Surrey, UK.

Jackes M, Lubell D, Meiklejohn C (1997) On physical anthropological aspects of the Mesolithic-Neolithic transition in the Iberian Peninsula. Current Anthropology 38:839–846.

Larsson NG, Clayton DA (1995) Molecular genetic aspects of human mitochondrial disorders. Annu Rev Genet 29:151–178.

Macaulay V, Richards M, Hickey E, Vega E, Cruciani F, Guida V, Scozzari R, Bonne.Tamir B, Sykes B, Torroni A (1999) The emerging tree of West Eurasian mtDNAs: A synthesis of control-region sequences and RFLPs. Am J Hum Genet 64:232–249.

Miller KWP, Dawson JL (1997&1998) Mitochondrial DNA Concordance. http://shelob.bioanth.cam.uk/mtDNA.

Richards M, Corte-Real H, Forster H, Macaulay V, Wilkinson-Hebbots H, Demaine A, Papiha S, Hedges R, Bandelt H-J, Sykes B (1996) Paleolithic and neolithic lineages in the European mitochondrial gene pool. Am J Hum Genet 59:185–203.

Richards MB, Macaulay VA, Bandelt H-J, Sykes BC (1998) Phylogeography of mitochondrial DNA in western Europe. Ann Hum Genet 62:241–260.

Torroni A, Lott MT, Cabell MF, Chen Y-S, Lavergne L, Wallace DC (1994) mtDNA and the origin of Caucasians: Identification of ancient Caucasian-specific haplogroups, one of which is prone to a recurrent somatic duplication in the D-loop region. Am J Hum Genet 55:760–776.

Torroni A, Houpanen K, Francalacci P, Petrozzi M, Morelli L, Scozzari R, Obinu D, Savontaus M-L, Wallace DC (1996) Classification of European mtDNA from an analysis of three European populations. Genetics 144:1835–1850.

Torroni A, Bandelt H-J, D'Urbano L, Lahermo P, Moral P, Sellitto D, Rengo C, Forster P, Savontaus M-L, Bonne-Tamir B, Scozzari R (1998) mtDNA analysis reveals a major Late Paleolithic population expansion from southwestern to northeastern Europe. Am J Hum Genet 62:1137–1152.

Wallace DC (1995) Mitochondrial DNA variation in human evolution, degenerative disease, and aging. Am J Hum Genet 57:201–223.

11

THE PLACE OF THE INDIAN MITOCHONDRIAL DNA VARIANTS IN THE GLOBAL NETWORK OF MATERNAL LINEAGES AND THE PEOPLING OF THE OLD WORLD

Toomas Kivisild[1], Katrin Kaldma[1], Mait Metspalu[1], Jüri Parik[1], Surinder Papiha[2], and Richard Villems[1]

[1]Institute of Molecular and Cell Biology
Tartu University and Estonian Biocentre
[2]Department of Biochemistry and Genetics
University of Newcastle upon Tyne, UK

ABSTRACT

The spectrum of mitochondrial DNA (mtDNA) variation in India clearly supports the African origin of modern humans. In their deepest branching points Indian mtDNA clusters share common combinations of mutations with clusters specific for eastern and western Eurasian populations. However, their further diversification appears to be specific to India. The sub-clusters of haplogroup U in India and in Europe overlap only at the basis of the multifurcation. Also the structure of Asian dominant haplogroup M in India differs profoundly from that of other Asian populations. Around 10% of Indian mtDNAs are found around a phylogenetically crucial node—R*—which is ancestral to more than 90% of typically European mtDNAs and also to a portion of Asian specific mtDNA lineages such as haplogroups B and F. Coalescence estimates, calculated separately for these three major clusters in India, indicate a major radiation of mtDNA lineages starting approximately 50,000 years ago. The fraction of lineages attributable to a recent influx of maternal genes either from eastern or western Eurasia is below 10%. Furthermore, their spread does not correlate with the spread of the Sanskrit- and Dravidian-based languages in India.

1. INTRODUCTION

Both archaeology and genetics suggest that modern humans originated 100,000 to 200,000 years ago in Africa (Cann *et al.* 1987; Stringer 1990). Their first skeletal remains outside Africa are about 100,000 years old but have been found so far only in the imme-

Genomic Diversity: Applications in Human Population Genetics
edited by Papiha, Deka, and Chakraborty, Kluwer Academic / Plenum Publishers, New York, 1999.

diate vicinity—from the caves in Near East (McDermott *et al.* 1993; Stringer 1992; Stringer *et al.* 1989; Aitken and Valladas 1992). There is no substantial evidence supporting further spatial dispersal of modern humans earlier than around 50,000 years ago. By that time they seem to have reached Papua New Guinea and Australia and soon after that they are found also in Europe. What happened during this 50,000 year long gap and where did the initial radiation of the Eurasian population take place remains largely an open question.

Western Asia and India stand geographically on the road early modern humans had almost inevitably pass to reach eastern Asia, New Guinea and Australia. Did some of the migrating waves of humans settle there instead of going *in corpore* further eastwards? Was it the place where the initial radiation of Eurasian mitochondrial DNA lineages took place? There is a lack of skeletal evidence of modern humans from East Asia older than the Upper Cave Zhoukoudian crania (Foley 1998) that are dated to around 30,000 years before present (BP). The earliest skeletal evidence from South Asia comes from Sri Lanka, where the Fa Hien Lena finds put forward 34,000 year old carbon datings (Deraniyagala 1998). These archaeological dates imply the approximate time line for the dispersal of modern humans in Asia.

Driven by fast mutation rate and lack of recombination, distinctive clusters of mtDNA lineages have emerged during the last tens of thousands of years. Low overall population density during Palaeolithic and vast geographic distances favoured the isolation of human populations and thus played an important role in secluding the differences arisen in DNA lineages. The present day mtDNA variability is highly continent-specific (Chen *et al.* 1995; Torroni *et al.* 1996; Wallace 1995). Therefore, already at the level of present day knowledge about the worldwide variation of mtDNA genome, one can reliably distinguish between mtDNAs of eastern Asian, European or sub-Saharan African origin (Fig. 1). Nevertheless, the same knowledge base shows that all mtDNA variants outside Africa, studied so far, derive from a single Pan-African mtDNA cluster L3a (Watson *et al.* 1997).

2. MITOCHONDRIAL DNA HAPLOGROUPS IN INDIA

Both eastern and western Eurasian mtDNA haplogroups have been found among modern Indian populations (Bamshad *et al.* 1997; Passarino *et al.* 1996; Passarino *et al.* 1996). Haplogroups M and U constitute the major portion (~75%) of the variation and pooled together, their spread is nearly uniform over India (Table 1). About 10% of the Indian mtDNAs belong to other haplogroups, geographically characteristic to Eurasia, while as many as 15% of the lineages do not belong to any of the continental clusters described before. The lack of African-specific variants in India should be considered with some precaution (Bamshad *et al.* 1997) because African-borne haplogroup L3 lacks any distinctive RFLP or HVS-I sequence motifs. However, the absence of *Hpa*I site at nucleotide position (np) 3592 among Indians seems clearly to negate any recent substantial gene flow from sub-Saharan Africa to the peninsula.

Comparing the frequencies of mtDNA haplogroups present among Indians with those among other Eurasian populations suggests (Table 2) that Indians are simultaneously part of two distinctive meta-populations: eastern and western Eurasians. In that respect Indians appear to be similar to the populations of Central Asia (Comas *et al.* 1998). This is not surprising when bearing in mind the geographical position of both regions. However, haplogroup frequency data alone are insufficient and can be sometimes

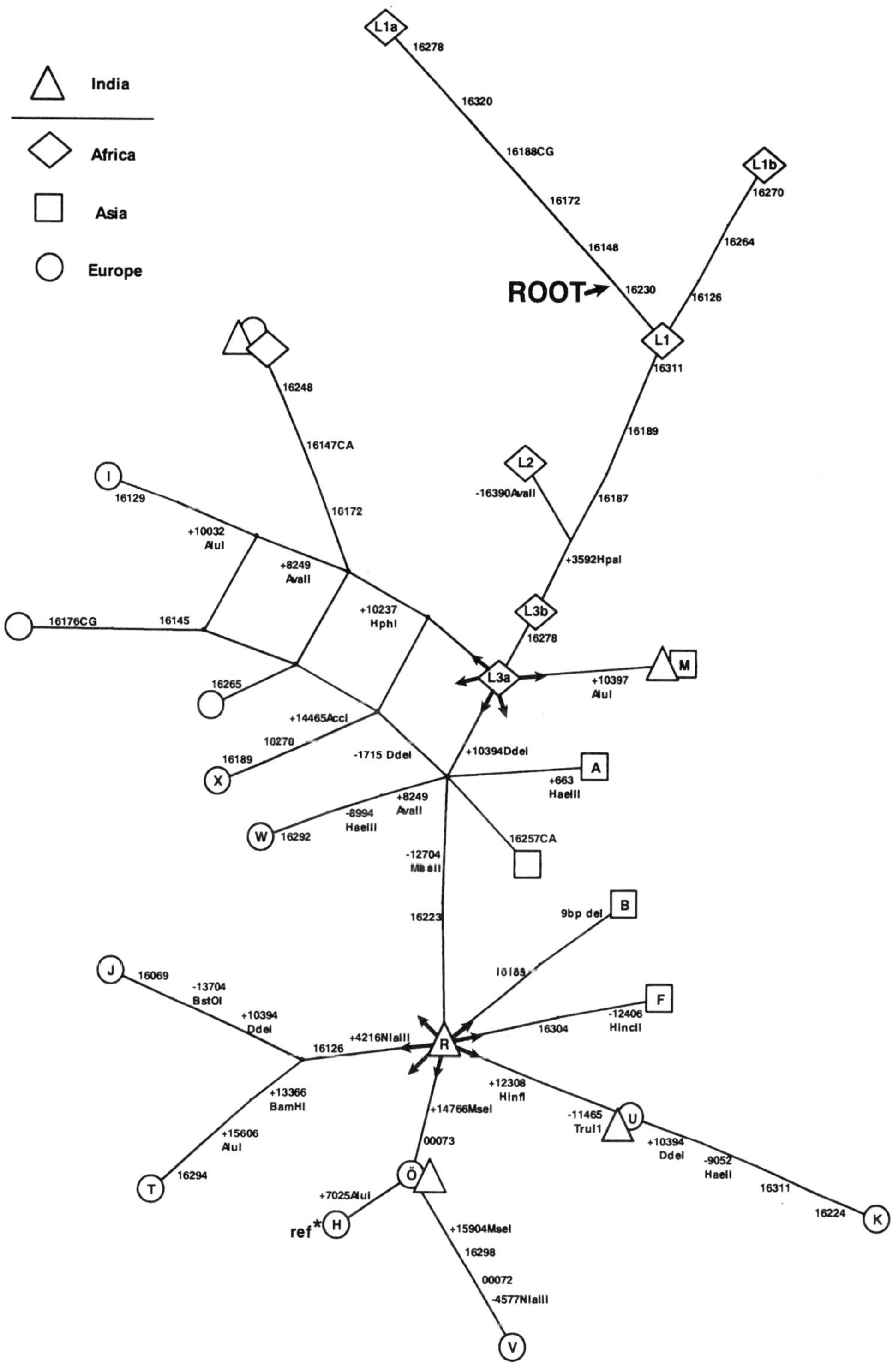

Figure 1. The skeleton network of mtDNA haplogroups and their geographic distribution. Haplogroup defining mutations are shown near lines connecting them, indicating base pair position relative to the Cambridge reference sequence (CRS; Anderson *et al.* 1981). Presence or absence of restriction enzyme recognition sites is shown also respective to the CRS, not the root, which is proposed using Neandertal sequence (Krings *et al.* 1997) as an outgroup. The definition and labelling of the haplogroups follows the scheme of earlier studies (Torroni *et al.* 1996; Richards *et al.* 1998; Watson *et al.* 1997; Macaulay *et al.* 1999). L3a is a super-haplogroup (indicated by arrows) encompassing Eurasian haplogroups. Another super-haplogroup is R, which on the other hand is a sub-cluster to L3a. Several so far undefined clusters are shown unlabelled while a new distinctive haplogroup Ö is defined inside a point previously labelled as HV* (Macaulay *et al.* 1999).

Table 1. mtDNA haplogroup frequencies in India

	Pakistan	Kashmir	Uttar Pradesh	Punjab (Lobanas)	Andhra Pradesh (Lambadis)	Other Localities	Total
African							
L	0	0	0	0	0	0	0
e. Eurasian							
A	0	0	1 (1%)	0	0	0	1 (0.3%)
B	0	0	0	0	0	0	0
F	0	4 (21%)	2 (2%)	1 (1%)	1 (1%)	0	8 (2.7%)
M	3 (38%)	5 (26%)	52 (51%)	34 (55%)	55 (64%)	9 (41%)	158 (52.7%)
w. Eurasian							
H	1 (12%)	1 (5%)	3 (3%)	0	2 (2%)	0	7 (2.3%)
I	0	0	2 (2%)	0	2 (2%)	0	4 (1.3%)
J	0	1 (5%)	0	0	1 (1%)	0	2 (0.7%)
K	0	0	0	0	0	0	0
T	0	0	1 (1%)	3 (5%)	1 (1%)	0	5 (1.7%)
U	2 (25%)	2 (11%)	24 (24%)	3 (5%)	11 (13%)	4 (18%)	46 (15.3%)
V	0	0	0	0	0	0	0
W	0	1 (5%)	0	8 (13%)	2 (2%)	0	11 (3.7%)
X	0	1 (5%)	0	0	1 (1%)	0	2 (0.7%)
Ö	1 (12%)	3 (16%)	1 (1%)	0	0	0	5 (1.7%)
Others	1 (12%)	1 (5%)	17 (17%)	13 (21%)	10 (12%)	9 (41%)	50 (16.7%)
n	8	19	103	62	86	22	299

misleading in choosing between alternative population genetic scenarios: whether the co-presence of these two sets of lineages is due to their initial diversification or because of a recent diffusion. To get deeper insight into the problem a detailed phylogenetic analysis of both eastern and western Eurasian lineages is needed.

2.1. Eastern Eurasian mtDNA Haplogroups in India

Although the data about Asian populations are still far from being representative, it is already apparent that the structure of Asian mtDNA phylogeny is different from that of European and African populations (Wallace 1995). A common feature for all the Asian populations studied so far is that the major fraction of their mtDNA pool is made up of haplogroups A, B, F and M.

Haplogroup M is a dominant mtDNA cluster among the populations of Mainland Asia as well as among Native Americans (Ballinger *et al.* 1992; Torroni *et al.* 1994; Torroni *et al.* 1994). It is defined by the presence of a *Dde*I site at np 10394 and an *Alu*I site at np 10397. The co-occurrence of these two polymorphisms is highly Asian-specific but Asians lack at the same time any possible progenitor of this haplogroup. Thus, the mtDNA coding region RFLP diversity-based coalescence age of 55,500–73,000 years for haplogroup M (Chen *et al.* 1995) could indicate the time of the initial colonisation of Asia by modern humans.

Haplogroup M has been sub-divided into discrete sub-clusters according to the accumulation of further synapomorphic mutations along Asian maternal lineages. Sub-clusters C, D, E and G, defined by certain RFLP and HVS-I sequence polymorphisms (Torroni *et al.* 1994; Torroni *et al.* 1993; Torroni *et al.* 1993), are spread over vast territories all over Mainland Asia (Fig. 2). These separate sub-clusters display surprisingly similar, approximately 40,000 year old, coalescence times (Table 3). The split of

Table 2. mtDNA haplogroup frequencies in some Eurasian populations

	n	WE[1]	H	I	J	K	T	U	V	X	W	EE[2]	A	B	F	$M^{C,D,E,G}$	M*	AF[3]	L1	L2	L3	Others
Indians[a]	299	27.4	2.3	1.3	0.7	0	1.7	15.3	0	0.7	3.7	55.7	0.3	0	2.7	1.7	51.0	0	0	0	0	16.7
Mongols[b]	103			0	0	1.0	1.0		0	0	0	77.7	3.9	9.7	5.8	38.9	19.4					
Tibetians[c]	54											94.5	11.1	5.6	14.8	33.4	29.6	0	0	0	0	5.5
Chinese[d]	66			0	0	0	0		0		0	94.0	6.0	20.0	22.0	22.0	24.0	0	0	0	0	6
Abor. Siberians[e]	259											97.3	23.6			49.8	23.9					2.7
Native Americans[f]	591											95.9	28.7	27.4		39.1		0	0	0	0	4.1
Central Asians[g]	205	30.5	14.0	1.0	2.5	0.5	3.5	8.0	0	0	1.0	61.0	6.8	6.8	5.4	29.2	9.3	0	0	0	0	8.5
Trans-Caucasians[a]	330	80.9	24.8	1.8	6.7	8.2	11.8	21.2	0	5.5	0.9	1.5	0	0.3	0	1.2	0	0.3	0	0	0.3	17.3
Italians[h]	99	93.9	33.3	4.0	7.1	8.1	9.1	22.2	5.1	3.0	2.0	0	0	0	0	0	0	0	0	0	0	6.1
Finno-Ugries[a]	149	97.9	45.6	1.4	12.1	3.3	6.0	22.8	2.0	2.0	2.7	0.7	0	0	0	0.7	0	0	0	0	0	1.4
Slavs[a]	324	94.7	41.4	2.8	10.5	3.7	12.3	19.4	3.1	0.6	0.9	0.9	0	0	0	0.9	0	0	0	0	0	4.4
Germans[i]	200	92.5	50	2.5	7.5	6.5	8.5	13.5	2.5	0.5	1.0	0	0	0	0	0	0	0	0	0	0	7.5
Ethiopians[j]	74	13.5	0	0	0	2.7	1.4	8.1	0	0	1.4	20.3						52.7				13.5
Sub-Saharan Africans[k]	407																		30.2	30.4	39.3	

[1]western Eurasian specific haplogroups.
[2]eastern Eurasian specific haplogroups.
[3]African specific haplogroups.
[a]Kivisild *et al.*, manuscript in preparation.
[b]Kolman *et al.* 1996.
[c]Torroni *et al.* 1994.
[d]deduced from sequence data of Horai *et al.* 1996.
[e]Torroni *et al.* 1993.
[f]Baillet *et al.* 1994; Torroni *et al.* 1992; Schurr *et al.* 1990.
[g]deduced from sequence data of Comas *et al.* 1998.
[h]Torroni *et al.* 1997.
[i]deduced from sequence data of Lutz *et al.* 1998.
[j]deduced from sequence data of Passarino *et al.* 1998.
[k]Watson *et al.* 1997.

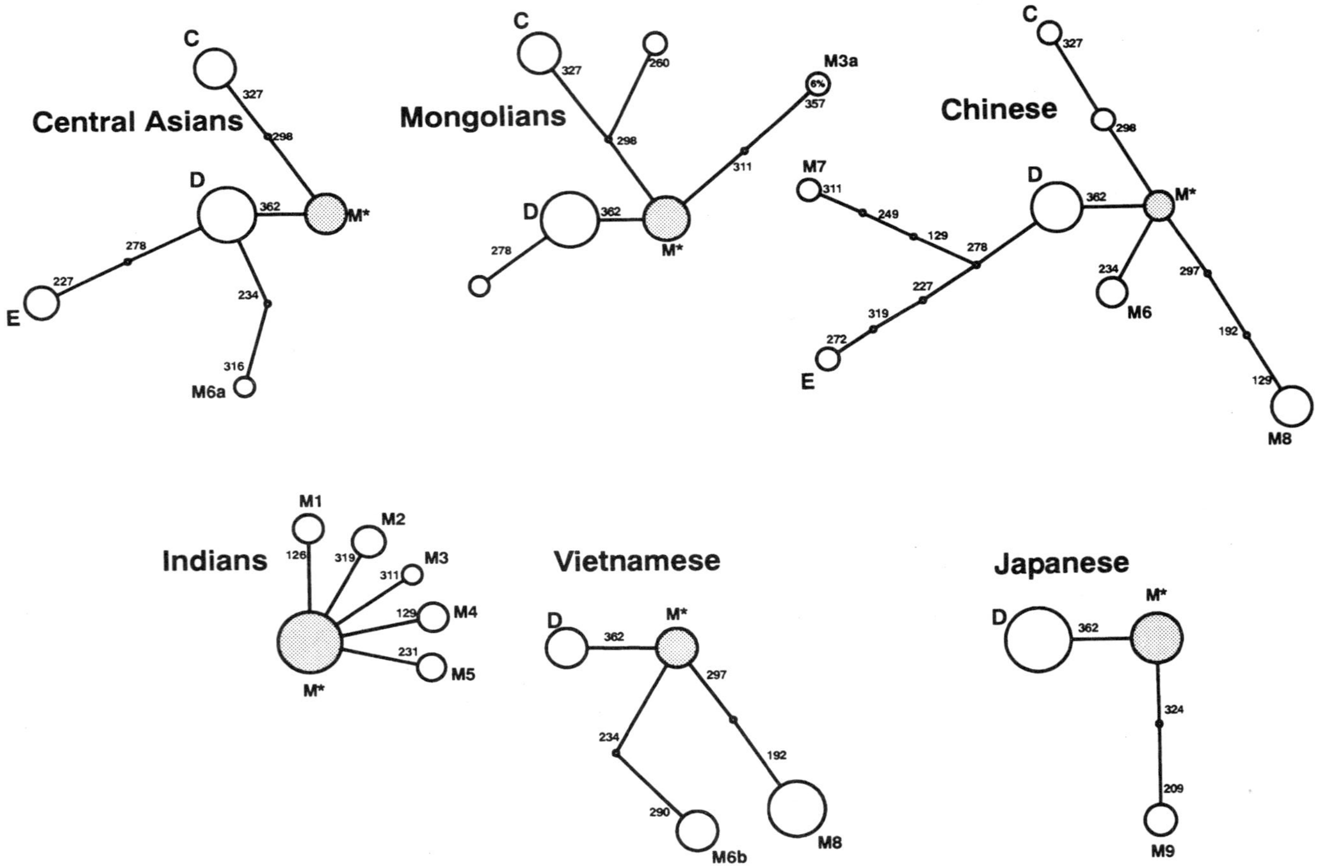

Figure 2. The structure of haplogroup M in some Asian populations. RFLP status of the sequences is known only for Indian (manuscript in preparation) and Mongolian (Kolman *et al.* 1996) sequences. Among Central Asian (Comas *et al.* 1998), Chinese, Japanese (Horai *et al.* 1996), and Vietnamese (Lum *et al.* 1998) populations the sequences were assumed to belong to haplogroup M according to indirect evidence—comparing them with sequences of known status. Only those haplogroup M sub-branches that constitute at least 5% of the total frequency of haplogroup M in the given population are shown. Shaded M* collects all the remaining branches that did not exceed the 5% criterion. The node area reflects the frequency of the sub-branch. Mutations at the given bp in HVS-I (less 16,000) are shown near lines connecting the nodes.

Table 3. Coalescence times of eastern Eurasian mtDNA haplogroups in Mongols[1]

Haplogroup[2]	Consensus	ρ[3]	Time[3]		Variance[3]
A	+663*Hae*III; 16223T; 16390T; 16,319G	1.75	35,500	±	13,000
B	9bp del; 16189C	2.5	50,500	±	10,000
B1	9bp del; 16189C; 16217C	2	40,500	±	11,000
B2	9bp del; 16189C; 16243C	1.7	33,500	±	15,000
F	(+12406*Hinc*II); 16304C	2.1	42,000	±	10,000
M	+10394*Dde*I; +10397*Alu*I; 16223T	3.7	74,000	±	5,000
C	+10394*Dde*I; +10397*Alu*I; 16223T; 16298C; 16327T	2.1	42,000	±	8,000
D	+10394*Dde*I; +10397*Alu*I; 16223T; 16362C	2.2	44,500	±	6,000

[1]data from Kolman *et al.* 1996.
[2]Haplogroups defined as in Torroni *et al.* 1994.
[3]Calculations as in Forster *et al.* 1996.

haplogroup M into these subclusters may therefore reflect a secondary expansion event, which led ultimately to the extension of modern Asian populations to northern and central Asia and to the Americas. Besides these haplogroup M sub-clusters, haplogroups A and two basic sub-clusters of haplogroup B display comparable diversity and coalescence ages (Table 3).

The sub-branching of haplogroup M in India is profoundly different from that described so far for any other Asian locality. Five novel sub-clusters can be defined and they form a half of the Indian haplogroup M lineages (Fig. 2, 3). The other half is scattered between multiple minor branches (Fig. 4). In contrast to that, typical 'Mongoloid' sub-clusters C, D, E and G are found at extremely low frequencies (1.2% all together, Table 2). Even though, these exceptions are mostly due to the sampling of borderline populations, such as Tharus, whose Oriental origins have been characterised earlier (Brega *et al.* 1986; Passarino *et al.* 1993; Passarino *et al.* 1992). Likewise, the other typically eastern Asian haplogroups A, B and F, are practically absent in India, although B and F are very common in neighbouring southeastern populations of Asia (Wallace 1995). Studies based on mtDNA RFLP polymorphisms in India have shown that haplogroup M divergence predates the separation of proto-Indians from proto-eastern Asians (Passarino *et al.* 1996; Passarino *et al.* 1996). Hence, an important conclusion can be drawn: during the last 50,000 years or so, there has been a very limited admixture of Indian populations with the Mongoloid populations living east and north of India. And the other way round: the five major Indian-specific sub-clusters of haplogroup M are not represented to any significant extent elsewhere in Asia. Thus, it appears that although Indian populations display haplogroup M frequency similar to or even higher than that in other Asian populations, the internal structure of haplogroup M lineages in India reflects their basically autochthonous development.

The spread of haplogroup M variants reveals some characteristic differences among different populations of the Indian Peninsula (Fig. 3). For example, sub-cluster M2 is predominantly found among populations living in the southern parts of the peninsula. Also, an interesting association of sub-cluster M1 with high caste Indians should be mentioned. Our search among the sequences from Kerala and Karnataka populations (Mountain *et al.* 1995) shows that only Brahmin caste Haviks, but not the lower caste Mukris, have a characteristic to M1 motif (transitions at nps 16,126 and 16,223). The distance between higher and lower caste Telugus was also found to be bigger than that between higher and middle or middle and lower castes; also no mtDNA haplotypes were shared exclusively between the upper and the lower castes (Bamshad *et al.* 1998). On the

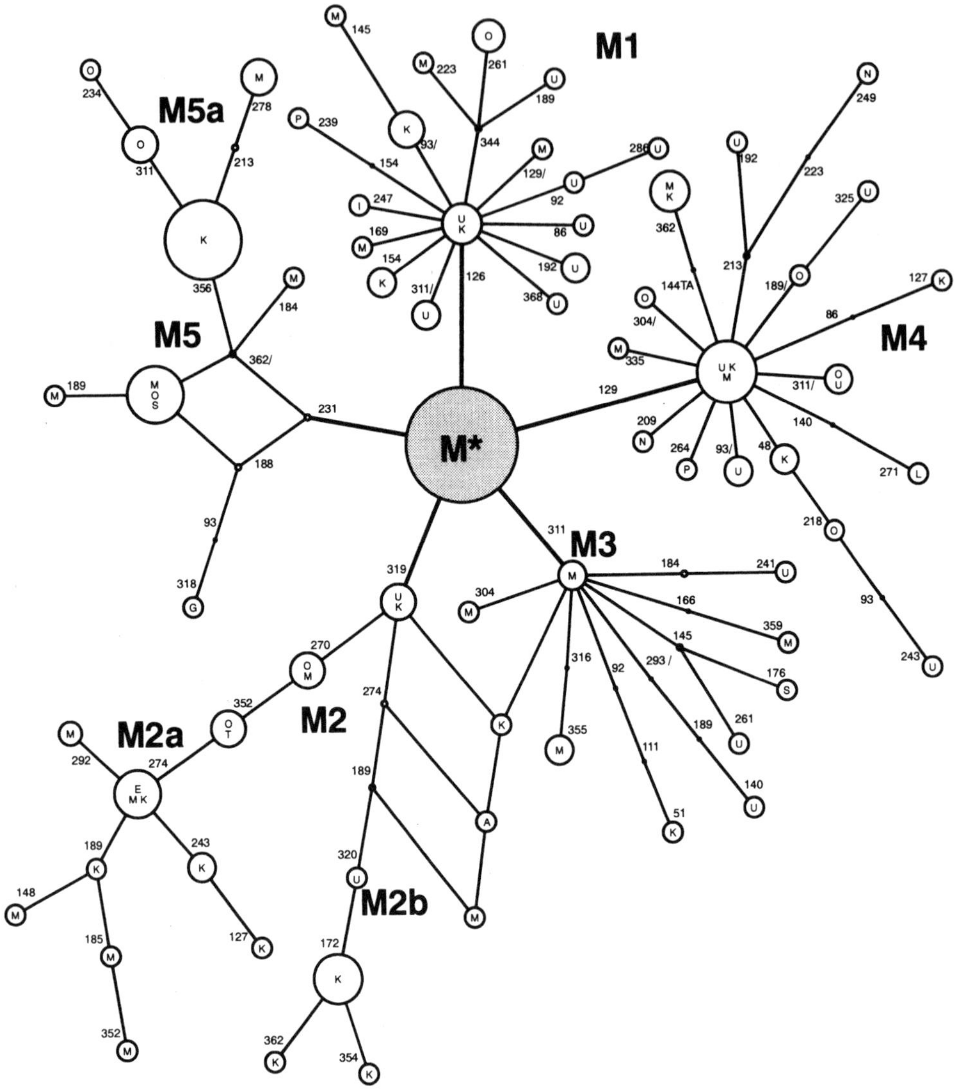

Figure 3. Reconstruction of 5 major haplogroup M subclusters in India. Node areas correspond to haplotype frequencies. Mutations at which two haplotypes differ are shown less 16,000 on lines connecting them. "/" indicates a broken reticulation. Central shaded node corresponds to the ancestral state of haplogroup M as in Fig. 1. U—Uttar Pradesh; O—Lobanas; M—Lambadis; S—Kashmir; K—Kerala; I—Bihar; N—Bangladesh; L—Bengal; A—Andhra Pradesh; P—Pakistan; G—Gujerat.

other hand, lineage clusters M3, M4, M5 and haplogroup M minor sub-clusters are spread over India without any evident population or caste-specific sequestration. The general intermixture of lineages of different caste groups at the level of major lineage clusters was shown by Mountain and colleagues (Mountain *et al.* 1995) by a HVR sequence inter-match analysis. These two findings can be explained by a common ancestry of distant caste groups of India, followed by a comparatively recent endogamy within the limits of caste system.

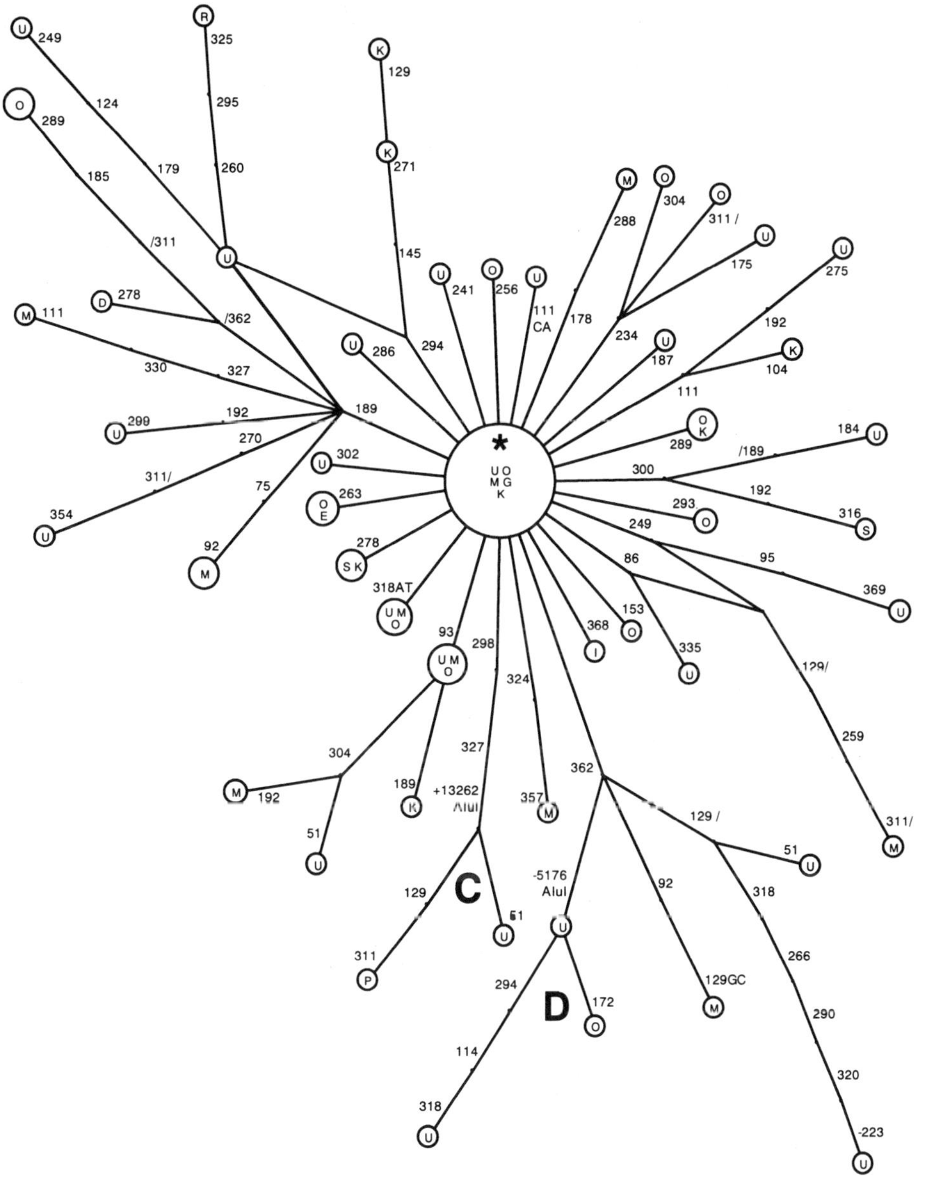

Figure 4. Reconstruction of the M* lineages in India. Populations: U—Uttar Pradesh; O—Lobanas; M—Lambadis; S—Kashmir; K—Kerala; I—Bihar; N—Bangladesh; L—Bengal; A—Andhra Pradesh; P—Pakistan; G—Gujerat; D—Tamil Nadu; R—Orissa. For other details, see Legend to Fig. 3.

We estimated the mean coalescence age of Indian M sequences as 47,000 ± 2,500 years (Table 4). This estimate is younger than the date of ~65,000 years proposed earlier (Mountain *et al.* 1995) for an expansion starting from South-Asia. However, the latter date derives from a 'non-African' cluster M2 (Fig. 5 in Mountain *et al.* 1995) that includes also African sequences and could therefore reflect (a possibility considered also by the authors) an expansion occurring in Africa just prior to the exodus. A support for this enigmatic cold and dry period expansion in eastern Africa (60,000–80,000 years ago) was

Table 4. The diversity and age of haplogroup M and its major subclusters in India

	n	ρ	t	±
M-total	158	2.33	47,000	2,500
sub-cluster				
M1	23	1.26	25,500	5,500
M2	34	3.12	63,000	6,000
M2a	13	0.85	17,000	5,000
M2b	11	1.1	22,000	6,500
M3	12	1.58	32,000	7,500
M4	35	0.91	18,500	3,500
M5	31	1.38	28,000	4,500

ρ, t, variance—estimated as in Forster *et al.* 1996.

given by a recent analysis of African mtDNAs (Watson *et al.* 1997). When we consider the possibility that more than just a single female walked out of Africa and gave enough progeny, any time calculation based on random sequences from a non-African population should reach the estimate of this initial radiation inside Africa. Therefore, the time for non-African expansions should be calculated from lineage clusters that show autochthonous development outside Africa. Although haplogroup M is predominantly found in Asian populations, a recent study showed also its presence in eastern Africa (Passarino *et al.* 1998). Within the Indian haplogroup M sequences averaging to the 47,000-year-old coalescence point, sub-cluster M2 and many long independent branches (Fig. 4) reveal divergence that exceeds that point. For example the coalescence age of all M2 sequences is 63,000 years. This estimate agrees with the expansion dates discussed above (Mountain *et al.* 1995; Watson *et al.* 1997) and also falls within the limits of the calculations of time depth based on the RFLP data of Asian M (Chen *et al.* 1995).

To conclude, two different scenarios can be drawn. First, the origin of haplogroup M lies in eastern Africa and multiple M lineages were present already at the onset of the population movement that finally colonised Asia. Second, the calculation of the initial

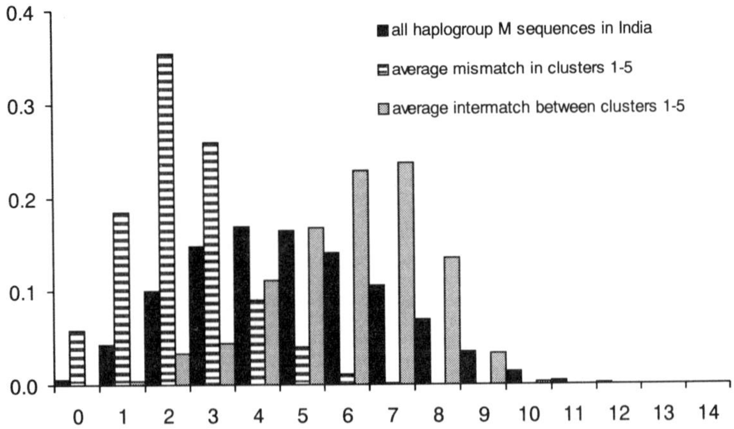

Figure 5. Mismatch distribution of haplogroup M in India. Mismatch and intermatch distribution of Indian-specific haplogroup M major subclusters 1–5 is shown alongside the mismatch distribution of all Indian M sequences.

expansion phase of haplogroup M may be hindered by cumulative demographic turbulences. Figure 5 shows intra- and inter-branch mismatch distribution patterns of five major sub-clusters of haplogroup M in India. Sub-clusters M1, M2a, M2b, M3-M5 all possess star-like topology (Fig. 3) and their expansion dates range from 17,000–32,000 years (Table 4). These star-like clusters thus reflect another influential demographic expansion in the history of Indian populations. In archaeology it roughly coincides with the proposed transition from Middle to Upper Palaeolithic in India (Joshi 1996). The inter-match distance of these clusters on the other hand exceeds the mean value of the total M sequence pool. There is also a significant portion of very divergent haplogroup M sub-lineages (Fig. 4) that do not reveal star-like traces of this recent expansion phase. The summary haplogroup M mismatch distribution and age derived from that could be therefore statistically elusive, affected by the sum of 'noise' from multiple founders and expansion phases. Below we use data from other mtDNA haplogroups to elucidate further the question when the first population expansion started in India.

2.2. mtDNA Haplogroups in India Which Are Considered Specific for Western Eurasia

Haplogroups H, I, J, K, T, U, V, W and X cover about 95% of the mtDNA variation in European populations (Torroni *et al.* 1996). Among them, haplogroup H is the most frequent, followed by U, T and J. Other minor haplogroups hardly ever exceed 5% in their frequency in Europe. Among Indian populations only haplogroup U displays frequency, comparable to that in European populations (Table 2). All other major western-Eurasian specific haplogroups are more than ten times less frequent. Haplogroup U is present also among some African populations (Torroni 1996, Passarino 1998). Hence, the possibility that the Indian haplogroup U varieties could be derived from a non European gene pool needs to be considered. Figure 6 represents a phylogenetical reconstruction (Kivisild *et al.* manuscript in preparation) of haplogroup U lineages in Indians. This phylogeny is dominated by a single but diverse sub-branch, which is defined by an A to G transition at np 16,051. The transition is rare in European haplogroup U sequences (sub-cluster U2) and is usually associated there with two additional mutations: an A to C transversion at np 16,129 and a T to C transition at np 16,189. The search in other world-wide mtDNA sequence data sets, including African ones, did not reveal sequences similar to those present in India. Thus, the extensive variation within the Indian mtDNA sub-cluster U2 has most likely developed locally in India. The coalescence age of the branches of U2 in India is 53,000 ± 4,000 years. This date is within the range of error limits of the haplogroup M coalescence time discussed above and synchronous with the oldest layer of haplogroup U found in Europe: sub-cluster U5 was dated 52,000 years old (Richards 1998). It is important to note, however, that these coalescence times predate the archeologically accepted time scale of peopling of Europe by modern humans.

There is still another fraction of haplogroup U, sub-cluster U7, which has likely developed in India from the Palaeolithic (Kivisild *et al.*, manuscript in preparation). It is occasionally found in populations of southern Europe and in the Trans-Caucasus but is present at much higher frequency and diversity in Indian populations. In contrast, other haplogroup U subclusters that are abundant in Europe are rare or missing in India. The same is true for the remaining European-specific mtDNA haplogroups (Table 2). Their combined frequency in India is below 8% and they are spread quite evenly: both in northern and southern regions. None of the extant western Eurasian populations display significantly higher affinity to the Indian "Caucasoid" lineages (Kivisild *et al*, manuscript

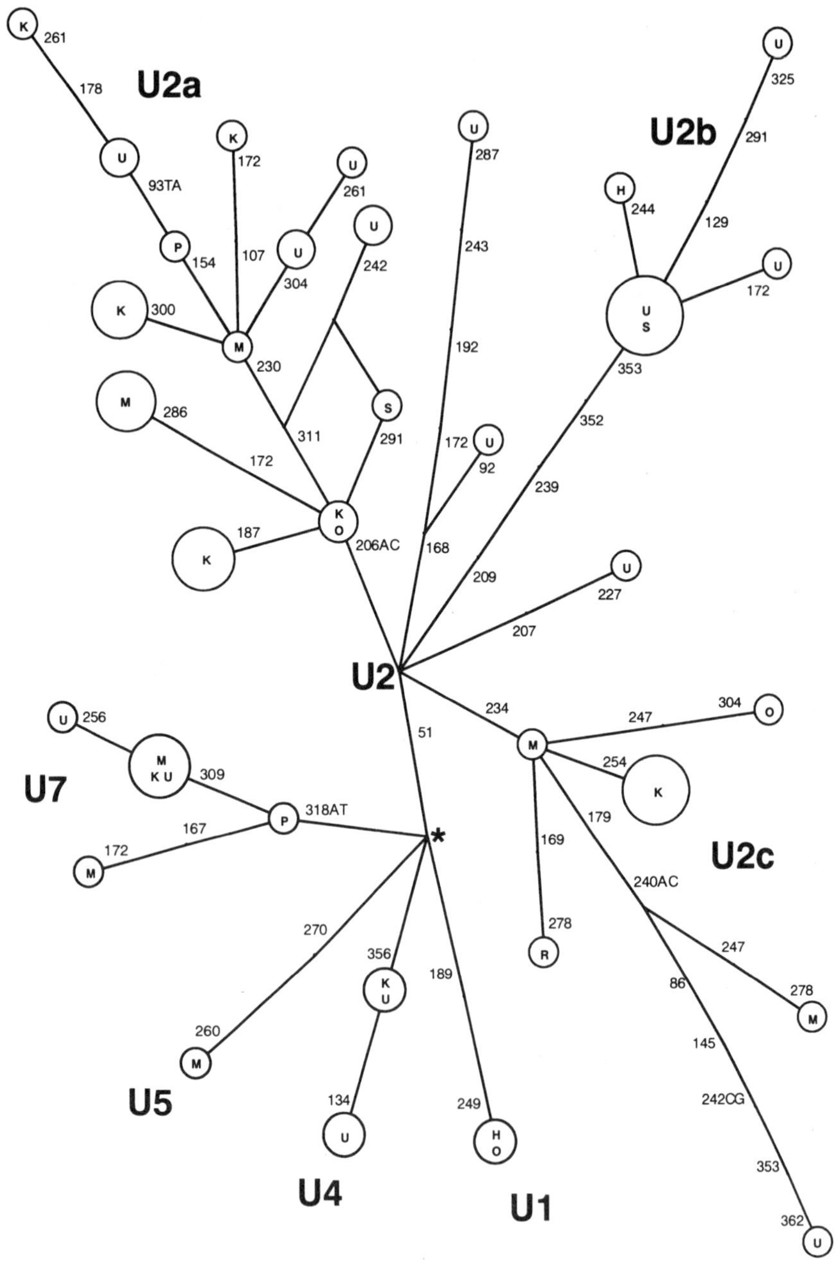

Figure 6. Reconstruction of haplogroup U in India. Populations as in Fig. 4. For other details, see Legend to Fig. 3.

in preparation). Thus, the putative approximately 4,000 year-old Indo-Aryan invasion can hardly be taken as an explanation for their spread in India. This 8% fraction of Indian maternal lineages that can be considered as a "loan" from Caucasoid populations is distant from the lineages shared with western Eurasian populations to an extent that suggests 9,000 years for their departure. This mean may and probably does reflect multiple minor immigration waves of nomadic people to India during a long time interval. Any

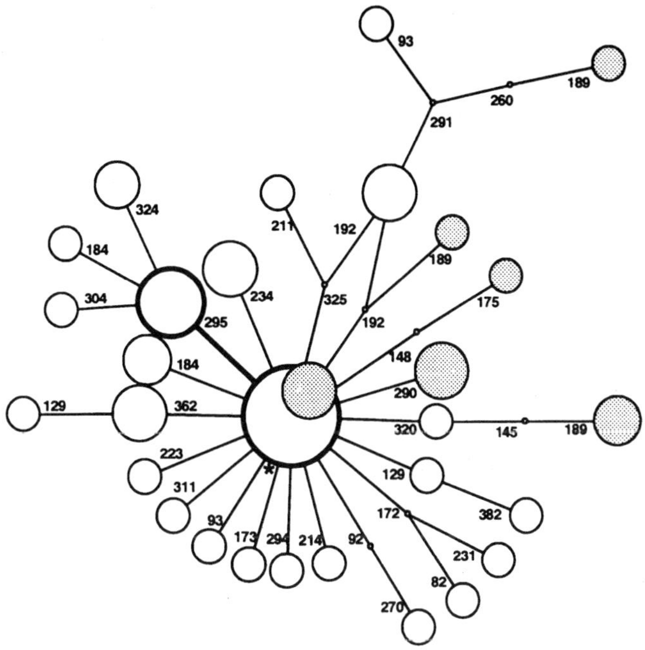

Figure 7. Reconstruction of haplogroup W in western Eurasia (open nodes) and in India (shaded nodes). For other details, see Legend to Fig. 3.

major recent immigration wave should be traceable in a form of frequency gradient. The only geographic variations detected in India are slight spatially ordered frequency changes in haplogroups U and M. However, both demonstrate variation by far older than the 4,000 years under discussion.

The presence of yet another western Eurasian-specific haplogroup in India deserves special attention. This is haplogroup W, which is a minor mtDNA variety in Europe (Torroni 1996) with a frequency of 3% and less. Haplogroup W was also found in India though with a rather limited and specific distribution (Kivisild *et al.*, manuscript in preparation). It is practically absent among most of the Indian populations but is present both in high frequency and diversity among the Lamani speaking Lobanas (13%) from Punjab. Another linguistically related Lamani population, the Lambadis from Andhra Pradesh, also possesses haplogroup W, but at several-fold lower frequency (3%). The European and Indian haplogroup W lineages overlap only at their ancestral node (Fig. 7). Coalescence ages calculated separately for Indian and European haplogroup W sequences are very similar: 22,000 and 25,000 years, respectively. The specific distribution of W in India suggests that it could have been imported there by a small number of populations, like the mobile Lobanas, who are known for their past inter-state salt trading occupation (Mastana *et al.* 1991). Nevertheless, as the lineages do not match those found so far in Europeans, their recent European admixture is unlikely. It is reasonable to guess that these lineages can be found also in Iran and/or Afghanistan.

Slightly less than one half of the European mtDNA-s belong to haplogroup H (Torroni *et al.* 1996). Haplogroup H has a minor sister-haplogroup, V, which is present in western Europe but can be found also at an exceptionally high frequency (>40%) among the Saamis (Torroni *et al.* 1998). These two haplogroups share a common

Table 5. Frequency of 00073 A in some Asian populations

	n/N	%
East Indonesia[1]	0/142	0
South China (Han)[1]	0/103	0
Java[1]	0/96	0
Malaysia[1]	0/81	0
Taiwan[1]	0/81	0
Bangladesh[1]	0/31	0
Philippines[1]	1/175	0.6
Borneo[1]	1/91	1.1
S. India and Sri Lanka[1]	2/73	2.7
S. India(Lambadis)[2]	2/86	2.3
N. India (Uttar Pradesh)[2]	4/70	5.7
N. India (Sikh)[1]	4/47	8.5
Pakistan[1]	11/73	15

[1]data from Melton and Stoneking 1996.
[2]Kivisild *et al.*, manuscript in preparation.

node in phylogeny (HV*in Macaulay *et al.* 1999), basic offshoots of which are rare in European populations. The derivatives of this ancestral node are, however, found at a remarkable frequency in the populations of Trans-Caucasus and also in northern India. This cluster, which we propose to define as haplogroup Ö, will be discussed in greater detail in our accompanying paper (E. Metspalu *et al*, this book). The populations of Kashmir and Pakistan are still insufficiently covered in this respect. Finding of four haplogroup Ö mtDNA genomes from a sample of 27 individuals from this region (see Fig. 1 in E. Metspalu *et al.*, this book) hints that the north-western part of the Indian Peninsula may be also rich in derivatives of this ancestral node, which has given rise to a half of European maternal inheritance. Nevertheless, haplogroup Ö has not penetrated deep into India. Taking the presence of nucleotide A at np 00073 as an indirect indicator (Table 5) one can see a clear diminishing south-eastward gradient of this lineage cluster in Asia. Haplogroup Ö reveals traces of a Palaeolithic demic expansion about 35,000 years ago (Kivisild *et al.*, manuscript in preparation). However, the lack of multiple direct matches in haplogroup Ö sequences between Indian samples and the Armenian, Ossetian and Georgian ones suggests that this link does not imply any recent admixture between these groups of populations. Instead, we can hypothesise that India, the Trans-Caucasus and the territories between them were the birthplace of the mitochondrial haplogroups (the place of their expansion) that eventually reached western as well as northern Europe.

2.3. Maternal Lineage Clusters, Which Are Highly Specific for Indian Populations While Being Rare Elsewhere

More than ten per cent of the Indian mtDNA sequences do not belong to any of the continent-specific mtDNA haplogroups characterised so far (Tables 1, 2). Nevertheless, the position of these lineages in the world-wide mtDNA phylogeny (Fig. 1) is not difficult to reveal: they all stem out of a node that occupies a crucial position in the human mtDNA phylogenetic tree (Kivisild *et al.*, manuscript in preparation). It coincides with a hypothetical branching point connecting a large number of distinct, well-characterised mtDNA haplogroups. Theoretically, the existence of such a node was

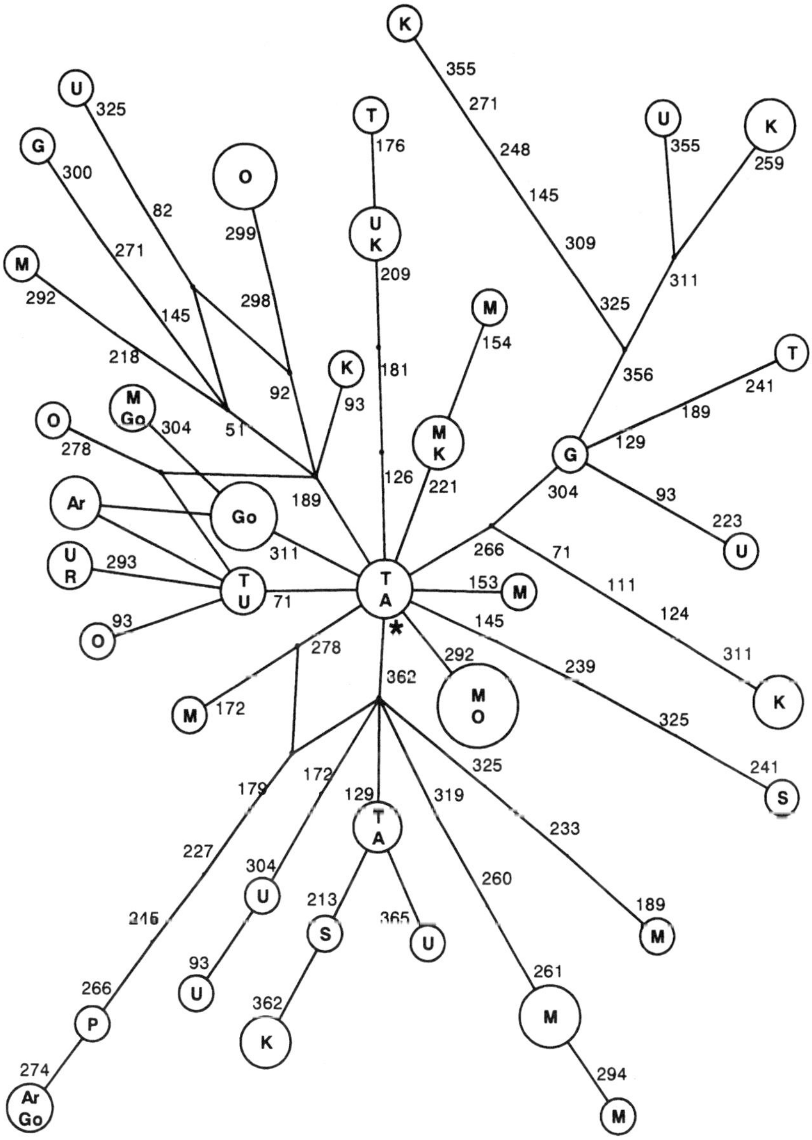

Figure 8. Reconstruction of super-haplogroup R lineages. Populations: U—Uttar Pradesh; O—Lobanas; M—Lambadis; S—Kashmir; K—Kerala; A—Andhra Pradesh; T—Tharus; P—Pakistan; G—Gujerat; R—Rajastan; Ar—Armenia; Go—Georgia. For other details, see Legend to Fig. 3.

obvious already earlier and in one of the schemes it has been defined as R* (Macaulay *et al.* 1999). However, thus far it has existed as an "empty node". Defining it as the founder of a super-haplogroup of mtDNA lineages allows one to say that it is an ancestral state of all western Eurasian sequences belonging to haplogroups H, V, J, T, U and K and has the same position relative to eastern Eurasian and Amerind sequences belonging to haplogroups F and B (Fig. 1). The western Eurasian haplogroups listed above constitute about 90% of mtDNA variation in Europe, whereas an Asian-specific haplogroup

B is close to fixation in some Polynesian populations (Lum *et al.* 1998; Sykes *et al.* 1995) and, together with haplogroup F, makes up a large portion of the mtDNA varieties found in southeastern Asian populations (Ballinger *et al.* 1992).

What is remarkable about the Indian populations is that here and so far only here, one can find at a significant frequency and diversity mtDNA sequences in the vicinity of this central node (Kivisild *et al.*, manuscript in preparation; Fig. 8). Calculation of the beginning of expansion time for R* in India resulted in estimate of 55,000 ± 5,000 years. It is noteworthy that this time estimate practically coincides with those we found for haplogroups M and U2.

3. CONCLUSIONS

Both western and eastern Eurasian-specific mtDNA haplogroups can be found in India together with strictly Indian-specific ones. However, in India the structure of the haplogroups shared either with western or eastern Eurasian populations is profoundly different. This indicates a local independent development over a very long time period. Minor overlaps with lineages described in other Eurasian populations clearly demonstrate that recent immigrations have had very little impact on the innate structure of the maternal gene pool of Indians. Despite the variations found within India, these populations stem from a limited number of founder lineages. These lineages were most likely introduced to the Indian subcontinent during the Middle Palaeolithic, before the peopling of Europe and perhaps the Old World in general. Our demographic analysis reveals at least two major expansion phases that have influenced the wide assortment of the Indian mtDNA lineages. The more recent phase, which according to our estimation started around 20,000–30,000 years ago, seems to correspond to the transition from the Middle to the Upper Palaeolithic. The first expansion phase may reflect a demographic burst immediately after the initial peopling of India around 50—60 thousand years ago. This wave of expansion brought forward also those maternal lineages that can rightfully claim the name of Eurasian Eves.

4. REFERENCES

Aitken MJ, Valladas H (1992) Luminescence dating relevant to human origins. Philos Trans R Soc Lond B Biol Sci 337:139–44.

Anderson S, Bankier AT, Barrell BG, de Bruijn MH, Coulson AR, Drouin J, Eperon IC, *et al.* (1981) Sequence and organization of the human mitochondrial genome. Nature 290:457–65.

Bailliet G, Rothhammer F, Carnese FR, Bravi CM, Bianchi NO (1994) Founder mitochondrial haplotypes in Amerindian populations. Am J Hum Genet 55:27–33.

Ballinger SW, Schurr TG, Torroni A, Gan YY, Hodge JA, Hassan K, Chen KH, *et al.* (1992) Southeast Asian mitochondrial DNA analysis reveals genetic continuity of ancient mongoloid migrations. Genetics 130:139–52.

Bamshad M, Rao BB, Naidu JM, Prasad BVR, Watkins S, Jorde LB (1997) Response to Spurdle *et al.* Human Biology 69:432–435.

Bamshad MJ, Watkins WS, Dixon ME, Jorde LB, Rao BB, Naidu JM, Prasad BV, *et al.* (1998) Female gene flow stratifies Hindu castes. Nature 395:651–2.

Brega A, Gardella R, Semino O, Morpurgo G, Astaldi Ricotti GB, Wallace DC, Santachiara Benerecetti AS (1986) Genetic studies on the Tharu population of Nepal: restriction endonuclease polymorphisms of mitochondrial DNA. Am J Hum Genet 39:502–12.

Cann RL, Stoneking M, Wilson AC (1987) Mitochondrial DNA and human evolution [see comments]. Nature 325:31–6.

Chen YS, Torroni A, Excoffier L, Santachiara-Benerecetti AS, Wallace DC (1995) Analysis of mtDNA variation in African populations reveals the most ancient of all human continent-specific haplogroups. Am J Hum Genet 57:133–49.

Comas D, Calafell F, Mateu E, Perez-Lezaun A, Bosch E, Martinez-Arias R, Clarimon J, *et al.* (1998) Trading genes along the silk road: mtDNA sequences and the origin of central Asian populations. Am J Hum Genet 63:1824–38.

Deraniyagala SU (1998) Pre- and protohistoric settlement in Sri Lanka. XIII U.I.S.P. Congress Proceedings. Vol. 5., A.B.A.C.O. s.r.l., Forli, pp. 277–285.

Foley R (1998) The context of human genetic evolution. Genome Res 8:339–47.

Forster P, Harding R, Torroni A, Bandelt HJ (1996) Origin and evolution of Native American mtDNA variation: a reappraisal. Am J Hum Genet 59:935–45.

Horai S, Murayama K, Hayasaka K, Matsubayashi S, Hattori Y, Fucharoen G, Harihara S, *et al.* (1996) mtDNA polymorphism in East Asian Populations, with special reference to the peopling of Japan. Am J Hum Genet 59:579–90.

Joshi RV (1996) SOUTH ASIA in the period of *Homo sapiens neanderthalensis* and contemporaries (Middle Palaeolithic) History of Humanity. Vol. I. UNESCO, pp 162–164.

Kolman C, Sambuughin N, Bermingham E (1996) Mitochondrial DNA analysis of Mongolian populations and implications for the origin of New World founders. Genetics 142:1321–34.

Krings M, Stone A, Schmitz RW, Krainitzki H, Stoneking M, Paabo S (1997) Neanderthal DNA sequences and the origin of modern humans. Cell 90:19–30.

Lum JK, Cann RL, Martinson JJ, Jorde LB (1998) Mitochondrial and nuclear genetic relationships among Pacific Island and Asian populations. Am J Hum Genet 63:613–24.

Lutz S, Weisser HJ, Heizmann J, Pollak S (1998) Location and frequency of polymorphic positions in the mtDNA control region of individuals from Germany. Int J Legal Med 111:67–77.

Macaulay VA, Richards MB, Hickey E, Vega E, Cruciani F, Guida V, Scozzari R, *et al.* (1999) The Emerging Tree of West Eurasian mtDNAs: A Synthesis of Control-Region Sequences and RFLPs. Am J Hum Genet 64.

Mastana SS, Garlick JP, Papiha SS (1991) Serogenetic studies of the Lobanas in north-west India. Anthropol Anz 49:177–86.

McDermott F, Grun R, Stringer CB, Hawkesworth CJ (1993) Mass-spectrometric U-series dates for Israeli Neanderthal/early modern hominid sites. Nature 363:252–5.

Melton T, Stoneking M (1996) Extent of heterogeneity in mitochondrial DNA of ethnic Asian populations. J Forensic Sci 41:591–602.

Mountain JL, Hebert JM, Bhattacharyya S, Underhill PA, Ottolenghi C, Gadgil M, Cavalli-Sforza LL (1995) Demographic history of India and mtDNA-sequence diversity. Am J Hum Genet 56:979–92.

Passarino G, Semino O, Bernini LF, Santachiara-Benerecetti AS (1996) Pre-Caucasoid and Caucasoid genetic features of the Indian population, revealed by mtDNA polymorphisms. Am J Hum Genet 59:927–34.

Passarino G, Semino O, Modiano G, Bernini LF, Santachiara Benerecetti AS (1996) mtDNA provides the first known marker distinguishing proto-Indians from the other Caucasoids; it probably predates the diversification between Indians and Orientals. Ann Hum Biol 23:121–6.

Passarino G, Semino O, Modiano G, Santachiara-Benerecetti AS (1993) COII/tRNA(Lys) intergenic 9-bp deletion and other mtDNA markers clearly reveal that the Tharus (southern Nepal) have Oriental affinities. Am J Hum Genet 53:609–18.

Passarino G, Semino O, Pepe G, Shrestha SL, Modiano G, Santachiara Benerecetti AS (1992) MtDNA polymorphisms among Tharus of eastern Terai (Nepal). Gene Geogr 6:139–47.

Passarino G, Semino O, Quintana-Murci L, Excoffier L, Hammer M, Santachiara-Benerecetti AS (1998) Different genetic components in the Ethiopian population, identified by mtDNA and Y-chromosome polymorphisms. Am J Hum Genet 62:420–34.

Richards MB, Macaulay VA, Bandelt HJ, Sykes BC (1998) Phylogeography of mitochondrial DNA in western Europe. Ann Hum Genet 62:241 60.

Schurr TG, Ballinger SW, Gan YY, Hodge JA, Merriwether DA, Lawrence DN, Knowler WC, *et al.* (1990) Amerindian mitochondrial DNAs have rare Asian mutations at high frequencies, suggesting they derived from four primary maternal lineages. Am J Hum Genet 46:613–23.

Stringer CB (1990) The emergence of modern humans. Sci Am 263:98–104.

Stringer CB (1992) Reconstructing recent human evolution. Philos Trans R Soc Lond B Biol Sci 337:217–24.

Stringer CB, Grun R, Schwarcz HP, Goldberg P (1989) ESR dates for the hominid burial site of Es Skhul in Israel. Nature 338:756–8.

Sykes B, Leiboff A, Low-Beer J, Tetzner S, Richards M (1995) The origins of the Polynesians: an interpretation from mitochondrial lineage analysis. Am J Hum Genet 57:1463–75.

Torroni A, Bandelt HJ, D'Urbano L, Lahermo P, Moral P, Sellitto D, Rengo C, *et al.* (1998) mtDNA analysis reveals a major late Paleolithic population expansion from southwestern to northeastern Europe. Am J Hum Genet 62:1137–52.

Torroni A, Huoponen K, Francalacci P, Petrozzi M, Morelli L, Scozzari R, Obinu D, *et al.* (1996) Classification of European mtDNAs from an analysis of three European populations. Genetics 144:1835–50.

Torroni A, Miller JA, Moore LG, Zamudio S, Zhuang J, Droma T, Wallace DC (1994) Mitochondrial DNA analysis in Tibet: implications for the origin of the Tibetan population and its adaptation to high altitude. Am J Phys Anthropol 93:189–99.

Torroni A, Neel JV, Barrantes R, Schurr TG, Wallace DC (1994) Mitochondrial DNA "clock" for the Amerinds and its implications for timing their entry into North America. Proc Natl Acad Sci U S A 91:1158–62.

Torroni A, Petrozzi M, D'Urbano L, Sellitto D, Zeviani M, Carrara F, Carducci C, *et al.* (1997) Haplotype and phylogenetic analyses suggest that one European-specific mtDNA background plays a role in the expression of Leber hereditary optic neuropathy by increasing the penetrance of the primary mutations 11778 and 14484. Am J Hum Genet 60:1107–21.

Torroni A, Schurr TG, Cabell MF, Brown MD, Neel JV, Larsen M, Smith DG, *et al.* (1993) Asian affinities and continental radiation of the four founding Native American mtDNAs. Am J Hum Genet 53:563–90.

Torroni A, Schurr TG, Yang CC, Szathmary EJ, Williams RC, Schanfield MS, Troup GA, *et al.* (1992) Native American mitochondrial DNA analysis indicates that the Amerind and the Nadene populations were founded by two independent migrations. Genetics 130:153–62.

Torroni A, Sukernik RI, Schurr TG, Starikorskaya YB, Cabell MF, Crawford MH, Comuzzie AG, *et al.* (1993) mtDNA variation of aboriginal Siberians reveals distinct genetic affinities with Native Americans. Am J Hum Genet 53:591–608.

Wallace D (1995) 1994 William Allan Award Address. Mitochondrial DNA variation in human evolution, degenerative disease, and aging. Am J Hum Genet 57:201–23.

Watson E, Forster P, Richards M, Bandelt HJ (1997) Mitochondrial footprints of human expansions in Africa. Am J Hum Genet 61:691–704.

12

MITOCHONDRIAL DNA VARIATION IN THE SOUTHWEST PACIFIC

D. Andrew Merriwether[1], Frederika A. Kaestle[1], Babette Zemel[2], George Koki[3], Charles Mgone[3], Michael Alpers[3], and Jonathan Friedlaender[4]

[1]Department of Anthropology
University of Michigan
Ann Arbor, Michigan, USA
[2]Krogman Growth Center
Children's Hospital of Philadelphia
Philadelphia, Pennsylvania, USA
[3]Institute of Medical Research
Goroka, Papua New Guinea
[4]Department of Anthropology
Temple University
Philadelphia, Pennsylvania, USA

ABSTRACT

Human populations in the southwest Pacific have a reputation of being particularly diverse genetically. In this preliminary survey of this region, we have identified five major clusters of mitochondrial haplotypes among 134 DNA samples from dispersed populations: From Papua New Guinea; Gende-speaking groups (Bundi) from Madang Province, groups belonging to the Torricelli language Phylum of West Sepik Province, groups near Minj in the highlands, villages near Popondetta and Garaina in Northern Province; the Baining of East New Britain Province, and Misima Island. We included, as well, samples from the Santa Cruz (see Fig. 1). Two of these clusters conform to clusters I and II identified by Lum *et al.* (1994). The distribution of Cluster I (a haplotype including the mtDNA Region V 9-bp deletion) is consistent with a recent introduction by Austronesians (presumably the ancestors of the Polynesians) approximately 3,000 years ago. Cluster I was detected among the Garaina, Misima Island, Popondetta, and Santa Cruz samples. We detected Cluster II in all of our mainland New Guinea populations except for Minj. We did not find any evidence of group III identified by Lum *et al.* but found three additional clusters (IV–VI) not found by Lum *et al.* (1994). Cluster IV was seen in the more easterly samples from Misima Island, Bundi,

Minj, and Popondetta. Cluster V was only found in some individuals from Garaina in our study (but 2 of the 6 mutations defining haplogroup V) have been seen by others in Borneo and the Philippines (Sykes *et al.* 1995). Cluster VI was found widely from the Sepik and Karawari River region to Popondetta, Baining, Minj, and Misima Island. Cluster VI contains the Cambridge reference sequence (Anderson *et al.* 1981) which is known to be common in Europeans, but is also widespread globally. We question whether this maternally-derived haplogroup found in widespread and remote regions of PNG and Island Melanesia is due to European admixture, as has been suggested elsewhere. Lastly, mismatch distributions of 9-bp deleted versus non-deleted PNG sequences were markedly different. The 9-bp deleted lineages appear to be quite recent in PNG (and Melanesia as a whole), The mean number of pairwise differences in 9-bp deleted lineages is 1.6, while the mean is nearly 5.8 for all the non-deleted lineages in PNG.

1. INTRODUCTION

Evidence exists placing modern humans in New Guinea by at least 45,000 years ago (Roberts *et al.* 1994; Groube *et al.* 1986), and on the neighboring islands to the east, including the interior of New Britain (Pavlides and Gosden 1994) and New Ireland (Allen 1996) by 35,000 years ago. Humans had even reached Buka Island and presumably Northern Bougainville by 29,000 years ago (Wickler and Spriggs 1988; Irwin 1992). Modern humans apparently spread no further than the Solomons Archipelago until about 3,000 years ago, when another demographic impulse originating in southeast Asia, and associated with a distinctive cultural tradition, is clearly detectable in the region. Populations in this relatively isolated corner of the planet have been evolving and diverging for over 40,000 years, and human variation there is often characterized linguistically and genetically as extreme (Attenborough and Alpers 1992; Bellwood 1979, 1989; Spriggs 1997; Kirch 1997). Over the years there have been many studies of variation in blood group markers (Simmons 1973; Mourant 1974; Booth *et al.* 1970; Salmon *et al.* 1988), enzyme and protein polymorphisms (Weitkamp *et al.* 1969; Wood *et al.* 1982; Serjeantson *et al.* 1973; Weisenfeld and Gadusek 1976; Booth *et al.* 1982; Steinberg and Larrick 1981; Schanfield *et al.* 1975; inter alia), and more recently mitochondrial DNA (Stoneking *et al.* 1990; Melton *et al.* 1995, 1999; Redd *et al.* 1995; Hagelberg *et al.* 1999; Lum *et al.* 1994, 1998; Lum and Cann 1998; Merriwether *et al.* in press), hemoglobin (Martinson 1996; Yenchitsomanus *et al.* 1985; Roberts-Thompson *et al.* 1996; Tsintsof *et al.* 1990), and the Y chromosome (Hammer 1995; Hammer *et al.* 1997; Forster *et al.* 1998; Breuil *et al.* 1987). The sampling in all of these studies is often haphazard, and the groups sampled in these different studies do not overlap, making comparisons between different systems of analysis difficult. Nonetheless, a picture of extreme diversity emerges from many different genetic systems (Tishkoff *et al.* 1996; Martinson 1996; Redd *et al.* 1995).

1.1. Language Families

There is a linguistic dichotomy commonly invoked in describing Oceania. Languages are generally classified as Austronesian (AN) or non-Austronesian (NAN). Austronesian is a well-recognized group of related languages, with well-defined sub-divisions (Ross 1988; Blust 1996a, 1996b; Kirch 1997; Pawley and Ross 1998), and geographically defined groupings. Non-Austronesian, on the other hand, is a catch-all for languages spoken in the region that are not related to Austronesian, and that very often cannot be linked to each other.

Most New Guinea populations (both Papua New Guinea and Irian Jaya) speak NAN languages. The populations sampled from the Sepik River area speak languages in the Torricelli language Phylum (Wurm 1982). Popondetta is surrounded by languages belonging to the Binandere Stock of the Trans-New Guinea Phylum (TNG), Garaina is surrounded by Guhu-Samane Family languages (same stock and phylum). Minj is also part of the TNG Phylum. The Baining of New Britain have been placed by Wurm in his proposed East Papuan Phylum, unrelated to either of the above groups. The Santa Cruz NAN languages are an anomaly. They are the easternmost NAN-speaking population in Oceania (and the only NAN population in remote Oceania, beyond the main Solomons archipelago), but could be remotely related to other East Papuan languages.

Austronesian languages in this region of New Guinea and Near Oceania are assumed to be intrusive, and are largely spoken along the coast of New Guinea and on most of the near-by islands. Misima Islanders speak a Milne Bay Family language of the Papuan Tip cluster of languages [one of Ross' (1988) four main clusters of Western Oceanic in the Near Oceanic region]. Intermixture with other populations, especially around Garaina and Popondetta, is likely to have occurred.

Despite the linguistic and genetic diversity, and perhaps because of it, very little is known about the origins of Melanesians in Near Oceania. While there have been numerous attempts to trace the origins of the Polynesian and Micronesian islanders (Lum and Cann 1998; Lum *et al.* 1994; Merriwether *et al.* 1999; Sykes *et al.* 1995; Martinson *et al.* 1996; Redd *et al.* 1995; Melton *et al.* 1995, 1999; Hagelberg *et al.* 1999; Hertzerg *et al.* 1989; Serjeantson *et al.* 1983), there have been no convincing arguments put forth as to the origins of the non-Austronesian speakers in Island Melanesia. In part, this is because proto-language reconstructions are not reliable using standard historical linguistic techniques beyond about 6–10,000 years. Similarly, many populations in this region are likely to be genetically equidistant from many other human populations because they have been diverging in relative isolation for so long. Here we present preliminary evidence for the types and distributions of mitochondrial DNA diversity in this region.

2. MATERIALS AND METHODS

We extracted DNA from 134 plasma and/or whole blood samples collected from Bundi (Gende—speakers—n = 8) in the 1980's by Babette Zemel (1989; Zemel and Jenkins 1989), and from Sepik area villages (n = 14. 7 from Momboyoto, 4 from Kambadwa, and 3 from Sagare villages), Garaina (n = 21. 9 from Garaina Village, 11 from Kasu Village, and 1 from Sopa Village), Popondetta (n = 13 from two un-named villages near Popondetta), Misima Island (n = 23 from Liak Village), and Minj (n = 2 from Uinba Village) collected by Lowell Weitkamp and colleagues between 1967 and 1969 (Weitkamp *et al.* 1969). See Fig. 1 for the map locations of these samples. We also present unpublished data from a paper in preparation on non-Austronesian speakers in the Santa Cruz Islands of Solomon Islands (n = 47, Green, *et al.* in preparation) and on the East Baining from East New Britain (n = 8, Mali dialect of the Baining language, Marambu Village, Sinivit District) collected by Friedlaender in 1998.

The protocol for new sample collection as well as old sample reanalysis was approved by the University of Michigan Human Subjects Committee Internal Review Board and by the Medical Research Advisory Committee of Papua New Guinea.

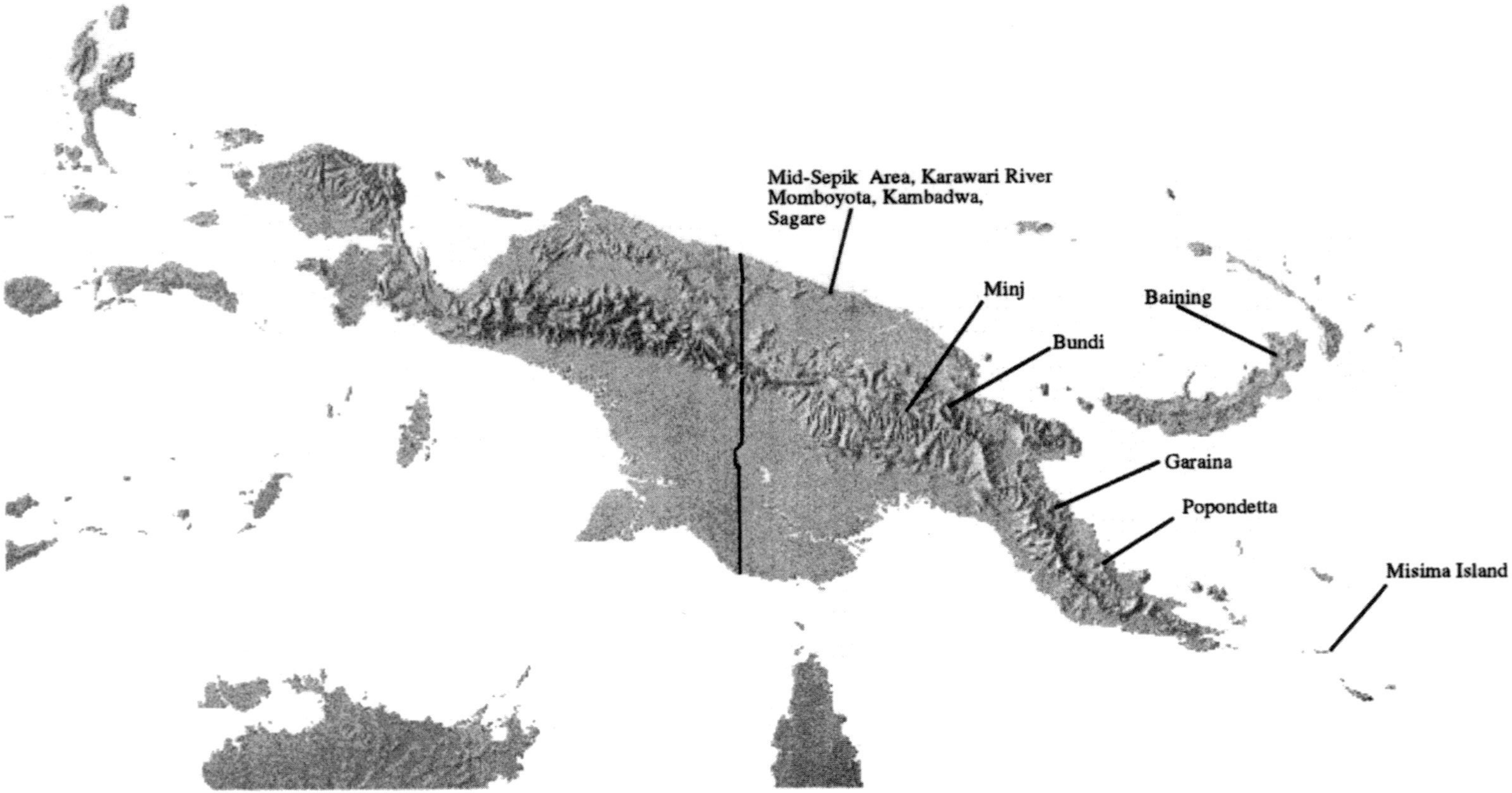

Figure 1. Locations of the populations sampled in this study.

2.1. Extraction and Sequencing

DNA was extracted using a modification of the IsoQuik protocol (Orca Scientific) using 100 microliters of plasma. All samples were then genomically amplified using the DOP procedure as described by Cheung and Nelson (1996). The DOP-DNA was then amplified using specific primers for the first hypervariable region of the D-loop (typically nts 15938–16315) as well as for primers spanning the Region V 9-bp deletion (spanning nts 8195–8317). All PCR products were electrophoresed through ethidium bromide stained agarose gels and visualized under ultraviolet light. Amplicons were purified for sequencing using Micron 100 microconcentrators (Amicon), then sequenced in both directions with the FS Dye Terminator kit (ABI) using half size reactions and manufacturers recommended conditions. Sequenced DNA was electrophoresed on an ABI 377XL automated sequencer.

2.2. Sequence Analysis

Sequences were aligned by hand using Sequence Navigator (ABI) and converted to sequential PHYLIP "infile" format using an Applescript program called Layout Converter (written by John McDonough and D. Andrew Merriwether, available upon request via email from D. A. Merriwether at andym@umich.edu). Additional sequences from PNG were added from the literature. These included 8 sequences described as "New Guineans from various parts of coastal and highland Papua New Guinea" by Vigilant *et al.* (1991: pp1503–1504) and herein labelled PNG. Also included were 23 Coastal PNG (CPNG) from Redd *et al.* (1995). Lastly, we included 5 African DNAs (4 !Kung and 1 Yoruba from Vigilant *et al.* 1991). Identical sequences were filtered and merged together using MaClade 3.06 (Maddison and Maddison 1992, Sinauer Associates). Phylogenetic trees were generated using two different methods: 1) Kimura 2-Parameter genetic distance matrices were made using PHYLIP 3.51′s DNADIST program (Felsenstein 1999), and a neighbor joining tree (abbreviated NJ hereafter, Saitou and Nei 1987) was generated using PHYLIP's NEIGHBOR program, and visualized using the TreeView PPC program (Page 1996) 2) Bandelt Median Networks (abbreviated as BMN hereafter, Bandelt *et al.* 1995) were generated using the RM and MJ programs (Bandelt and 1995, 1999).

3. RESULTS

Using the non-Austronesian/Austronesian dichotomy as a starting point, it is useful to look at the distribution of one particularly informative mtDNA marker: the Region V 9-bp deletion. This 9-bp deletion is found in all Polynesian and Micronesian populations, and its frequency varies from 85–100% in Polynesia. It also exists in lower frequencies in Taiwan and southeast Asia, where Austronesians are thought to have originated, yet early studies failed to detect it in non-Austronesian and native Australian populations. Therefore, it is assumed that all or most of the founders of Polynesia carried the 9-bp deletion. This makes it an excellent marker to trace the expansion that led to the peopling of Polynesia (See Fig. 2). The frequencies of the 9-bp deletion in more complete surveys in Near and Remote Oceania show that the 9-bp deletion is quite frequent in Near Oceania in NAN populations as well. However, more remote, highland, populations such as the Aita and Rotokas in North Bougainville, the Baining in East New

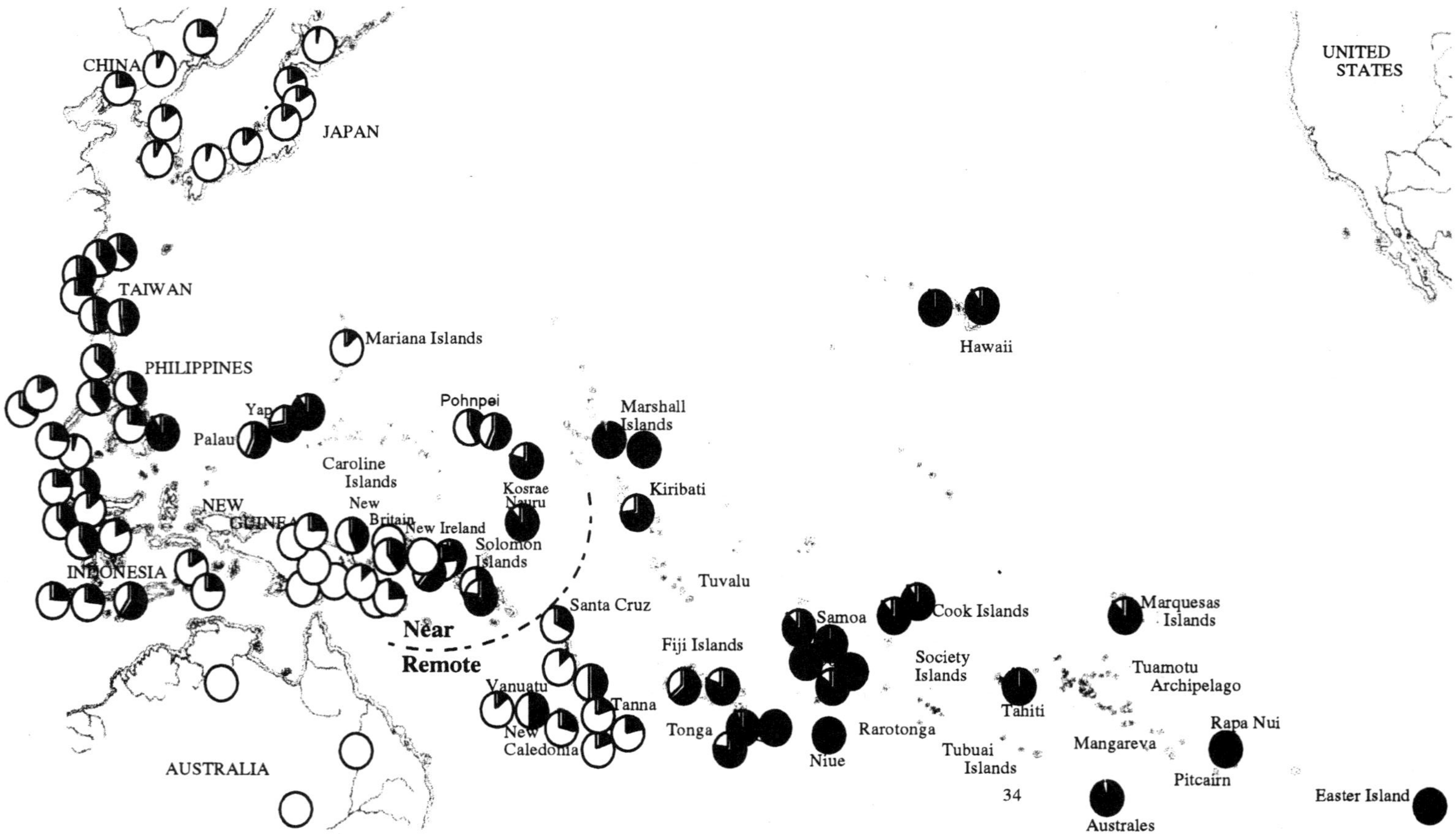

Figure 2. Frequencies of the 9-bp Region V deletion in the Pacific. The proportion of the pie-chart that is filled (black) is equal to the frequency of the 9-bp deletion in that population. Data from the literature is from: Ballinger *et al.* 1992; Betty *et al.* 1996; Green *et al.* in preparation; Harihara *et al.* 1992; Hertzberg *et al.* 1989; Lum *et al.* 1994, 1998; Lum and Cann 1998; Melton *et al.* 1995; Merriwether *et al.* in review; Redd *et al.* 1995; Stoneking *et al.* 1989, 1990; Sykes *et al.* 1995; Vigilant *et al.* 1991.

Table 1. Frequencies of the 9-bp deletion in the samples in this study

Group	Non-deleted	Deleted	Total	Deletion Freq.
Popondetta	11	2	13	0.15
Garaina	15	8	23	0.35
Bundi	8	0	8	0
Minj	2	0	2	0
Momboyoto	7	0	7	0
Kambadwa	4	0	4	0
Sagare	3	0	3	0
Mid-Sepik total	14	0	14	0
PNG total	50	10	60	0.17

Britain, and highland Papua New Guineans all uniformly lack the deletion. This absence in more isolated areas has been once more interpreted by Merriwether *et al.* (in press) as representing the condition of the inhabitants of the region prior to the intrusion of the Austronesian expansion through the region (see Table 1 for the frequencies of the 9-bp deletion in the populations from this study). Also note that a significant proportion of individuals in Polynesia carry non-deleted haplotypes (15% in American Samoa, Merriwether *et al.* in press). It remains unclear whether these were part of the initial migration into remote Oceania or if they represent admixture with non-Austronesians as the founders moved through Near Oceania.

The distribution of Haplogroups I-V from our study and from the literature is listed in Table 2. Since the sample sizes to date are very small, the distributions are at best approximate. Figure 3 shows the distribution of Haplogroups I-V on a map of Oceania. Haplogroups I and II were found in Polynesia, Micronesia, and Melanesia. Haplogroup III was found in Indonesia, western Micronesia, and Samoa (although only one point mutation links just one Samoan individual to Haplogroup III). Haplogroup IV has been found only in New Guinea, Near Oceania and northwestern-most Micronesia. Haplogroup V appears in New Guinea, and at least two of the six mutations defining Haplogroup V are seen in the Philippines and Borneo (the 16157 and 16256 sites are detected there in Sykes *et al.* 1995). Haplogroup VI is found in the mid-Sepik River area, Minj, Popondetta, CPNG, Misima, and Bundi, and includes the presumed Caucasian European Anderson Reference sequence (Anderson *et al.* 1981).

Figure 4 is a Bandelt Median Network representation (BMN) of nts 16024–16259 for 186 individuals from this study and from the literature. The six haplogroups are circled and their members identified. Cluster I, corresponding to individuals with the Region V 9-bp deletion, was seen in Garaina, Misima, Popondetta, and Santa Cruz. Cluster II was found in all of our PNG populations except Minj. Cluster IV was seen in Misima, Bundi, Minj, and Popondetta. Cluster V was only found in some individuals from Garaina. Cluster VI was found in the Sepik region, Popondetta, Baining, Minj, and Misima.

The polymorphic sites that define these haplotypes are given in Table 3, not including those sites identical to the Anderson *et al.* (1981) published reference sequence. Figure 5 is a neighbor-joining tree (NJ) of nts 16024–16310. We were able to analyze a longer fragment for the NJ tree because the NJ method can accommodate missing data, whereas the BMN method cannot. Since some sequences were longer or shorter than others, there are clearly missing data at the ends of sequences. This does not particularly

Table 2. Frequencies of the Haplogroups found in our study and from the literature

Group	Region	Language*	Haplogroups I	II	III	IV	V	N	References
Australs	Pol.	OcAN	0	0	0	0	0	1	1
Aoteoroa (NZ)	Pol.	OcAN	0	0	0	0	0	1	2
American Samoa	Pol.	OcAN	9 (100%)	0	0	0	0	9	3
Samoa	Pol.	OcAN	11 (73%)	0	1 (7%)	0	0	15	1, 2, 4
Cooks Island	Pol.	OcAN	3 (30%)	1 (10%)	0	0	0	10	2
Hawai'i	Pol.	OcAN	10 (91%)	0	0	0	0	11	1, 4
Kapingamarangi	Pol.	OcAN	25 (89%)	0	0	0	0	28	1
Marquesas	Pol.	OcAN	0	0	0	0	0	5	2
Rapa Nui (Easter Is.)	Pol.	OcAN	10 (100%)	0	0	0	0	10	1
Tonga	Pol.	OcAN	0	2 (67%)	0	0	0	3	2
Tahiti	Pol.	OcAN	0	0	0	0	0	2	2
Kiribati	Mic.	OcAN	3 (38%)	0	0	1 (13%)	0	8	1
Kosrae	Mic.	OcAN	27 (79%)	0	0	0	0	34	1
Marianas	Mic.	AN	6 (12%)	0	0	0	0	49	1
Micronesia	Mic.	AN	0	0	0	0	0	1	4
Marshall Isl.	Mic.	OcAN	25 (93%)	0	0	0	0	27	1, 2
Nauru	Mic.	OcAN	21 (84%)	0	0	1 (4%)	0	25	1
Outer Island Yap	Mic.	OcAN	109 (87%)	8 (6%)	0	3 (2%)	0	126	1
Palau	Mic.	AN	46 (40%)	0	2 (2%)	34 (29%)	0	116	1
Southwest Palau	Mic.	AN	12 (100%)	0	0	0	0	12	1
Pohnpei	Mic.	OcAN	13 (59%)	2 (9%)	0	0	0	22	1
Yap	Mic.	AN	36 (62%)	0	0	0	0	58	1
Micronesia total	Mic.	mult.	366 (64%)	13 (2%)	3 (<1%)	39 (7%)	0	576	1, 2, 3, 4
Garaina (PNG)	Mel.	NAN	10 (48%)	5 (24%)	0	1 (4%)	5 (24%)	21	8
Popondetta (PNG)	Mel.	NAN	2 (15%)	6 (46%)	0	3 (23%)	0	13	8
Bundi (PNG)	Mel.	NAN	0	3 (38%)	0	5 (62%)	0	8	8
Minj (PNG)	Mel.	NAN	1 (50%)	0	0	0	0	2	8
Momboyoto (m-SPNG)	Mel.	NAN	0	5 (71%)	0	0	0	7	8

Sagare (m-SPNG)	Mel.	NAN	1 (33%)	0	0	0	0	3	8
Kamadwa (m-SPNG)	Mel.	NAN	0	2 (50%)	0	0	0	4	8
Mid-Sepik total (PNG)	Mel.	NAN	1 (7%)	7 (50%)	0	0	0	14	8
total our PNG	Mel.	NAN	14 (24%)	21 (36%)	0	8 (14%)	5 (9%)	58	8
lit PNG	Mel.	NAN	14 (30%)	17 (36%)	0	7 (15%)	0	47	1, 2, 3, 4, 6
Total PNG	Mel.	NAN	23 (27%)	38 (36%)	0	15 (14%)	5 (5%)	105	1, 2, 3, 4, 6, 8
Baining	Mel.	NAN	0	5 (83%)	0	0	0	6	8
Misima	Mel.	OcAN	3 (13%)	1 (4%)	0	18 (75%)	0	24	8
Santa Cruz	Mel.	NAN	7 (12%)	30 (53%)	0	5 (9%)	0	57	9
Vanuatu	Mel.	OcAN	3 (5%)	21 (38%)	0	7 (13%)	0	56	1, 2
Fiji	Mel.	OcAN	9 (69%)	1 (8%)	0	1 (8%)	0	13	1
Melanesia total	Mel.	mult.	50 (19%)	96 (37%)	0	46 (18%)	5 (2%)	261	1, 2, 3, 4, 5, 8, 9
Multiple	mult.	mult.	7 (50%)	2 (14%)	0	0	0	14	2, 4
Australia	Aus.	NAN	0	2 poss. (3%)	0	0	0	58	4, 5, 6, 7
Borneo	ISEAsia	AN	3 (12%)	0	0	0	0	26	1, 2
Indonesia	ISEAsia	AN	20 (67%)	0	0	0	0	30	3, 4
Java	ISEAsia	AN	1 (6%)	0	0	0	0	18	1
Philippines	ISEAsia	AN	7 (18%)	0	2 (5%)	0	0	40	1, 2
Taiwan	ISEAsia	AN	5 (31%)	0	0	0	0	16	2
Thailand	Main. Asia	other	0	0	0	0	0	9	1
Vietnam	Main. Asia	other	4 (17%)	0	0	1 (4%)	0	24	1, 2
Moken	Ind. Ocean	other	0	0	0	0	0	8	1

*language abbreviations are as follows: Austronesian other than Oceanic branch—AN; Oceanic Branch of Austronesian—OcAN; Non-Austronesian, NAN; languages outside Oceania—other.

[1] Lum *et al.* (1998)
[2] Sykes *et al.* (1995)
[3] Redd *et al.* (1995)
[4] Lum *et al.* (1994)
[5] Betty *et al.* (1996)
[6] Vigilant *et al.* (1991)
[7] von Holst Pellekaan *et al.* (1998)
[8] This study
[9] Green *et al.* (in prep.)

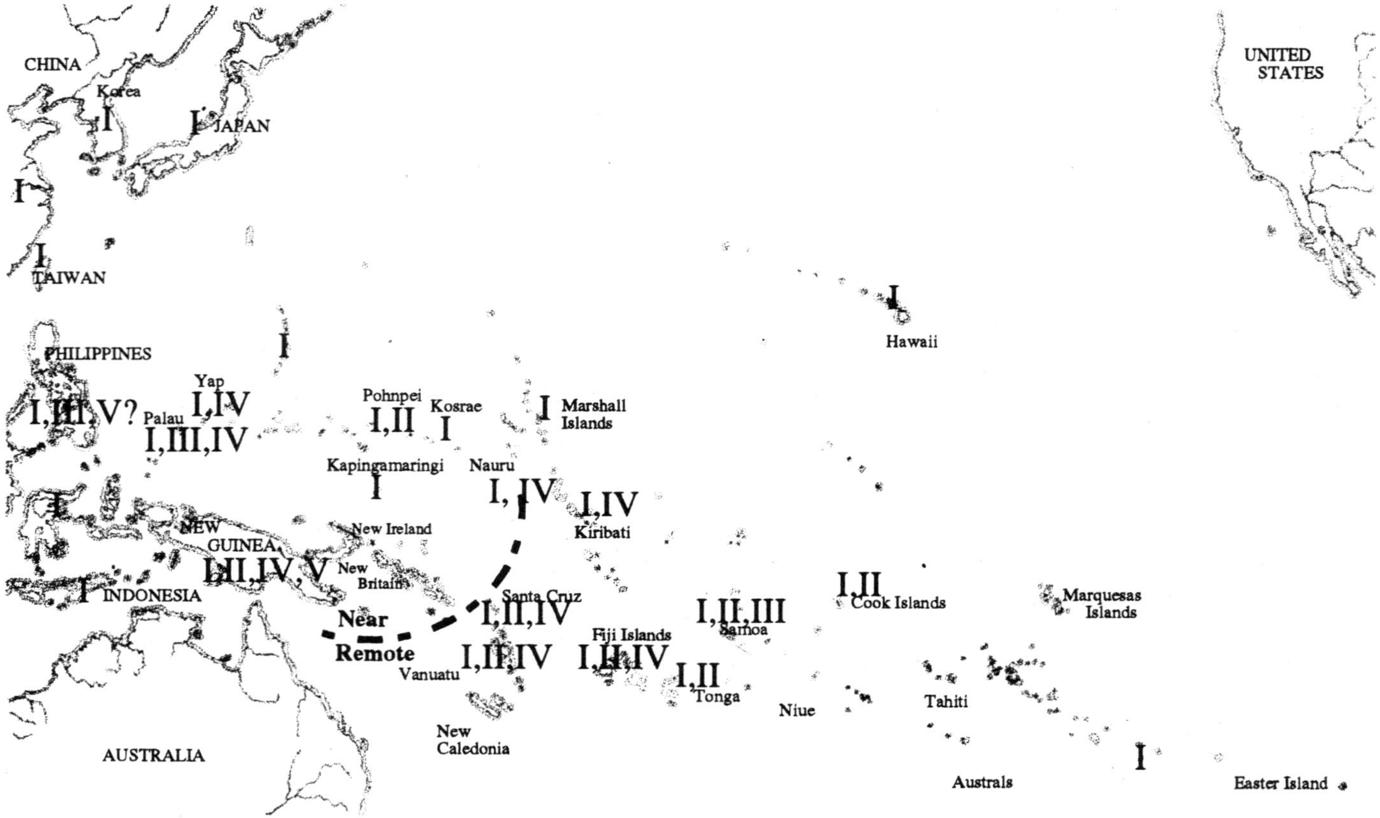

Figure 3. Distribution of Haplogroups I-V in the Pacific. Roman numeral indicates presence of that haplogroup in that location. Locations with no numbers were not surveyed.

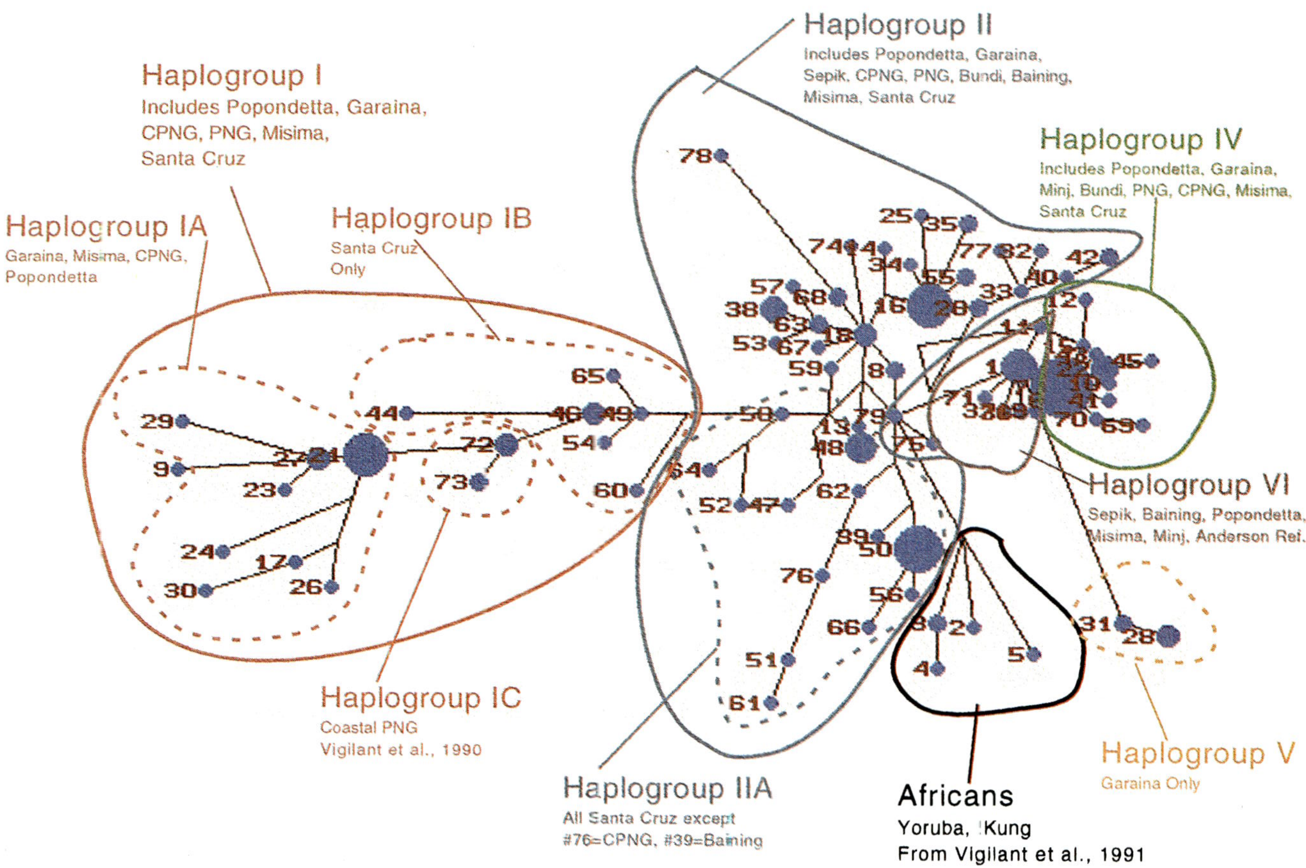

Figure 4. Bandelt Median Joined Network of HVSI sequences from this study and the literature. Haplogroups are circled in colored solid lines. Sub-groups are circled in colored dotted lines. The size of the circles representing each haplotype is proportional to the number of individuals possessing that haplotype. The length of the line connecting two haplotypes is proportional to the number of nt differences between the two sequences (with transversions weighted 2 : 1 over transitions). The numbers assigned to the haplotypes correspond to the haplotype numbers in Table 3.

effect NJ trees, but causes the BMN programs (entitled RM and MJ by the authors) to freeze, crash, or generate nonsensical trees. We therefore cropped the BMN data set shorter than the NJ data set. Both were cropped to conform in length to sequences from the literature.

4. DISCUSSION

4.1. Haplogroup Definitions

We identified six distinct clusters and four population-specific sub-clusters in this data set. The first two haplogroups follow definitions from Lum *et al.* (1994). We did not observe Lum *et al.*'s (1994) haplogroup III.

Haplogroup I is found very widely in Polynesia, Micronesia, and Melanesia, and is widely represented in this study on Papua New Guinea. The mutations Lum *et al.* (1994) identified to define this haplogroup were 16217 (T → C), 16247 (A → G), and 16261 (C → T). We add 16189 (T → C) in the poly C region of the HVS1 to this definition. The Region V 9-bp deletion is also found in virtually all Haplogroup I cases. We found a few exceptions to the association between the 9-bp deletion and Haplogroup I, namely two individuals from Garaina (Gar010 and Gar019) who lack the 9-bp deletion, but possess the D-loop mutations characteristic of Haplogroup I.

Haplogroup I is further divided into population-specific clusters.

IA contains individuals from Garaina, Popondetta, Misima, and Coastal PNG. IA is defined by 16182 (A → C) and 16183 (A → C) in all cases, and 16247 (A → G) in most.

IB only contains individuals from the Santa Cruz Islands. IB is defined by 16234 (C → T) and 16236 (C → T) in all cases.

IC contains individuals listed as Coastal PNG by Redd *et al.* 1995. It is defined by the absence of markers for subgoups A and B (and all have 16261 C → T). It is possible these may belong to haplogroup IA, but as they were reported in the literature with question marks at nts 16182 and 16183, we were unable to assign them with confidence. Some classify this as a deletion of two adenines and then an insertion of two cytosines, rather than two A → C transversions. Either way, the two changes occur together in our samples thus far. When some authors are unsure how to code this, they code them as missing data for their analyses.

Haplogroup II is defined by Lum *et al.* (1994) to include the mutations 16223 (C → T) in all individuals and the mutations 16241 (A → G) and 16265 (A → C) in many. We also found the mutations 16148 (C → T) and 16129 (G → A) in most and 16144 (T → C) in many members of this haplogroup. In addition, a small subset possessed the mutation 16222 (C → T) instead of the 16223 mutation. Haplogroup II is also widely found in Micronesia and Melanesia, less so in Polynesia. Haplogroup II contains one major population-specific subgroup which we label IIA.

IIA includes only Santa Cruz Islanders with the exceptions of haplotype #76 (from Coastal PNG, Redd *et al.* 1995) and haplotype #39 (a Baining from Northeast New Britain). IIA is defined by the lack of certain mutations present in the rest of Haplogroup II, including lack of 16241 (A → G) and 16265 (A → C).

Table 3. Variable sites in the control region nucleotide positions 16024–10310 (as numbered in Anderson *et al.* 1981). Dots indicate identity with the reference sequence (Anderson sequence, Anderson *et al.* 1981). Lineages are identified by the name of one of the samples possessing that lineage, and the distribution of each lineage among the groups sampled for this study is indicated in the last 12 columns of the table (with the abbreviations Popondetta, Pop; Garaina, Gar; Momboyoto, Mom; Kambadwa, Kam; Sagare, Sag; Baining, Bain; Misima, Mis; Santa Cruz, SCz; Coastal Papua New Guinea, CPNG; Papua New Guinea, PNG). *The number assigned to each lineage in the Bandelt Network (see fig. 4) is noted in parenthesis after the lineage identification.*

Haplogroup	Lineage	16025	16029	16051	16058	16066	16077	16086	16092	16093	16110	16114	16126	16129	16144	16145	16148	16150	16157	16169	16172	16176
VI	Andersen+ (1)	T	T	A	A	A	A	T	T	T	G	C	T	G	T	G	C	C	T	C	T	C
	Pop004 (1)	.	.	.	.	.	.	.	.	.	.	.	.	.	.	.	.	.	.	.	.	.
	Bain059 (1)	.	.	.	.	.	.	.	.	.	.	.	.	.	.	.	.	.	.	.	.	.
	Kam004 (1)	.	.	.	.	.	.	.	.	.	.	.	.	.	.	.	.	.	.	.	.	.
	Mis012 (19)	.	.	.	.	.	.	.	.	.	.	.	.	.	.	.	.	.	.	.	.	.
	Mis048 (19)	.	.	.	.	.	.	.	.	.	.	.	.	.	.	.	.	.	.	.	.	.
	Sag004 (36)	.	.	.	.	.	.	.	.	.	.	.	.	.	.	.	.	.	.	.	.	.
	Minj114 (6)	.	.	.	.	.	.	C	.	.	.	.	.	.	.	.	.	.	.	.	.	.
	Sag007 (37)	A	.	.	.	.	.	.	.	.	.	.	.	.	.	.	.	.	.	.	.	.
	Pop005 (11)	.	.	.	.	.	.	.	.	.	.	.	.	.	C	.	.	.	.	.	.	.
	Cpng5410 (71)	?	.	.	.	.	.	.	.	.	.	.	.	.	.	A	.	.	.	.	.	.
IA	Mis016+ (21)	.	.	.	.	.	.	.	.	.	.	.	.	.	.	.	.	.	.	.	.	.
	Mis023+ (21)	.	.	.	.	.	.	.	.	.	.	.	.	.	.	.	.	.	.	.	.	.
	Cpng5377 (21)	.	.	.	.	.	.	.	.	.	.	.	.	.	.	.	.	.	.	.	.	.
	Pop002 (9)	.	.	.	.	.	.	.	.	.	.	.	.	.	.	.	.	.	.	.	.	.
	Gar024+ (27)	.	.	.	.	.	.	.	.	.	.	.	.	.	.	.	.	.	.	.	.	.
	Gar005 (23)	.	.	.	.	.	.	.	.	.	.	.	.	.	.	.	.	.	C	.	.	.
	Gar031 (29)	?	.	.	.	.	.	.	.	.	C	.	.	.	.	.	.	.	.	.	.	T
	Gar010 (24)	.	.	.	.	.	.	.	.	.	.	.	.	A	C	.	T	.	.	.	.	.
	Pop021 (17)	.	.	.	.	.	.	.	.	.	.	.	.	.	.	.	.	.	.	.	.	.
	Gar019 (26)	.	.	.	.	.	.	.	.	.	.	.	.	.	.	.	.	.	.	.	.	T
	Gar037 (30)	.	.	.	.	.	.	.	.	C	.	.	.	.	.	.	.	.	.	.	.	.
IB	Scz15d68 (49)	.	.	.	.	.	.	.	.	.	.	.	.	.	.	.	.	.	.	.	.	.
	Scz28d38+ (46)	.	.	.	.	.	.	.	.	.	.	.	.	.	.	.	.	.	.	.	.	.
	Scz140d197 (65)	.	.	.	.	.	.	.	.	.	.	.	.	.	.	.	.	.	.	.	.	.
	Scz19d26 (54)	.	.	.	.	.	.	.	.	.	.	.	.	.	.	.	.	.	.	.	.	.
	Scz87d133 (60)	.	.	.	.	.	.	.	.	.	.	.	.	.	.	.	.	.	.	.	.	.
	Scz18d31 (44)	.	.	.	.	.	.	.	.	.	.	.	.	A	C	.	T	.	.	T	.	.
IC	Cpng5391+ (72)	.	.	.	.	.	.	.	.	.	.	.	.	.	.	.	.	.	.	.	.	.
	Cpng5418 (72)	.	.	.	.	.	.	.	.	.	.	.	.	.	.	.	.	.	.	.	.	.
	Cpng5388+ (73)	.	.	.	.	.	.	.	.	C	.	.	.	.	.	.	.	.	.	.	.	.
II	Pop008 (13)	.	.	.	.	.	.	.	.	.	.	.	.	A	.	.	.	.	.	.	.	.
	Pop001+ (8)	.	.	.	.	.	.	.	.	.	.	.	.	.	.	.	.	.	.	.	.	.
	Pop022+ (18)	.	.	.	.	.	.	.	.	.	.	.	.	A	.	.	.	.	.	.	.	.
	Cpng5403 (74)	?	?	?	?	.	.	.	.	.	.	.	.	A	.	.	T	.	.	.	.	.
	Cpng5405 (67)	?	.	.	.	.	.	.	.	.	.	.	.	A	.	.	.	.	.	.	.	.
	Scz69d110 (63)	.	.	.	.	G	.	.	.	.	.	.	.	A	.	.	.	.	.	.	.	.
	Scz124d177 (63)	.	.	.	.	G	.	.	.	.	.	.	.	A	.	.	.	.	.	.	.	.
	Scz121d173 (53)	.	.	.	.	G	.	.	.	.	.	.	.	A	.	.	.	.	.	.	.	.
	Scz99d148 (57)	.	.	.	.	G	.	.	.	.	.	.	.	A	.	.	.	.	.	.	.	.
	Bain053+ (38)	.	.	.	.	G	.	.	.	.	.	.	.	A	.	.	.	.	.	.	.	T
	Bain054 (38)	.	.	.	.	G.	.	.	.	.	.	.	.	A	.	.	.	.	.	.	.	T
	Cpng5402+ (68)	.	.	.	.	.	.	.	.	.	.	.	.	A	.	.	.	.	.	.	.	.
	Cpng5403 (74)	.	.	.	.	.	.	.	.	.	.	.	.	A	.	A	.	T	.	.	.	.
	Cpng5395 (78)	.	.	.	.	.	.	.	.	C	.	.	.	A	C	.	T	.	.	.	.	.

16179	16182	16183	16184	16185	16187	16188	16189	16192	16197	16203	16209	16217	16219	16221	16222	16223	16224	16230	16233	16234	16235	16236	16239	16241	16242
C	A	A	C	C	C	C	T	C	C	A	T	T	A	C	C	C	T	A	A	C	A	C	C	A	C
.	.	.	.	.	.	.	.	.	.	.	.	.	.	.	.	.	.	.	.	.	.	.	.	.	.
.	.	.	.	.	.	.	.	.	.	.	.	.	.	.	.	.	.	.	.	.	.	.	.	.	.
.	.	.	.	.	.	.	.	.	.	.	.	.	.	.	.	.	.	.	.	.	.	.	.	.	.
.	.	.	.	.	.	.	.	.	.	.	.	.	.	.	.	.	.	.	.	.	.	.	.	.	.
.	.	.	.	.	.	T	.	.	.	.	.	.	.	.	.	.	.	.	.	.	.	.	.	.	.
.	.	.	.	.	.	.	.	.	.	.	.	.	.	.	.	.	.	.	.	.	.	.	.	.	.
.	.	.	.	.	.	.	.	.	.	.	.	.	.	.	.	.	.	.	.	.	.	.	.	.	.
.	.	.	.	.	.	.	.	.	.	.	.	.	.	.	.	.	.	.	.	.	.	.	.	.	?
.	.	.	.	.	.	.	.	.	.	.	.	.	.	.	.	.	.	.	.	.	.	.	.	.	.
.	C	C	.	.	.	.	C	.	.	.	.	C	.	.	.	.	.	.	.	.	.	.	.	.	.
.	C	C	.	.	.	.	C	.	.	.	.	C	.	.	.	.	.	.	.	.	.	.	.	.	.
.	?	?	.	.	.	.	C	.	.	.	.	C	.	.	.	.	.	.	.	.	.	.	.	.	.
.	C	C	.	.	.	.	C	.	.	.	.	C	.	.	.	T	.	G	.	.	.	.	.	.	T
.	C	C	.	.	.	.	C	.	.	.	.	C	.	.	.	.	.	.	.	.	.	.	.	.	T
.	C	C	.	.	.	.	C	.	.	.	.	C	.	.	.	.	.	.	.	.	.	.	.	.	T
.	C	C	.	.	.	.	C	.	.	.	.	C	.	.	.	.	.	.	.	.	.	.	.	.	T
.	C	C	.	.	.	.	C	.	.	.	.	.	.	.	.	.	.	.	.	.	.	.	.	.	.
.	C	C	.	.	.	.	C	.	.	.	.	.	.	.	.	.	.	.	.	.	.	.	.	.	.
.	C	C	.	.	.	.	C	.	.	.	.	.	.	.	.	.	.	.	.	.	.	.	.	.	.
.	C	C	.	.	.	.	C	.	.	.	.	.	.	.	?	.	.	.	.	.	.	T	.	.	?
.	.	.	.	.	.	.	C	.	.	.	.	C	.	.	.	T	.	.	.	T	.	T	.	.	.
.	.	.	.	.	.	.	C	.	.	.	.	C	.	.	.	.	.	.	.	T	.	T	.	.	.
.	.	.	.	.	.	.	C	.	.	.	.	C	.	.	.	T	.	.	.	T	.	T	.	G	.
.	.	.	.	.	.	.	C	.	T	.	.	C	.	.	.	T	.	.	.	T	.	T	.	.	.
.	C	C	.	.	.	.	C	.	.	.	.	.	.	.	.	T	.	.	.	T	.	T	.	.	.
.	.	.	.	.	.	.	C	.	.	.	.	C	.	.	.	.	.	.	.	T	.	T	.	.	.
.	?	?	.	.	.	.	C	.	.	.	.	C	.	.	.	.	.	.	.	.	.	.	.	.	.
.	?	?	.	.	.	.	C	.	.	.	.	C	.	.	.	.	.	.	.	.	.	.	.	.	.
.	?	?	.	.	.	.	C	.	.	.	.	C	.	.	.	.	.	.	.	.	.	.	.	.	.
.	.	.	.	.	.	.	.	.	.	G	.	.	.	.	.	T	.	.	.	.	.	.	.	.	?
.	.	.	.	.	.	.	.	.	.	.	.	.	.	.	.	T	.	.	.	.	.	.	.	G	.
.	.	.	.	.	.	.	.	.	.	.	.	.	.	.	.	T	.	.	.	.	.	.	.	G	.
.	.	.	.	.	.	.	.	.	.	.	.	.	.	.	.	T	.	.	.	.	.	.	.	G	.
.	.	.	.	.	.	.	.	.	.	.	.	.	.	.	.	T	.	.	.	.	.	.	.	G	A
.	.	.	.	.	.	.	.	.	.	.	.	.	.	.	.	T	.	.	.	.	.	.	.	G	.
.	.	.	.	.	.	.	.	.	.	.	.	.	.	.	.	T	.	.	.	.	.	.	.	G	.
.	.	.	.	.	.	.	.	.	.	.	.	.	.	.	.	T	.	.	.	T	.	.	.	G	.
.	.	.	.	.	.	.	.	T	.	.	.	.	.	.	.	T	.	.	.	.	.	.	.	G	.
.	.	.	.	.	.	.	.	.	.	.	.	.	.	.	.	T	.	.	.	.	.	.	.	G	.
.	.	.	.	.	.	.	.	.	.	.	.	.	.	.	.	T	.	.	.	.	.	.	.	G	?
.	.	.	.	.	.	.	.	.	.	.	C	.	.	.	.	T	.	.	.	.	.	.	.	G	.
.	.	.	.	.	.	.	.	.	.	.	.	.	.	.	.	T	.	.	.	.	.	.	.	G	.
.	.	.	.	.	.	.	.	.	.	.	C	.	.	A	.	T	.	.	.	.	.	.	.	G	.

Table 3. *Continued*

Haplogroup	Lineage	16025	16029	16051	16058	16066	16077	16086	16092	16093	16110	16114	16126	16129	16144	16145	16148	16150	16157	16169	16172	16176
	Pop020+ (16)	.	.	.	.	.	.	.	.	.	.	.	.	A	C	.	T	.	.	.	.	.
	Mom014 (16)	.	.	.	.	.	.	.	.	.	.	.	.	A	C	.	T	.	.	.	.	.
	Kam009 (16)	.	.	.	.	.	.	.	.	.	.	.	.	A	C	.	T	.	.	.	.	.
	Sag005 (16)	.	.	.	.	.	.	.	.	.	.	.	.	A	C	.	T	.	.	.	.	.
	Cpng5400 (16)	?	.	.	.	.	.	.	.	.	.	.	.	A	C	.	T	.	.	.	.	.
	Pop10 (14)	.	.	.	.	.	.	.	.	.	.	.	.	A	C	.	.	.	.	.	.	.
	Mom025 (34)	.	.	.	.	.	.	.	.	.	.	.	.	A	C	.	T	.	.	.	.	.
	Gar013 (25)	.	.	.	.	.	.	.	.	.	.	.	.	A	C	.	T	.	.	.	.	.
	Kam012+ (35)	.	.	.	.	.	.	.	.	.	.	.	.	A	C	.	T	.	.	C	.	.
	Scz129d184 (55)	.	.	.	.	.	.	.	.	.	.	.	.	A	C	.	T	.	.	.	.	T
	Scz44d59 (55)	.	.	.	.	.	.	.	.	.	.	.	.	A	C	.	T	.	.	.	.	T
	Mis013+ (20)	.	.	.	.	.	.	.	.	.	.	.	.	A	C	.	T	.	.	.	.	.
	Mom021 (33)	.	.	.	.	.	.	.	.	.	.	.	.	A	C	.	T	.	.	.	.	.
	Cpng5396 (77)	?	.	.	.	.	.	.	.	.	.	.	.	A	C	.	T	.	.	.	.	.
	Mom018 (32)	.	.	.	.	.	.	C	.	.	.	.	.	A	C	.	T	.	.	.	.	.
	Bun043 (40)	.	.	.	.	.	.	.	.	.	.	.	.	A	C	.	T	.	.	.	.	.
	Bun082 (42)	.	.	.	.	.	.	.	.	C	.	.	.	A	C	.	T	.	.	.	.	.
	Bun083 (42)	.	.	.	.	.	.	.	.	C	.	.	.	A	C	.	T	.	.	.	.	.
	Scz91d139 (59)	.	.	.	.	.	.	.	.	.	.	.	.	A	.	.	.	.	.	.	.	.
IIA	Scz63d86+ (48)	.	.	.	.	.	.	.	.	.	.	.	.	.	.	.	.	.	.	.	.	.
	Scz67d108 (58)	.	.	.	.	.	.	.	.	.	.	.	.	A	.	.	.	.	.	.	.	.
	Scz8d8 (64)	.	.	.	.	.	.	.	.	.	.	.	.	.	.	.	.	.	.	.	.	T
	Scz13d15 (52)	.	.	.	.	.	.	.	.	.	.	.	.	A	C	.	T	.	.	.	.	T
	Scz37d52 (47)	.	.	.	.	.	.	.	.	.	.	.	.	A	C	.	T	.	.	.	.	T
	Scz9d12 (62)	.	.	.	.	.	.	.	.	C	.	.	.	.	.	.	T	.	.	.	.	.
	Cpng5399 (76)	.	.	.	.	.	.	.	.	C	.	.	.	A	C	.	T	.	.	.	.	.
	Scz112d161 (51)	.	.	.	.	.	.	.	.	.	.	.	.	A	C	.	?	.	.	T	.	.
	Scz7d6 (61)	.	.	.	.	.	.	.	.	.	.	.	.	A	C	.	T	.	.	T	.	.
	png188795 (80)	.	.	.	.	.	.	.	.	.	.	.	.	A	C	.	T	.	.	.	.	.
	Bain057 (39)	.	.	.	.	.	.	C	.	.	.	.	.	.	.	.	T	.	.	.	.	.
	Scz57d90+ (50)	.	.	.	.	.	.	C	.	.	.	.	.	A	.	.	T	.	.	.	.	.
	Scz15d20 (50)	.	.	.	.	.	.	C	.	.	.	.	.	A	.	.	T	.	.	.	.	.
	Scz59d92 (50)	.	.	.	.	.	.	C	.	.	.	.	.	A	.	.	T	.	.	.	.	.
	Scz64d105 (56)	.	.	.	.	.	.	C	.	.	A	.	.	A	.	.	T	.	.	.	.	.
	Scz25d35 (66)	.	.	.	.	.	G	C	.	.	.	G	.	A	.	.	T	.	.	.	.	.
IV	Minj116+ (7)	.	.	.	.	.	.	.	.	.	.	.	.	.	.	.	.	.	.	.	.	T
	Gar014+ (7)	.	.	.	.	.	.	.	.	.	.	.	.	.	.	.	.	.	.	.	.	T
	Bun064 (7)	.	.	.	.	.	.	.	.	.	.	.	.	.	.	.	.	.	.	.	.	T
	Mis006+ (7)	.	.	.	.	.	.	.	.	.	.	.	.	.	.	.	.	.	.	.	.	T
	Mis014 (7)	.	.	.	.	.	.	.	.	.	.	.	.	.	.	.	.	.	.	.	.	T
	Mis031+ (7)	.	.	.	.	.	.	.	.	.	.	.	.	.	.	.	.	.	.	.	.	T
	Cpng5406 (70)	.	.	.	.	.	.	.	.	.	.	.	.	.	.	.	.	.	.	.	.	T
	Cpng5407 (69)	.	.	.	.	.	.	.	.	.	.	.	.	.	.	.	.	.	.	.	.	T
	Bun066 (41)	C	.	.	.	.	.	.	.	.	.	.	.	.	.	.	.	.	.	T	.	.
	Pop003 (10)	.	.	.	.	.	.	.	.	.	.	.	.	.	.	.	.	.	.	.	.	T
	Mis049+ (22)	.	.	.	.	.	.	.	.	.	.	.	.	.	.	.	.	.	.	.	.	T
	Bun087 (43)	.	.	G	.	.	.	.	.	.	.	.	.	.	.	.	.	.	.	.	T	.
	Scz38d61 (45)	.	C	.	.	.	.	.	.	.	.	.	.	.	.	.	.	.	.	.	.	T
	Pop015 (15)	.	.	.	.	.	.	.	.	.	.	.	.	.	C	.	.	.	.	.	.	T
	Pop006 (12)	.	.	.	.	.	.	.	.	.	.	.	.	.	C	.	.	.	.	.	.	T
V	Gar038+ (31)	.	.	.	.	.	.	.	.	.	.	.	C	.	.	.	.	.	C	.	.	.
	Gar028+ (28)	.	.	.	.	.	.	.	.	.	.	.	C	.	.	.	.	.	C	.	.	.
	Cpng5411 (75)	.	.	.	.	.	.	.	.	.	.	.	.	.	.	.	.	.	.	.	.	.
	png188763 (79)	.	.	.	.	.	.	.	.	.	.	.	.	.	.	.	.	.	.	.	.	.

16179	16182	16183	16184	16185	16187	16188	16189	16192	16197	16203	16209	16217	16219	16221	16222	16223	16224	16230	16233	16234	16235	16236	16239	16241	16242
.	.	.	.	.	.	.	.	.	.	.	.	.	.	.	.	T	.	.	.	.	.	.	.	G	.
.	.	.	.	.	.	.	.	.	.	.	.	.	.	.	.	T	.	.	.	.	.	.	.	G	.
.	.	.	.	.	.	.	.	.	.	.	.	.	.	.	.	T	.	.	.	.	.	.	.	G	.
.	.	.	.	.	.	.	.	.	.	.	.	.	.	.	.	T	.	.	.	.	.	.	.	G	.
.	.	.	.	.	.	.	.	.	.	.	.	.	.	.	.	T	.	.	.	.	.	.	.	G	.
.	.	.	.	.	.	.	.	.	.	.	.	.	.	.	.	T	.	.	.	.	.	.	.	G	.
.	.	.	.	.	.	.	.	T	.	.	.	.	.	.	.	T	.	.	.	.	.	.	.	G	.
.	.	.	.	.	.	.	.	.	.	.	.	.	.	.	.	T	.	.	.	.	.	.	T	G	.
.	.	.	.	.	.	.	.	.	.	.	.	.	.	.	.	T	.	.	.	.	.	.	.	G	.
.	.	.	.	.	.	.	.	.	.	.	.	.	.	.	.	T	.	.	.	.	.	.	.	G	.
.	.	.	.	.	.	.	.	.	.	.	.	.	.	.	.	T	.	.	.	.	.	.	.	G	.
.	.	.	.	.	.	.	.	.	.	.	.	.	.	.	.	.	.	.	.	.	.	.	.	G	.
.	.	.	.	.	.	.	.	.	.	.	.	.	.	.	T	.	.	.	.	.	.	.	.	G	.
.	.	.	.	.	T	.	.	.	.	.	.	.	.	.	T	.	.	.	.	.	.	.	.	G	.
.	.	.	.	.	.	.	.	.	.	.	.	.	.	.	T	.	.	.	.	.	.	.	.	G	.
.	.	.	.	.	.	.	.	.	.	.	.	.	.	.	T	.	.	.	.	.	.	.	.	G	.
.	.	.	.	.	.	.	.	.	.	.	.	.	.	.	T	.	.	.	.	.	.	.	.	G	.
.	.	.	.	.	.	.	.	.	.	.	.	.	.	.	T	.	.	.	.	.	.	.	.	G	.
.	.	.	.	.	.	.	.	.	.	.	.	.	.	.	.	T	.	.	.	.	.	T	.	G	.
.	.	.	.	.	.	.	.	.	.	.	.	.	.	.	.	T	.	.	.	.	.	T	.	.	.
.	.	.	.	.	.	.	.	.	.	.	.	.	.	.	.	T	.	.	.	T	.	T	.	.	.
.	.	.	.	.	.	.	.	.	.	.	.	.	.	.	.	T	.	.	.	T	.	T	.	.	.
.	.	.	.	.	.	.	.	.	.	.	.	.	.	.	.	T	.	.	.	T	.	T	.	.	.
.	.	.	.	.	.	.	.	.	.	.	.	.	.	.	.	T	.	.	.	.	.	T	.	.	.
.	.	.	.	.	.	.	.	.	.	.	.	.	.	.	.	T	.	.	.	.	.	.	.	.	.
.	.	.	.	.	.	.	.	.	.	.	.	.	.	.	.	T	.	.	.	.	.	.	.	.	.
.	.	.	.	.	.	.	.	.	.	.	.	.	.	.	.	T	.	.	.	.	.	.	.	.	.
.	.	.	.	.	.	.	.	.	.	.	.	.	.	.	.	T	.	.	.	.	.	T	.	.	.
.	.	.	.	?	?	?	.	.	.	.	.	.	.	.	.	T	.	.	.	.	.	.	.	.	.
T	.	.	.	.	.	.	.	.	.	.	.	.	.	.	.	T	.	.	.	.	.	.	.	.	.
.	.	.	.	.	.	.	.	.	.	.	.	.	.	.	.	T	.	.	.	.	.	.	.	.	.
.	.	.	.	.	.	.	.	.	.	.	.	.	.	.	.	T	.	.	.	.	.	.	.	.	.
.	.	.	.	.	.	.	.	.	.	.	.	.	.	.	.	T	.	.	.	.	.	.	.	.	.
.	.	.	.	.	.	.	.	.	.	.	.	.	.	.	.	T	.	.	.	.	.	.	.	.	.
.	.	.	.	.	.	.	.	.	.	.	.	.	.	.	.	T	.	.	.	.	.	.	.	.	.
.	.	.	.	.	.	.	.	.	.	.	.	.	.	.	.	.	.	.	.	.	.	.	.	.	.
.	.	.	.	.	.	.	.	.	.	.	.	.	.	.	.	.	.	.	.	.	.	.	.	.	.
.	.	.	.	.	.	.	.	.	.	.	.	.	.	.	.	.	.	.	.	.	.	.	.	.	.
.	.	.	.	.	.	.	.	.	.	.	.	.	.	.	.	.	.	.	.	.	.	.	.	.	.
.	.	.	.	.	.	.	.	.	.	.	.	.	.	.	.	.	.	.	.	.	.	.	.	.	.
.	.	.	.	.	.	.	.	.	.	.	.	.	.	.	.	.	.	.	.	.	.	.	.	.	.
.	.	.	.	.	.	.	.	.	.	.	.	.	.	.	.	.	.	.	.	.	.	.	.	.	.
.	.	.	.	.	.	.	.	.	.	.	.	.	.	.	.	.	.	.	.	.	G	.	.	.	.
.	.	.	.	.	.	.	.	.	.	.	.	.	.	.	.	.	.	.	.	.	.	.	.	.	.
.	.	.	.	.	.	.	.	.	.	.	.	.	.	.	.	.	.	.	.	.	.	.	.	.	T
.	.	.	T	.	.	.	.	.	.	.	.	.	.	.	.	.	.	.	.	.	.	.	.	.	.
.	.	.	.	.	.	.	.	.	.	.	.	.	.	.	.	.	.	.	.	.	.	.	.	.	.
.	.	T	.	.	.	.	.	.	.	.	.	.	.	.	.	.	.	.	.	.	.	.	.	.	.
.	.	.	.	.	.	.	.	.	.	.	.	.	.	.	.	.	.	.	.	.	.	.	.	.	.
.	.	.	.	.	.	.	.	.	.	.	.	.	.	.	.	.	.	.	.	.	G	.	.	.	.
.	.	.	T	T	.	.	C	.	.	.	.	.	.	.	.	.	.	.	.	.	.	.	.	.	.
.	.	.	T	T	.	.	C	.	.	.	.	.	.	.	.	.	.	.	.	.	.	.	.	.	T
.	.	.	A	.	.	.	.	.	.	.	.	.	.	.	.	T	.	.	.	.	.	.	.	.	.
.	.	.	.	?	?	?	?	?	.	.	.	.	.	.	.	T	.	.	.	.	.	.	.	.	.

Table 3. *Continued*

Haplogroup	Lineage	16246	16247	16248	16255	16256	16257	16259	16260	16261	16264	16265	16266	16270	16273	16274	16278	16280	16287	16289	16290	16291
VI	Andersen+ (1)	A	A	C	G	C	C	C	C	C	C	A	C	C	G	G	C	A	C	A	C	C
	Pop004 (1)	.	.	.	.	.	.	.	.	.	.	.	T	.	.	.	T	.	.	.	.	.
	Bain059 (1)	.	.	.	.	.	.	.	.	.	.	.	T	.	.	.	.	.	.	.	.	.
	Kam004 (1)	.	.	.	.	.	.	.	.	.	.	.	.	.	.	.	.	G	.	.	.	.
	Mis012 (19)	.	.	.	.	T	.	.	.	.	.	.	.	.	.	.	.	.	.	.	.	.
	Mis048 (19)	.	.	.	.	T	.	.	.	.	.	.	.	T	.	.	.	.	.	.	.	.
	Sag004 (36)	.	.	.	.	.	.	.	.	.	.	.	.	.	.	.	T	.	.	.	.	.
	Minj114 (6)	.	.	.	.	.	.	.	.	.	.	.	.	.	.	.	T	.	.	.	.	.
	Sag007 (37)	.	.	.	.	.	.	.	.	.	.	.	.	.	.	.	.	G	.	.	.	.
	Pop005 (11)	.	.	.	.	.	.	.	.	.	.	.	T	.	?	A	T	.	.	.	.	.
	Cpng5410 (71)	.	.	.	.	.	.	.	.	.	.	.	.	.	.	.	.	.	.	.	.	.
IA	Mis016+ (21)	.	G	.	.	.	.	.	.	T	.	.	.	.	.	.	.	.	.	.	.	.
	Mis023+ (21)	.	G	.	.	.	.	.	.	T	.	.	.	T	.	.	.	.	.	.	.	?
	Cpng5377 (21)	.	G	.	.	.	.	.	.	T	.	.	.	.	.	.	.	.	.	.	.	.
	Pop002 (9)	.	.	.	.	.	.	.	.	?	?	?	?	?	?	?	?	?	?	?	?	?
	Gar024+ (27)	.	G	.	.	.	.	.	.	T	.	.	.	.	.	.	.	.	.	.	.	.
	Gar005 (23)	.	G	.	.	.	.	.	?	?	?	?	?	?	?	?	?	?	?	?	?	?
	Gar031 (29)	.	G	T	.	.	.	.	T	T	.	.	.	?	?	?	?	?	?	?	?	?
	Gar010 (24)	.	G	.	.	.	.	.	.	.	.	.	.	.	.	.	.	.	.	.	.	.
	Pop021 (17)	G	.	.	.	.	.	.	?	?	?	?	?	?	?	?	?	?	?	?	?	?
	Gar019 (26)	.	.	.	.	.	.	.	.	.	.	.	.	.	.	.	?	?	?	?	?	?
	Gar037 (30)	?	?	?	?	?	?	?	?	?	?	?	?	?	?	?	?	?	?	?	?	?
IB	Scz15d68 (49)	.	.	.	.	.	.	.	.	.	.	.	.	.	.	.	.	.	.	.	.	.
	Scz28d38+ (46)	.	.	.	.	.	.	.	.	.	.	.	.	.	.	.	.	.	.	.	.	.
	Scz140d197 (65)	.	.	.	.	.	.	.	.	T	.	.	.	.	.	.	.	.	.	.	.	.
	Scz19d26 (54)	.	.	.	.	.	.	.	.	.	.	.	.	.	.	.	.	.	.	.	.	.
	Scz87d133 (60)	.	.	.	.	.	.	.	.	.	.	.	.	.	.	.	.	.	.	.	.	.
	Scz18d31 (44)	.	.	.	.	.	.	.	.	.	.	.	.	.	.	.	.	.	.	.	.	.
IC	Cpng5391+ (72)	.	.	.	.	.	.	.	.	T	.	.	.	.	.	.	.	.	.	.	.	.
	Cpng5418 (72)	.	.	.	.	.	.	.	.	T	.	.	.	.	.	.	.	.	.	.	.	.
	Cpng5388+ (73)	.	.	.	.	.	.	.	.	T	.	.	.	.	.	.	.	.	.	.	.	.
II	Pop008 (13)	.	.	.	.	.	.	.	?	?	?	?	?	?	?	?	?	?	?	?	?	?
	Pop001+ (8)	.	.	.	.	.	.	.	.	.	.	.	.	.	.	.	.	.	.	.	.	.
	Pop022+ (18)	.	.	.	.	.	.	.	.	.	.	.	.	.	.	.	.	.	.	.	.	.
	Cpng5403 (74)	.	.	.	.	.	.	.	.	.	.	C	.	.	.	.	.	.	.	.	.	.
	Cpng5405 (67)	.	.	.	.	.	.	.	.	.	.	.	.	.	.	.	.	.	.	.	.	.
	Scz69d110 (63)	.	.	.	.	.	.	.	.	T	.	.	.	.	.	.	.	.	T	.	.	.
	Scz124d177 (63)	.	.	.	.	.	.	.	.	T	.	.	.	.	.	.	.	.	.	.	.	.
	Scz121d173 (53)	.	.	.	.	.	.	.	.	T	.	.	.	.	.	.	.	.	.	.	.	.
	Scz99d148 (57)	.	.	.	.	.	.	.	.	T	.	.	.	.	.	.	.	.	.	.	.	.
	Bain053+ (38)	.	.	.	.	.	.	.	.	.	.	.	.	.	.	.	.	.	.	.	.	.
	Bain054 (38)	.	.	?	.	.	.	.	.	.	.	.	.	.	T	.	.	.	.	.	.	.
	Cpng5402+ (68)	.	.	.	.	.	.	.	.	.	.	.	.	.	.	.	.	.	.	.	.	.
	Cpng5403 (74)	.	.	.	.	.	.	.	.	.	.	.	.	.	.	.	.	.	.	.	.	.
	Cpng5395 (78)	.	.	.	.	.	.	.	.	.	.	C	.	.	.	.	.	.	.	.	.	.
	Pop020+ (16)	.	.	.	.	.	.	.	.	.	.	C	.	.	.	.	.	.	.	.	.	.
	Mom014 (16)	.	.	.	.	.	.	.	.	.	.	C	.	.	.	.	.	.	.	.	.	.
	Kam009 (16)	.	.	.	.	.	.	.	.	.	.	C	.	G	?	?	?	?	?	?	?	?
	Sag005 (16)	.	.	.	.	.	.	.	.	.	.	C	.	.	.	.	.	.	.	.	T	.
	Cpng5400 (16)	.	.	.	.	.	.	.	.	.	.	C	.	.	.	.	.	.	.	.	.	T

16292	16294	16296	16298	16302	16304	16305	16310	Pop	Gar	Mom	Kam	Sag	Minj	Bundi	Bain	Mis	SCz	CPNG	PNG
C	C	C	T	A	T	A	G	—	—	2	1	—	—	—	—	—	—	—	—
.	.	.	.	.	.	.	.	1	—	—	—	—	—	—	—	—	—	—	—
.	.	.	.	.	C	.	.	—	—	—	—	—	—	—	1	—	—	—	—
.	.	.	.	.	.	.	.	—	—	—	1	—	—	—	—	—	—	—	—
.	.	.	.	.	.	.	.	—	—	—	—	—	—	—	—	1	—	—	—
.	.	.	.	.	.	.	.	—	—	—	—	—	—	—	—	1	—	—	—
.	.	.	.	.	.	.	.	—	—	—	—	1	—	—	—	—	—	—	—
.	.	.	.	.	.	.	.	—	—	—	—	—	1	—	—	—	—	—	—
.	.	.	.	.	.	.	.	—	—	—	—	1	—	—	—	—	—	—	—
.	.	.	.	.	.	.	?	1	—	—	—	—	—	—	—	—	—	—	—
.	.	.	.	.	.	.	.	—	—	—	—	—	—	—	—	—	—	1	—
.	.	.	.	.	.	.	.	—	2	—	—	—	—	—	—	1	—	3	2
?	?	?	?	?	?	?	?	—	—	—	—	—	—	—	—	2	—	—	—
.	.	.	C	.	.	.	.	—	—	—	—	—	—	—	—	—	—	1	—
?	?	?	?	?	?	?	?	1	—	—	—	—	—	—	—	—	—	—	—
?	?	?	?	?	?	?	?	—	3	—	—	—	—	—	—	—	—	—	—
?	?	?	?	?	?	?	?	—	1	—	—	—	—	—	—	—	—	—	—
?	?	?	?	?	?	?	?	—	1	—	—	—	—	—	—	—	—	—	—
.	.	.	.	.	.	.	.	—	1	—	—	—	—	—	—	—	—	—	—
?	?	?	?	?	?	?	?	1	—	—	—	—	—	—	—	—	—	—	—
?	?	?	?	?	?	?	?	—	1	—	—	—	—	—	—	—	—	—	—
?	?	?	?	?	?	?	?	—	1	—	—	—	—	—	—	—	—	—	—
.	.	.	.	.	.	.	.	—	—	—	—	—	—	—	—	—	1	—	—
.	.	.	.	.	.	.	.	—	—	—	—	—	—	—	—	—	3	—	—
.	.	.	.	.	.	.	.	—	—	—	—	—	—	—	—	—	1	—	—
.	.	.	.	.	.	.	.	—	—	—	—	—	—	—	—	—	1	—	—
.	.	.	.	.	.	.	.	—	—	—	—	—	—	—	—	—	1	—	—
.	.	.	.	.	.	.	.	—	—	—	—	—	—	—	—	—	1	—	—
.	.	.	.	.	.	.	.	—	—	—	—	—	—	—	—	—	—	2	—
.	.	.	.	G	.	.	.	—	—	—	—	—	—	—	—	—	—	1	—
.	.	.	.	.	.	.	.	—	—	—	—	—	—	—	—	—	—	2	—
?	?	?	?	?	?	?	?	1	—	—	—	—	—	—	—	—	—	—	—
.	.	.	.	.	.	.	.	2	—	—	—	—	—	—	—	—	—	—	—
.	.	.	.	.	.	.	.	1	—	—	—	—	—	—	—	—	—	1	1
.	.	.	.	.	.	.	.	—	—	—	—	—	—	—	—	—	—	1	—
.	.	.	.	.	.	.	.	—	—	—	—	—	—	—	—	—	—	1	—
T	.	.	.	.	.	.	.	—	—	—	—	—	—	—	—	—	1	—	—
T	.	.	.	.	.	.	.	—	—	—	—	—	—	—	—	—	1	—	—
T	.	.	.	.	.	.	.	—	—	—	—	—	—	—	—	—	1	—	—
.	.	.	.	.	.	.	.	—	—	—	—	—	—	—	—	—	1	—	—
.	.	.	.	.	.	.	.	—	—	—	—	—	—	—	3	—	—	—	—
.	.	.	.	.	.	.	.	—	—	—	—	—	—	—	1	—	—	—	—
.	.	.	.	.	.	.	.	—	—	—	—	—	—	—	—	—	—	2	—
.	.	.	.	.	.	.	.	—	—	—	—	—	—	—	—	—	—	1	—
.	.	.	.	.	.	.	.	—	—	—	—	—	—	—	—	—	—	1	—
.	.	.	.	.	.	.	.	1	4	1	—	—	—	—	—	—	—	—	1
.	.	.	.	.	.	.	A	—	—	1	—	—	—	—	—	—	—	—	—
?	?	?	?	?	?	?	?	—	—	—	1	—	—	—	—	—	—	—	—
.	.	.	.	.	.	.	?	—	—	—	—	1	—	—	—	—	—	—	—
.	.	.	.	.	.	.	.	—	—	—	—	—	—	—	—	—	—	1	—

Table 3. *Continued*

Haplogroup	Lineage	16246	16247	16248	16255	16256	16257	16259	16260	16261	16264	16265	16266	16270	16273	16274	16278	16280	16287	16289	16290	16291
	Pop010 (14)	G	.	.	.	.	.	.	.	.	.	.	.	.	.	.	.	.	.	.	.	.
	Mom025 (34)	.	.	.	.	.	.	.	.	.	.	C	.	.	.	.	.	.	.	.	.	.
	Gar013 (25)	.	.	.	?	.	.	A	.	.	.	C	.	.	.	.	.	.	.	G	.	.
	Kam012+ (35)	.	.	.	A	.	.	.	.	.	.	C	.	.	.	.	.	.	.	.	.	.
	Scz129d184 (55)	.	.	.	.	.	.	.	.	.	.	.	.	.	.	.	.	.	.	.	.	.
	Scz44d59 (55)	.	.	.	.	.	.	.	.	.	.	C	.	.	.	.	.	.	.	.	.	.
	Mis013+ (20)	.	.	.	.	.	.	.	.	.	.	C	.	.	.	.	.	.	.	.	.	.
	Mom021 (33)	.	.	.	.	.	.	.	.	.	.	C	.	.	.	.	.	.	.	.	.	.
	Cpng5396 (77)	.	.	.	.	.	.	.	.	.	.	C	.	.	.	.	.	.	.	.	.	.
	Mom018 (32)	.	.	.	.	.	.	.	.	.	.	C	.	.	.	.	.	.	.	.	.	.
	Bun043 (40)	.	.	.	.	.	T	.	.	.	.	C	.	.	.	.	.	.	.	.	.	.
	Bun082 (42)	.	.	.	.	.	T	.	.	.	.	C	.	G	.	.	.	.	.	.	.	.
	Bun083 (42)	.	.	.	.	.	T	.	.	.	.	.	.	.	.	.	.	.	.	.	.	.
	Scz91d139 (59)	.	.	.	.	.	.	.	.	.	.	C	.	.	.	.	.	.	T	.	.	.
IIA	Scz63d86+ (48)	.	.	.	.	.	.	.	.	.	.	.	.	.	.	.	.	.	.	.	.	.
	Scz67d108 (58)	.	.	.	.	.	.	.	.	.	.	.	.	.	.	.	.	.	.	.	.	.
	Scz8d8 (64)	.	.	.	.	.	.	.	.	.	.	.	.	.	.	.	.	.	.	.	.	T
	Scz13d15 (52)	.	.	.	.	.	.	.	.	.	.	.	.	.	.	.	.	.	.	.	.	.
	Scz37d52 (47)	.	.	.	.	.	.	.	.	.	.	.	.	.	.	.	.	.	T	.	.	.
	Scz9d12 (62)	.	.	.	.	.	.	.	.	.	.	.	.	.	.	.	.	.	.	.	.	T
	Cpng5399 (76)	.	.	.	.	.	.	.	.	.	.	C	.	.	.	.	.	.	.	.	.	.
	Scz112d161 (51)	.	.	.	.	.	.	.	.	.	.	.	.	.	.	.	.	.	.	.	.	.
	Scz7d6 (61)	.	.	.	.	.	.	.	.	.	.	.	.	.	.	.	.	.	.	.	.	.
	png188795 (80)	.	.	.	.	.	.	.	.	.	.	C	.	.	.	.	.	.	.	.	.	.
	Bain057 (39)	.	.	.	.	.	.	.	.	.	.	.	.	.	.	.	.	.	.	.	.	.
	Scz57d90+ (50)	.	.	.	.	.	.	.	.	.	.	.	.	.	.	.	.	.	.	.	.	.
	Scz15d20 (50)	.	.	.	.	.	.	.	.	T	.	.	.	.	.	.	.	.	.	.	.	.
	Scz59d92 (50)	.	.	.	.	.	.	.	.	.	.	.	.	.	.	.	.	.	T	.	.	.
	Scz64d105 (56)	.	.	.	.	.	.	.	.	.	.	.	.	.	.	.	.	.	.	.	.	.
	Scz25d35 (66)	.	.	.	.	.	.	.	.	.	.	.	.	.	.	.	.	.	.	.	.	.
IV	Minj116+ (7)	.	.	.	.	.	.	.	.	.	.	.	T	.	.	.	.	.	T	.	.	.
	Gar014+ (7)	.	.	.	.	.	.	.	.	.	.	.	.	.	.	.	.	.	T	.	.	.
	Bun064 (7)	.	.	.	.	.	.	.	.	.	.	.	T	.	.	.	.	.	T	.	.	.
	Mis006+ (7)	.	.	.	.	.	.	.	.	.	.	.	T	T	.	.	.	.	.	.	.	.
	Mis014 (7)	.	.	.	.	.	.	.	.	.	.	.	T	.	.	.	.	.	.	.	.	.
	Mis031+ (7)	.	.	.	.	.	.	.	.	.	.	.	T	.	.	.	.	.	.	.	.	.
	Cpng5406 (70)	.	.	.	.	.	T	.	.	.	.	.	T	T	.	.	.	.	.	.	.	.
	Cpng5407 (69)	.	.	.	.	.	T	.	.	.	T	.	T	T	.	.	.	.	.	.	.	.
	Bun066 (41)	.	.	.	.	.	.	.	.	.	.	.	T	.	.	.	.	.	T	.	.	.
	Pop003 (10)	.	.	.	.	.	.	.	.	.	.	.	.	.	?	.	.	.	.	.	.	.
	Mis049+ (22)	.	.	.	.	.	.	.	.	.	.	.	T	.	.	.	.	.	.	.	.	.
	Bun087 (43)	.	.	.	.	.	.	.	.	.	.	.	.	.	.	.	.	.	T	.	.	.
	Scz38d61 (45)	.	.	.	.	.	.	.	.	.	.	.	T	.	.	.	.	?	?	.	.	.
	Pop015 (15)	.	.	.	.	.	.	.	.	.	.	.	.	.	.	.	.	.	T	.	.	.
	Pop006 (12)	.	.	.	.	.	.	.	.	.	.	.	T	T	.	.	.	.	.	.	.	.
V	Gar038+ (31)	.	.	.	.	T	.	.	.	.	.	.	.	.	.	.	.	.	.	.	.	.
	Gar028+ (28)	.	.	.	.	T	.	.	.	.	.	.	.	.	.	.	.	.	.	.	.	.
	Cpng5411 (75)	.	.	.	.	.	.	.	.	.	.	.	.	.	.	.	T	.	.	.	.	T
	png188763 (79)	.	.	.	.	.	.	.	.	.	.	.	.	.	.	.	.	.	.	.	.	.

16292	16294	16296	16298	16302	16304	16305	16310	Pop	Gar	Mom	Kam	Sag	Minj	Bundi	Bain	Mis	SCz	CPNG	PNG
.	.	.	.	.	.	.	.	1	—	—	—	—	—	—	—	—	—	—	—
.	.	.	.	.	.	.	.	—	—	1	—	—	—	—	—	—	—	—	—
.	?	.	.	.	.	.	A	—	1	—	—	—	—	—	—	—	—	—	—
.	.	.	.	.	.	.	.	—	—	—	1	—	—	—	—	—	—	1	—
.	.	.	.	.	.	.	.	—	—	—	—	—	—	—	—	—	1	—	—
.	.	.	.	.	.	.	.	—	—	—	—	—	—	—	—	—	1	—	—
.	.	.	.	.	.	.	.	—	—	—	—	—	—	—	—	1	—	1	—
.	.	.	.	.	.	.	.	—	—	1	—	—	—	—	—	—	—	—	—
.	.	.	.	.	.	.	.	—	—	—	—	—	—	—	—	—	—	1	—
.	.	.	.	.	.	.	.	—	—	1	—	—	—	—	—	—	—	—	—
.	.	.	.	.	.	.	.	—	—	—	—	—	—	1	—	—	—	—	—
.	.	.	.	.	.	.	?	—	—	—	—	—	—	1	—	—	—	—	—
.	.	.	.	.	.	.	.	—	—	—	—	—	—	1	—	—	—	—	—
.	.	.	.	.	.	.	.	—	—	—	—	—	—	—	—	—	1	—	—
.	.	.	.	.	.	.	.	—	—	—	—	—	—	—	—	—	6	—	—
.	.	.	.	.	.	.	.	—	—	—	—	—	—	—	—	—	1	—	—
.	.	.	.	.	.	.	.	—	—	—	—	—	—	—	—	—	1	—	
.	.	.	.	.	.	.	.	—	—	—	—	—	—	—	—	—	1	—	—
.	.	T	.	.	.	.	.	—	—	—			—	—	—	—	1	—	—
.	.	.	.	.	.	.	.	—	—	—	—	—	—	—	—	—	1	—	—
.	.	.	.	.	.	.	.	—	—	—	—	—	—	—	—	—	—	1	—
.	.	.	.	.	.	.	.	—	—	—	—	—	—	—	—	—	1	—	—
.	.	.	.	.	.	.	.	—	—	—	—	—	—	—	—	—	1	—	—
.	.	.	.	.	.	.		—	—		—	—			—	—	—	—	1
.	.	.	.	.	.	.	.	—	—	—	—	—	—	—	1	—	—	—	—
.	.	.	.	.	.	.	.	—	—	—	—	—	—	—	—	—	11	—	—
.	.	.	.	.	.	.	.	—	—	—	—	—	—	—	—	—	1	—	—
.	.	.	.	.	.	.	.	—	—	—	—	—	—	—			1	—	—
.	.	.	.	.	.	.	.	—	—	—	—	—	—	—	—	—	1	—	—
.	.	.	.	.	.	.	.	—	—	—	—	—	—	—	—	—	1	—	—
.	.	.	.	.	.	.	A	—	—	—	—	—	1	1	—	—	—		—
.	.	.	.	.	.	.	.	—	1	—	—	—	—	1	—	—	—	—	—
.	.	.	.	.	.	.	.	—	—	—	—	—	—	1	—	—	—	—	—
.	.	.	.	.	.	.	.	—	—	—	—	—	—	—	—	9	—	—	—
.	.	.	.	.	.	G	.	—	—	—	—	—	—	—	—	1	—	—	—
.	.	.	.	.	.	.	.	—	—	—	—	—	—	—	—	7	—	—	1
.	.	.	.	.	.	.	.	—	—	—	—	—	—	—	—	—	—	1	—
.	.	.	.	.	C	.	.	—	—	—	—	—	—	—	—	—	—	1	—
.	.	.	.	.	.	.	.	—	—	—	—	—	—	1	—	—	—	—	—
.	.	.	.	.	.	.	.	1	—	—	—	—	—	—	—	—	—	—	—
.	.	.	.	.	.	.	.	—	—	—	—	—	—	—	—	1	3	—	—
.	.	.	.	.	.	.	.	—	—	—	—	—	—	1			—	—	—
.	.	.	.	.	.	.	.	—	—	—	—	—	—	—	—	—	1	—	—
.	T	.	.	.	.	.	.	1	—	—	—	—	—	—	—	—	—	—	—
.	?	.	.	.	C	.	.	1	—	—	—	—	—	—	—	—	—	—	—
.	.	.	.	.	.	.	?	—	2	—	—	—	—	—	—	—	—	—	—
.	.	.	.	.	.	.	.	—	3	—	—	—	—	—	—	—	—	—	—
.	.	.	.	.	.	.	.	—	—	—	—	—	—	—	—	—	—	1	—
.	.	.	.	.	.	.	.	—	—	—	—	—	—	—	—	—	—	—	1

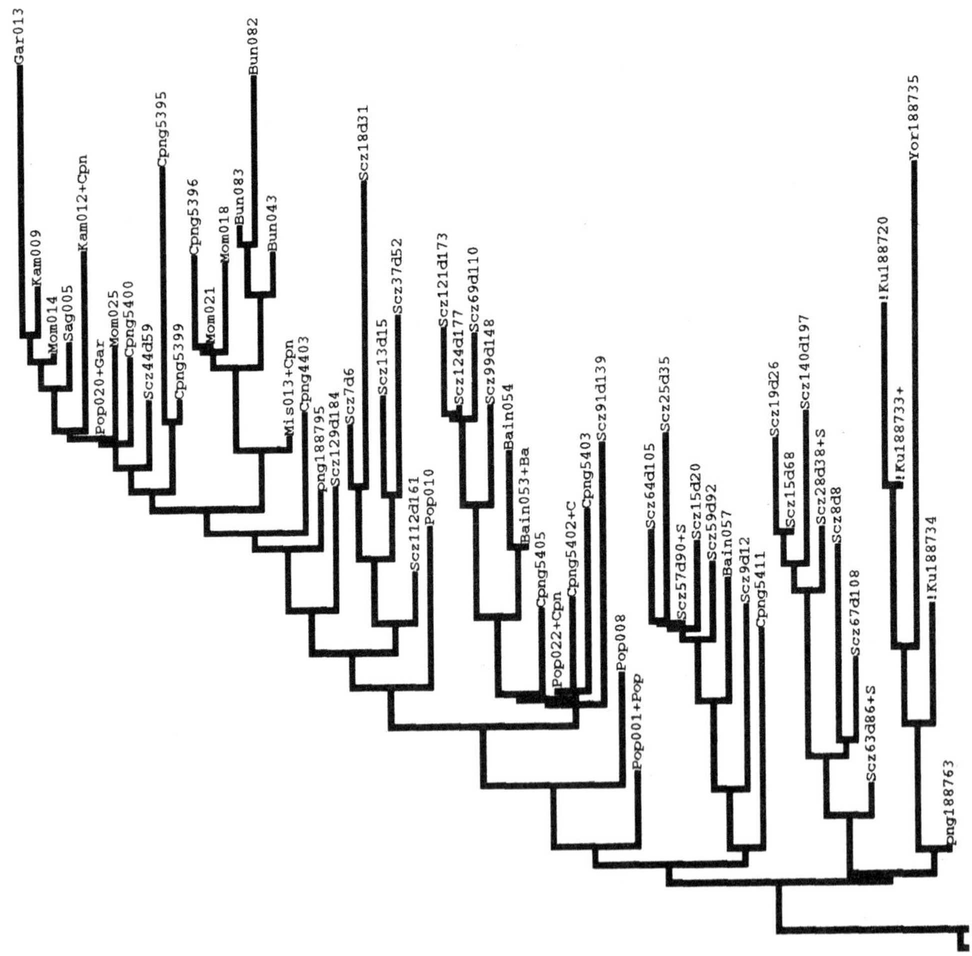
Gar013
Kam009
Mom014
Sag005
Kam012+Cpn
Pop020+Gar
Mom025
Cpng5400
Scz44d59
Cpng5395
Cpng5399
Cpng5396
Mom021
Mom018
Bun083
Bun082
Bun043
Mis013+Cpn
Cpng4403
png188795
Scz129d184
Scz7d6
Scz18d31
Scz13d15
Scz37d52
Scz112d161
Pop010
Scz121d173
Scz124d177
Scz69d110
Scz99d148
Bain054
Bain053+Ba
Cpng5405
Pop022+Cpn
Cpng5402+C
Cpng5403
Scz91d139
Pop008
Pop001+Pop
Scz64d105
Scz25d35
Scz57d90+S
Scz15d20
Scz59d92
Bain057
Scz9d12
Cpng5411
Scz19d26
Scz15d68
Scz140d197
Scz28d38+S
Scz8d8
Scz67d108
Scz63d86+S
!Ku188720
!Ku188733+
Yor188735
!Ku188734
png188763

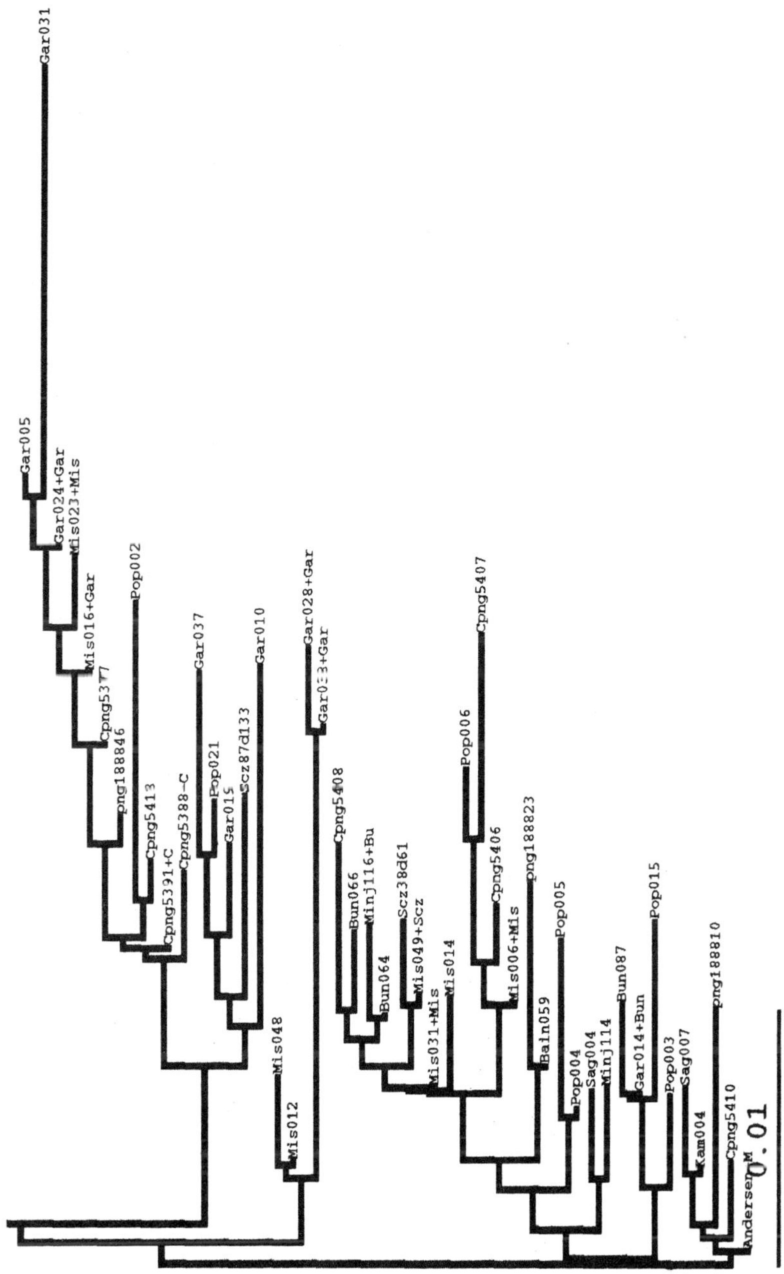

Figure 5. Neighbor-joining tree of the sequences from this study and from PNG from the literature. Populations are abbreviated as: Popondetta, Pop; Garaina, Gar; Mom-boyoto, Mom; Kambadwa, Kam; Sagare, Sag; Baining, Bain; Misima, Mis; Santa Cruz, SCz; Coastal Papua New Guinea, CPNG; Papua New Guinea, PNG.

Haplogroup IV is seen in northwest Micronesia, Indonesia, and Melanesia. Haplogroup IV is defined by us as having 16176 (C → T) and 16266 (C → T). We also see 16287 (C → T) in many group IV haplotypes.

BLAST searches (using NCBI's ENTREZ BLAST search program: http://www.ncbi.nlm.nih.gov/BLAST/) of GENBANK and a search of the HvrBase mitochondrial sequence database (Handt *et al.* 1998) found no matches in Genbank to Pop003 or Pop015.

Haplogroup V was found in 5 individuals (of 21) from Garaina, PNG. Haplogroup V is defined by us as 16126 (T → C), 16157 (T → C), 16184 (C → T), 16185 (C → T), 16189 (T → C), and 16256 (C → T).

ENTREZ® BLAST searches of GENBANK and HvrBase searches (Handt *et al.* 1998) found no matches to Gar038 or Gar039.

Haplogroup VI is seen in the Sepik, Popondetta, Baining, Minj, and Misima samples, and includes the Anderson Reference sequence (Anderson *et al.* 1981, haplotype #1 in the BMN in Fig. 4). Members of haplogroup VI are all only 0 to 2 mutations away from the Anderson reference sequence (Anderson *et al.* 1981).

ENTREZ® BLAST searches of GENBANK and HvrBase searches (Handt *et al.* 1998) using each of the Anderson-like sequences from Haplogroup VI revealed the following exact matches (at least for the nucleotides sequenced in common between the different studies):

1) Pop004 was identical to a Turkish sequence (GENBANK accession number U59024, Comas *et al.* 1996), and a Caucasian sequence (individual 84) (GENBANK accession number M58140, DiRienzo and Wilson 1991).
2) Mis012 was identical to a Swedish sequence (GENBANK accession number U67634, Sajantila *et al.* 1996), a Danish sequence (HVRBase isolate GER232, Richards *et al.* 1996) and an Italian sequence from Tuscany (HVRBase isolate 51, Francalacci *et al.* 1996).
3) Sag004 was identical to a PNG sequence (GENBANK accession number M76342, Huli (Indo Pacific) speaker, Vigilant *et al.* 1991).
4) Mis048 was identical to a Finnish sequence (GENBANK accession number X90120 Sajantila *et al.* 1995), four Karelian sequences (Finno-Ugric language group, inhabitants of Russia, GENBANK accession numbers X90167, X90192, X90197 and X90213, Sajantila *et al.* 1995) and an Unterwallis sequence (French-speaking inhabitant of Switzerland, GENBANK accession number X83183, Pult *et al.* 1994).
5) Bain059, Kam004, Minj114, Sag007, Pop005 were all unique and no matches were found.

4.2. Distribution of Haplogroups

At this preliminary stage, we can only be sure we have identified certain haplotypes in the region. Their actual distributions are unclear. Also, we cannot be sure we have identified all haplotypes occurring in the region. Nevertheless, we have identified two haplogroups (V and VI) that seem to be absent in Polynesia, and rare in Southeast Asia and Indonesia. Haplogroup V was only seen in Garaina (in five individuals out of 21) in our study. Sykes *et al.* (1995) found two of the six defining mutations for group V (16157 and 16256) in Borneo and the Philippines. However, no sequences possessing all six defining Haplogroup V mutations have been reported outside of our study. Searches of all Australian and Oceanic mtDNA HVSI sequences in HvrBase (Handt *et al.* 1998) found no matches to any of the haplogroup VI lineages other than the PNG individual reported above by Vigilant *et al.* (1991). Haplogroup IV seems to be limited to Melanesia, Westernmost Micronesia, and Indonesia, but absent in Polynesia and most of Micronesia. The Polynesian expansion has been associated (Lum *et al.* 1994; Melton *et al.* 1995; Redd *et al.* 1995; Sykes *et al.* 1995) with Haplogroup I. The series of mutations defining Haplogroup I has been termed the "Polynesian Motif"(Redd *et al.* 1995; Melton *et al.* 1995). This motif is absent in our Highland PNG samples, as well as those reported thus far in the literature (Hagelberg *et al.* 1999). This confirms earlier RFLP studies by Stoneking *et al.* (1990) that showed the 9-bp deletion was absent in the PNG highlands and in Australia.

4.3. Andersons in the Pacific

The apparently wide distribution in Island Melanesia and New Guinea of the published reference sequence, often called the Anderson sequence or the Cambridge reference sequence (Anderson *et al.* 1981), and many sequences closely related to the Anderson sequence (Haplogroup VI) has two possible explanations:

1) This represents European admixture in all cases. Since this is mtDNA, it requires that the admixture be European women who settled in these myriad locations and bore children.
2) Haplogroup VI was one of many clusters to initially people New Guinea and Island Melanesia. It is widely found around the world.

The first case, European admixture, is plausible in the sense that there are now, and certainly have been Europeans interacting with indigenous populations throughout the southwest Pacific for the past century. However, almost all of these interactions involved European males rather than females. Nevertheless, the exact identity of these haplogroups with Finnish, Danish, French, Italian, Turkish, Russian, and Swedish mtDNA sequences suggests strongly that these are instances of European admixture.

The second case, that they represent founding lineages in the region, would argue that the Anderson sequence is central to the entire human mtDNA phylogeny. This combined with the apparently unlikely scenario of European women settling and reproducing in so many remote regions on New Guinea (and Misima Island), argues in favor of it being a founding lineage. If European admixture were responsible, we would expect to find haplogroup VI widely distributed in Polynesia (Remote Oceania generally), rather than in New Guinea and Island Melanesia.

Since the number of Haplogroup VI individuals is still rather small at the present time, we cannot hope to differentiate between these two hypotheses in this paper. However, larger sample sizes and a more thorough sampling distribution may help to resolve the issue.

A third possibility, sequencing errors, is very unlikely because every strand was sequenced in the forward and reverse directions for each amplicon that was sequenced. In addition, many of the sequences, despite being similar to the Anderson reference sequence, had unique mutations never before reported in the literature or Genbank.

4.4. Origins and Ages of the Highlanders

It has been suggested that human population sizes have expanded in the highlands fairly recently following the advent of agriculture (Bayliss-Smith 1996). Our sample sizes in this study are small for the highland populations, and Hagelberg *et al.* (1999) have yet to publish their sequences. Hagelberg *et al.* (1999) suggested that their Y STR data showed a reduction in haplotype diversity for the Y chromosome in PNG (0.85 highland, 0.86 coastal) versus Polynesia (0.93 in Western Samoa) and Island Melanesia (0.98 in the Tolais of New Britain and in the Trobriand Islanders). Mitochondrial DNA did not show any reduction in haplotype diversity in PNG Highlanders (0.99) but did show reduction in the coastal PNG population (0.73 in the Roro) and far more diversity than the Western Samoans (0.59). From these data it would appear that the mtDNA lineages are much more diverse (and therefore older), than the Y lineages. When we have completed sequencing the 200–300 highlanders in our data set we should be able to get more accurate estimates of the ages of these populations, and their expansion times.

4.5. Mismatch Distributions and the Age of the 9-bp deletion in Melanesia

Green *et al.* (in preparation) and Merriwether (1999) showed that the 9-bp deleted lineages in Santa Cruz were much younger than the non-9-bp deleted lineages. The mismatch distribution (Rogers and Harpending 1992) for the D-loop sequences of the deleted lineages was only 1–3 differences (and only one pair of sequences had three differences) with a mean of 1 difference. The non-deleted lineages averaged over 7 pairwise differences. This indicated that in Santa Cruz, the deleted lineage individuals had arrived relatively recently compared to the other lineages. Sample sizes of 9-bp deleted lineages within-populations are too small to recreate this with the data in this study. Therefore, we pooled all the mainland PNG samples and generated mismatch distributions for all the lineages combined, for just the non-9-bp-deleted lineages, and for just the 9-bp deleted lineages (See Figs. 6a, 6b, and 6c respectively). The mean number of pairwise differences for all PNG samples in our study was very weakly bimodal, with means near 8 and 3, and a very clear central mode at 6 differences. The mean number of pairwise differences for the non-9-bp deleted individuals was strongly bimodal, with means near 7 and 1–2, and modes at 7 and 3. The mean number of pairwise differences for the 9-bp deleted samples was 1.6 (with a unimodal distribution). This closely mirrors what Green *et al.* (in preparation) found in Santa Cruz, except that the range of the 9-bp deleted mismatches extended up to 6 pairwise differences. We take this as additional evidence that the 9-bp deletion is a relatively recent addition in Melanesia as well as Polynesia, while it is believed to be considerably older in Asia and Indonesia (Redd *et al.* 1995).

Our preliminary study is consistent with the absence of the 9-bp deletion, and Haplogroup I, in the Highlands (Minj and Bundi), although only 10 of the 134 sequences first reported in this paper are from those two regions. Analyses of larger sample sizes from each population need to be completed before any mismatch distributions or population genetic statistics would be valid. Additional sequencing is already underway.

5. ONGOING RESEARCH

We are currently sequencing the mtDNA HVSI of more individuals from each of the locations in this study, plus populations from the West Coast of Irian Jaya, Riga (west coast PNG), Markham Valley (PNG), Torricelli Mountains (PNG), Rossel Island, Fergusson Island, New Britain, New Ireland, Bougainville, Ontong Java and Samoa. We have just begun Y-chromosome STR studies of the Solomon Islanders and New Britain natives, and if it proves valuable, we hope to extend the Y-chromosome analyses to all

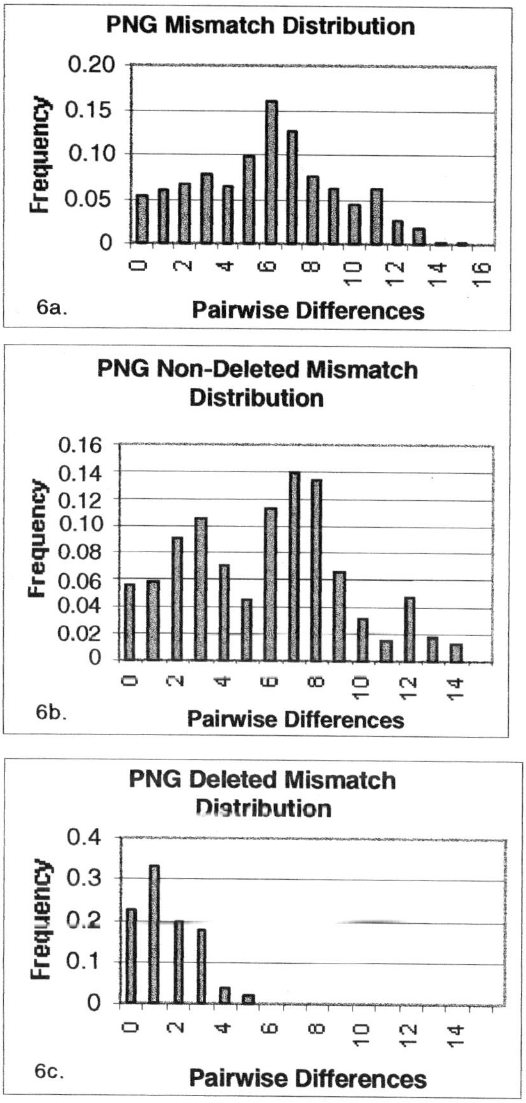

Figure 6. Mismatch distributions of the PNG samples in this study. 6a) Mismatch distribution of all the PNG mainland HVSI sequences from this study. 6b) Mismatch distribution of the non-9-bp deleted HVSI sequences from this study. 6c) Mismatch distribution of the 9-bp deleted HVSI sequences from this study.

our (male) samples. The work reported here is a preliminary progress report which defines many of the key haplogroups found on and around PNG.

6. ACKNOWLEDGMENTS

We would like to thank the following people: Robert Ferrell and Lowell Weitkamp for providing most of the samples; Baruch Blumberg for providing the Santa Cruz samples; Jose Mediavilla and Dan Shraeger for help aliquoting samples and Kirsten

Green for seqeuencing and aligning them; numerous University of Michigan Undergraduate Research Opportunity Program students for extracting the DNAs; Sarah Telfer, Wilson Long, and Saumil Shah for help in sequence preparation; Keith Hunley for help with the Mismatch distributions; Peter Forster for providing the BMN program and advice running it; Fred Gentz for help with references; and the people of Papua New Guinea, New Britain, Misima Island and the Santa Cruz Islands for their participation in this study. This research was supported by NSF grant #9601020 and Sloan-NSF Grant DBI-9803896, the University of Michigan, and Temple University.

7. REFERENCES

Allen J (1996) The Pre-Austronesian settlement of Island Melanesia. In WH Goodenough (ed.): Prehistoric Settlement of the Pacific. Philadelphia: American Philosophical Society, pp. 11–27.

Anderson S, Bankier AT, Barrel BG, DeBulin MHL, Coulson AR, Drouin J, Eperon IC, Nierlich DP, Roe BA, Sanger F, Schreier PH, Smith AJH, Staden R, Young IG (1981) Sequence and organization of the human mitochondrial genome. Nature 290:457–465.

Attenborough RD, Alpers MP, editors (1992) Human Biology in Papua New Guinea: The Small Cosmos. New York: Oxford University Press. 427 pp.

Ballinger SW, Schurr TG, Torroni A, Gan YY, Hodge JA, Hassan K, Chen K-H, Wallace DC (1992) Southeast Asian mitochondrial DNA analysis reveals genetic continuity of ancient Mongoloid migrations. Genetics 130:139–152.

Bandelt HJ, Forster P, Sykes BC, Richards MB (1995) Mitochondrial portraits of human populations using median networks. Genetics 1995 Oct;141(2):743–53

Bandelt H-J, Forster P, Rohl A (1999) Median-joining Networks for inferring Intraspecific Phylogenies. Molecular Biology and Evolution 16(1):37–48.

Bayliss-Smith T (1986) People-plant interactions in the New Guinea highlands: agricultural heartland or horticultural backwater? In The origins and spread of agriculture and pastoralism in Eurasia (ed. D. R. Harris), pp 499–523. London, UCL Press.

Bellwood P (1979) Man's conquest of the Pacific: The prehistory of Southeast Asia and Oceania, Oxford University Press, New York.

Bellwood P (1989) The colonization of the Pacific: some current hypotheses. In AVS Hill, SW Serjeantson (eds.): The Colonization of the Pacific: a Genetic Trail. Oxford University Press, pp. 1–59.

Betty DA, Chin-Atkins AN, Croft L, Sraml M, Easteal S (1996) Multiple independent origins of the COII/tRNA^{Lys} intergenic 9-bp mtDNA deletion in aboriginal Australians. American Journal of Human Genetics 58:428–433.

Blust R (1996a) Austronesian culture history: the windows of language. In WH Goodenough (ed.): Prehistoric Settlement of the Pacific. Philadelphia: American Philosophical Society, pp. 28–37.

Blust R (1996b) Beyond the Austronesian homeland: the Austric hypothesis. In WH Goodenough (ed.): Prehistoric Settlement of the Pacific. Philadelphia: American Philosophical Society, pp. 117–140.

Booth PB, Albery JA, Whittaker J, Sanger R (1970) Gerbich blood group system: a useful genetic marker in certain Melanesians of Papua and New Guinea. *Nature* 228, 462.

Booth PB, Tills D, Warlow A, Kopec AC, Mourant AE, Teesdale P, Hornabrook RW (1982) Red cell antigen, serum protein and red cell enzyme polymorphisms in Karkar Islanders and inhabitants of the adjacent North Coast of New Guinea. *Human Heredity* 32, 385–403.

Bowcock AM, Ruiz-Linares A, Thofohrde J, Minch E, Kidd JR, Cavalli-Sforza LL (1994) High resolution of human evolutionary trees with polymorphic microsattelites. Nature 368:455–457.

Breuil S, Halle L, Fuffie J, Lucotte G (1987) Polymorphism of p49 Y-specific probe in Papuas Baruyas of Papua New Guinea. *Ann Genet* 30, 209–212.

Cheung VG, Nelson SF (1996) Whole genome amplification using a degenerate oligonucleotide primer allows hundreds of genotypes to be performed on less than one nanogram of genomic DNA. Proc Natl Acad Sci USA 93(25):14676–14679.

Comas D, Calafell F, Mateu E, Perez-Lezaun A, Bertranpetit J (1996) Geographic variation nin human mitochondrial DNA control region sequence: The population history of Turkey and its relationship to the European populations. Molecular Biology and Evolution 13(8):1067–1077.

Curtain CC, van Loghem E, Baumgarten A, Golab T, Gorman J, Rutgers CF, Kidson C (1971) The ethnological significance of the gamma-globulin (Gm) factors in Melanesia. American Journal of Physical Anthropology 34:257–271.

Curtain CC, van Loghem E, Fudenberg HH, TIndale NB, Simmons RT, Doherty RL, Vos G (1972) Distribution of the immunoglobulin markers at the IgG1, IgG2, IgG3, IgA2 and K-chain loci in Australian Aborigines: Comparison with New Guinea populations. American Journal of Human Genetics 24:145–55.

Davenport W (1962) Reply to Arthur Cappell: Oceanic Linguistics Today. Current Anthropology 3:400–402.

DiRienzo A, Wilson AC (1991) Branching pattern in the evolutionary tree for human mitochondrial DNA. Proceedings of the National Academy of Sciences USA 88(5):1597–1601.

Felsenstein J (1999) PHYLIP (Phylogenetic Inference Package), Department of Genetics, University of Washington, Seattle.

Foley WA (1986) The Papuan Languages of New Guinea. Cambridge: Cambridge University Press.

Forster P, Kayser M, Meyer E, Roewer L, Pfeiffer H, Benkmann H, Brinkmann B (1998) Phylogenetic resolution of complex mutational features at Y-STR DYS390 in aboriginal Australians and Papuans. *Molecular Biology and Evolution* 15, 1108–1114.

Francalacci P, Bertranpetit J, Calafell F, Underhill PA (1996) Sequence diversity of the control region of mitochondrial DNA in Tuscany and its implications for the peopling of Europe. Americna Journal of Physical Anthropology 100(4):443–460.

Green KA, Friedlaender J, Blumberg B, Merriwether DA (in preparation) mtDNA variation in the Santa Cruz Islanders. American Journal of Human Genetics.

Groube L, Chappell J, Muke J, Price D (1986) A 40,000 year-old human occupation site at Huon Peninsula, Papua New Guinea. *Nature* 324, 453–455.

Hagelberg E, Kayser M, Nagy M, Roewer L, Zimdahl H, Krawczak M, Lio P, Schiefenhovel W (1999) Molecular genetic evidence for the human settlement of the Pacific: analysis of mitochondrial DNA, Y chromosome adn HLA markers. *Philosphical Transactions of the Royal Society London B* 354, 141–152.

Hammer MH (1995) A recent common ancestry for human Y chromosomes. *Nature* 378, 376–378.

Hammer MF, Spurdle AB, Karafet T, Bonner MR, Wood ET, Novelletto A, Malaspina P, Mitchell RJ, Horai S, Jenkins T, Zegura SL (1997) The geographic distribution of human Y chromosome variation. *Genetics* 145, 787–805.

Handt O, Meyer S, von Haeseler A (1998) Compilation of human mtDNA control region sequences. Nucleic Acids Research 26(1):126–129.

Harihara S, Momoki H, Suutou Y, Shimizu K, Omoto K (1992) Frequency of the 9-bp deletion of mitochondrial DNA among Asian populations. Human Biology 64:161–166.

Hertzberg M, Mickleson KNP, Serjeantson SW, Prior JF, Trent RJ (1989) An Asian specific 9-bp deletion of mitochondrial DNA is frequently found in Polynesians. American Journal of Human Genetics 44:504–510.

Hill AVS, O'Shaughnessy DF, Clegg JB (1989) Haemoglobin and globin gene variants in the Pacific. In AV Hill, SW Serjeantson (eds.): The colonization of the Pacific: a genetic trail. Oxford: Clarendon, pp. 246–285.

Irwin GJ (1992) The prehistoric exploration and colonization of the Pacific. Cambridge: Cambridge University Press.

Kelly KM (1990) Gm polymorphism, linguistic affinities, and natural selection in Melanesia. Current Anthropology 31:201–219.

Kidd KK, Kidd JR (1996) A nuclear perspective on human evolution. In AJ Boyce, CGN Mascie-Taylor (eds.): Society for the Study of Human Biology Symposium 38: Molecular biology and Human Diversity. Cambridge: Cambridge U Press, pp. 242–264.

Kirch PV (1997) The Lapita People: Ancestors of the Oceanic World. Cambridge: Blackwell.

Kirk RL (1982) Linguistic, ecological and genetic differentiation in New Guinea and the Western Pacific. In M Crawford, J Mielke (eds.): Current Developments in Anthropological Genetics vol II. Plenum, New York, pp. 229–254.

Long J, Naidu J, Mohrenweiser H, Gershowitz H, Johnson P, Wood J, Smouse P (1986) Genetic characterization of the Gainj and Kalam speaking peoples of Papua New Guinea. American Journal of Physical Anthropology 70:75–96.

Lum K, Richards O, Change C, Cann RL (1994) Polynesian mitochondrial DNA's reveal three deep maternal lineage clusters. Human Biology 66:567–590.

Lum JK, Cann RL, Martinson JJ, Jorde LB (1998) Mitochondrial and nuclear genetic relationships among Pacific Island and Asian populations. American Journal of Human Genetics 63(2):613–624.

Lum JK, Cann RL (1998) mtDNA and language support a common origin of Micronesians and Polynesians in Island Southeast Asia. American Journal of Physical Anthropology 105(2):109–119.

Maddison WP, Maddison DR (1992) MacClade 3.0.1: Analysis of phylogeny and character evolution., Sinauer Associates, Sunderland, Massachusetts.

Martinson JJ (1996) Molecular perspectives on the colonization of the Pacific. In AJ Boyce, CGN Mascie-Taylor (eds.): Society for the Study of Human Biology Symposium 38: Molecular Biology and Human Diversity. Cambridge: Cambridge University Press, pp. 171–195.

Mazzur S, Burgert S, Blumberg BS (1973a) Geographic distribution of Australia Antigen determinants d, y, and "w". Nature 247:38–40.

Mazzur S, Falker D, Blumberg BS (1973b) Geographical variation of the "w" subtype of Australia Antigen. Nature New Biology 243.

Melton T, Peterson R, Redd AJ, Saha N, Sofrσ ASM, Martinson J, Stoneking M (1995) Polynesian genetic affinities with Southeast Asian populations identified by mtDNA analysis. American Journal of Human Genetics 57:403–414.

Melton T, Clifford S, Batzer M, Stoneking M (1999) Genetic evidence for the proto-Austronesian homeland in Asia: mtDNA and nuclear DNA variation in Taiwanese aboriginal tribes. American Journal of Human Genetics 63(6):1807–1823.

Merriwether DA (1999) Freezer Anthropology: New Uses for Old DNA Philosophical Transactions of the Royal Society London B 354:121–129.

Merriwether DA, Friedlaender JS, Mediavilla J, Mgone C, Ferrell RE (in press): Mitochondrial DNA variation is an indicator of Austronesian influence in Island Melanesia. *American Journal of Physical Anthropology*.

Mourant AE (1974) The hereditary blood factors of the peoples of New Guinea and the surrounding regions. *Philosophical Transactions of the Royal Society of London B Biological Science* 268, 251–255.

Page RDM (1996) TREEVIEW: An application to display phylogenetic trees on personal computers. Computer Applications in the Biosciences 12:357–358.

Pavlides C, Gosden C (1994) 35,000 year-old sites in the rain forests of West New Britain, Papua New Guinea. Antiquity 69:604–610.

Pawley AK, Ross M (1995) The prehistory of the Oceanic languages: a current view. In P Bellwood, JJ Fox, D Tryon (eds.): The Austronesians: Historical and Comparative Perspectives. Canberra: Australian National University (Department of Anthropology), pp. 39–95.

Pult I, Sajantila A, Simanainen J, Georgiev O, Schaffner W, Paabo S (1994) Mitochondrial DNA sequences from Switzerland reveal striking homogeneity of European populations. Biol. Chem. Hoppe Seyler 375(12):837–840.

Redd AJ, Takezaki N, Sherry ST, McGarvey ST, Sofro ASM, Stoneking M (1995) Evolutionary History of the COII/tRNALys Intergenic 9 Base Pair Deletion in Human Mitochondrial DNAs from the Pacific. Molecular Biology & Evolution 12:604–615.

Richards M, Corte-Real H, Forster P, Macaulay V, Wilkinson-Herbots H, Demaine A, Papiha S, Hedges R, Bandelt HJ, Sykes B (1996) Paleolithic and neolithic lineages in the European mitochondrial gene pool. American Journal of Human Genetics 59(1):185–203.

Roberts RG, Jones R, Smith MA (1994) Beyond the radiocarbon barrier in Australian prehistory. Antiquity 69:611–616.

Roberts-Thomson JM, Martinson JJ, Norwich JT, Harding RM, Clegg JB, Boettcher B (1996) An ancient common origin of aboriginal Australians and New Guinea highlanders is supported by alphapglobin analysis. *American Journal of Human Genetics* 58, 1017–1024.

Rogers A, Harpending H (1992) Population growth makes waves in the distribution of pairwise differences. Molecular Biology and Evolution, 9:552–569.

Ross MD (1988) Proto Oceanic and the Austronesian Languages of Western Melanesia. Canberra: Department of Linguistics, Research School of Pacific Studies, Australian National University. Pacific Linguistics C-98.

Saiki RK, Glefland DH, Stoffel S, Scharf SJ, Higuchi R, Horn GT, Mullis KB, Erlich HA (1988) Primer-directed enzymatic amplification of DNA with a thermostable DNA polymerase. Science 239:487–491.

Saitou N, Nei M (1987) The neighbor-joining method: A new method for reconstructing phylogenetic trees. Molecular Biology and Evolution 4(4):406–425.

Sajantila A, Lahermo P, Anttinen T, Lukka M, sistonen P, Savontaus ML, Aula P, Beckman L, Tranebjaerg L, Gedde-Dahl T, Issel-Tarver L, DiRienzo A, Paabo S (1995) Genes and languages in Europe: An analysis of mitochondrial lineages. Genome Research 5(1):42–52.

Sajantila A, Salem AH, Savolainen P, Bauer K, Gierig C, an dPaabo S (1996) Paternal and maternal DNA lineages ereveal a bottleneck in the founding of the Finnish population. Proceedings of the National Academy of Sciences USA 93(21):12035–12039.

Schanfield MS, Giles E, Gershowitz H (1975) Genetic studies in the Markham Valley, northeastern Papua New Guinea: gamma globulin (Gm and Inv), group specific component (Gc) adn ceruloplasmin (Cp) typing. *American Journal of Physical Anthropology* 42, 1–7.

Salmon D, Godelier M, Halle L, Lemonnier P, Lory JL, Rouger P, Ruffie J, Salmon C (1988) Blood groups in Papua New Guinea Eastern Highlands. *Gene Geography* 2, 89–98.

Serjeantson S, Lai LY, Baumgarten A, Curtain CC (1973) Blood Group, erythrocyte acid phosphatase, serum group specific component, haptoglobin and transferin gene frequencies amongst the Gogodara (Balimo), Kuman (Minj), and Enga (Laiagam) and peoples of New Guinea. *Hum Biol Coeania* 2, 139–146.

Serjeantson S, Kirk RL, Booth PB (1983) Linguistic and genetic differentiation in New Guinea. Journal of Human Evolution 12(1):77–92.

Simmons RT (1973) Blood group genetic patterns and heterogeneity in New Guinea. *Hum Biol Oceania* 2, 63–71.

Spriggs MJT (1997) The Island Melanesians. Blackwell Publishers. Cambridge, MA.

Steinberg AG, Larrick JW (1981) Gm and Inv (Km) studies of Melanesian people of the Huon Peninsula in Northeast Papua New Guinea: polymorphism for a Gm1,5,10,11,13,14,17,21,26 Haplotype. *American Journal of Physical Anthropology* 55, 89–94.

Stoncking M, Jorde LB, Bhatia K, Wilson AC (1990) Geographic variation in human mitochondrial DNA from Papua New Guinea. Genetics 124:717–733.

Stoneking M, Wilson AC (1989) Mitochondrial DNA. In A Hill (ed.): The colonization of the Pacific: a genetic trail. Oxford: Clarendon, pp. 215–245.

Sykes B, Lieboff A, Low-Beer J, Tetznere S, Richards M (1995) The origins of the Polynesians: an interpretation from mitochondrial lineage analysis. American Journal of Human Genetics 57:1463–1475.

Tishkoff SA, Dietzsch E, Speed W, Pakstis AJ, Cheung K, Kidd JR, Bonne-Tamir B, Santachiara-Benercetti AS, Moral P, Watson E, Krings M, Paabo S, Risch N, Jenkins T, Kidd KK (1996) Global Patterns of Linkage Disequilibrium at the CD4 Locus and Modern Human Origins. Science 271:1380–1387.

Tsintof AS, Hertzberg MS, Prior JF, Mickleson KN, Trent RJ (1990) Alpha-globin gene markers identify genetic differences between Australian aborigines and Melanesians. *American Journal of Human Genetics* 46, 138–143.

Vigilant L, Stoneking M, Harpending H, Hawkes K, Wilson AC (1991) African populations and the evolution of human mitochondrial DNA. *Science* 253, 1503–1507.

Wiesenfeld SL, Gajdusek C (1976) Genetic structure and heterozygosity in the kuru region, eastern highlands of New Guinea. *American Journal of Physical Anthropology* 45, 177–190.

Weitkamp LR, Shreffler DC, Saave JJ (1969) Serum Albumin variants in New Guinea Indigenes. *Vox Sang* 17, 237–240.

Wickler S, Spriggs M (1988) Pleistocene human occupation of the Solomon Islands, Melanesia Antiquity 62:703–706.

Wood JW, Johnson PL, Kirk RL, McLoughlin K, Blake NM, Matheson FA (1982) The genetic demography of the Gainj of Papua New Guinea. I. Local differentiation fo blood group, red cell enzyme, and serum protein allele frequencies. *American Journal of Physical Anthropology* 57, 15–25.

Wurm |S| (1982) Languages of Oceania. Gunter Narr Berlag, Tubingen.

Yenchitsomanus PT, Summers KM, Bhatia KK, Cattani J, Board PG (1985) Extremely high frequencies of alpha-globin gene deletion in Madang and on Kar Kar Island, Papua New Guinea. *American Journal of Human Genetics* 37, 778–784.

Zemel BS, Jenkins C (1989) Urban-rural differences in growth and maturation among the Bundi (Gende speakers) of Papua New Guinea. American Journal of Human Biology (1), pp. 44–66.

Zemel BM (1989) "Dietary Change and Adolescent Growth Among the Bundi (Gende-speaking) People of Papua New Guinea," Ph.D. Dissertation, University of Pennsylvania, 1989.

13

MOLECULAR ANTHROPOLOGY

Progress and Perspectives on Ancient DNA Technology

Connie J. Kolman

Smithsonian Center for Materials Research and Education
Suitland, MD 20746 and National Institute of Alcohol Abuse and Alcoholism, National Institutes of Health, Rockville, MD 20852*

ABSTRACT

The use of ancient DNA (aDNA) in the reconstruction of population origins and evolution is becoming increasingly common. However, contamination of ancient specimens and DNA extracts with modern DNA is more widespread than is generally acknowledged and remains a significant problem in aDNA analysis. Furthermore, studies of human aDNA are uniquely sensitive to contamination due to the continual presence of potential contamination sources and these studies require novel solutions for identification of contamination. Careful selection of polymorphic markers capable of discriminating between ancient DNA and probable DNA contaminants is critical. A promising new application of aDNA technology is the analysis of ancient human pathogens, which can be conducted with reduced risk of contamination. Fascinating questions to be addressed through ancient pathogen studies include the evolutionary history of pathogens and the migration patterns and evolutionary relationship of human populations.

1. INTRODUCTION

Ancient DNA (aDNA) technology has emerged within the past decade as one of the breakthroughs borne of the revolution in molecular biology that originated with the polymerase chain reaction (PCR). The enormous potential of PCR is based on a deceptively simple methodology that enables the exponential amplification of a specific region of DNA or RNA from only a few molecules, even in the midst of a significant quantity of non-target nucleic acid. This capability has stimulated the study of large

* Current address: National Institutes of Health, NIAAA/DICBR/LNG, 12420 Parklawn Dr., Park Bldg Rm 451 MSC 8110, Rockville, MD 20852, E-mail: *ckolman@mail.nih.gov*

Genomic Diversity: Applications in Human Population Genetics
edited by Papiha, Deka, and Chakraborty, Kluwer Academic / Plenum Publishers, New York, 1999.

numbers of individuals that is fundamental to population genetics and has also permitted the investigation of DNA samples that are too degraded or damaged for analysis by traditional cloning methods. Thus, the specialized field of ancient DNA analysis of populations was created. "Ancient" samples are generally those that were not collected for the purpose of immediate DNA or RNA analysis and include archaeological, clinical, and natural history specimens. Since these specimens were not originally collected or preserved for nucleic acid analysis, endogenous DNA or RNA is typically damaged to an extent that enzymatic amplification can be quite difficult, if not impossible, to achieve. The types of DNA damage that are primarily encountered include modifications of pyrimidines and sugar residues as well as baseless sites and intermolecular crosslinks (Pääbo 1989). Only limited research has been conducted on the chemistry of aDNA damage and possible methods of *in vitro* repair. Various protocols have been developed to determine the utility of specific ancient specimens for aDNA analysis (Handt *et al.* 1994, 1996; Richards *et al.* 1995; Poinar *et al.* 1996) and these methods have recently been evaluated in a comparative study (Kolman and Tuross, in press).

Damage of aDNA increases the potential for another, more widely publicized characteristic of aDNA PCR, that of contamination. Since PCR analysis involves the exponential generation of new, synthetic DNA products from a small number of molecules, contamination with exogenous DNA in one of the initial PCR cycles can result in exclusive amplification of the contaminating DNA. This possibility is increased in aDNA analysis where the contaminant is likely to be undamaged DNA which will be amplified preferentially over the damaged, endogenous DNA. The growing number of aDNA studies published and number of samples and polymorphic sites assayed may give the impression that all technological hurdles associated with aDNA technology have been overcome. However, identification of contamination remains the single most critical issue in aDNA methodology. Early, spectacular claims of successful DNA extraction from extremely old specimens, such as 17–20 million year old (Myr old) *Magnolia* leaf fossils (Golenberg *et al.* 1990), 25–135 Myr old specimens preserved in amber (DeSalle *et al.* 1992; Cano *et al.* 1993), and 80 Myr old dinosaur bones (Woodward *et al.* 1994), have generally been disproved or cast into serious doubt (Sidow *et al.* 1991; DeSalle *et al.* 1993; Hedges and Schweitzer 1995; Austin *et al.* 1997).

Human specimens, although relatively recent in origin and of particular relevance for anthropological studies, are uniquely sensitive to contamination simply because every researcher involved in the study represents a potential source of contaminating DNA. Recent analysis of DNA extracted from the Neandertal type specimen (Krings *et al.* 1997) revealed two distinct sets of mitochondrial D-loop sequences, one significantly different from modern humans and proposed to be Neandertal in origin and one identical to the human reference sequence (Anderson *et al.* 1981) and presumed to reflect modern human contamination. In a study of animal bones excavated from a site in England, Richards *et al.* (1995) reported that approximately 50% of the bones exhibited contamination with human DNA sequences. Similar contamination should be assumed for all human bones and measures to identify contaminants should be integrated into the research design of human aDNA studies.

A promising area for future aDNA research is the analysis of pathogenic DNA extracted from ancient human remains. The issue of contamination is resolved because experiments can be conducted in a laboratory in which the pathogen has never been studied. A fascinating range of questions can be addressed including the evolutionary history of pathogens and the migration patterns and evolutionary relationships of human populations. For example, research on the distribution of human T-cell lymophotrophic retrovirus type II in contemporary populations has been used to propose that the first migrants to the New World originated in east Central Asia (Neel *et al.* 1994). Study of ancient specimens could shed light on such long-standing controversies as the origin of venereal syphilis. The relevant question is whether venereal syphilis was endemic to the New World prior to European contact or was present, but undiagnosed, in the Old World at that time. Nucleic acid-based identification of the causative agent in pre-contact skeletons from either the New or Old World would quickly put the debate to rest.

This paper reports results on ancient human and pathogen studies and presents technical considerations for these types of analyses with a particular emphasis on identification of contamination. Specific methods for enhancing amplification of aDNA and for detecting contaminants are summarized. The presented research is also used as a means to assess previously published aDNA protocols and results and to suggest improvements for future studies. In general, the results demonstrate that the analysis of aDNA is still an excruciatingly slow and meticulous process. Current strategies that assume contamination can be eliminated are inadvisably applied to human aDNA studies. Instead, research strategies must be designed with a goal of identifying all DNA contaminants in order to differentiate convincingly contamination from endogenous DNA. Pathogen studies are free from the problem of contamination only if no modern DNA is introduced into the laboratory environment, including positive PCR control DNA. The ability of ancient DNA analyses to provide unique insights into anthropological questions assures the merit of such projects, but only if extensive precautions have been taken to ensure the accuracy and reliability of the results.

2. ANCIENT DNA ANALYSIS OF HUMAN POPULATIONS

2.1. Analysis of New World Pre-Contact Skeletal Specimens

Noreen Tuross (Smithsonian Center for Materials Research and Education) and I conducted a study of specimens excavated in the Americas in the 1950s. More than 100 skeletons were recovered from approximately 30 sites that dated from 0 AD to the 1800s. The geologic condition of the excavation sites was cool and dry and the skeletal material was in uniformly excellent condition. The skeletal material was generally a pale color similar to fresh bone, with little of the dark discoloration found in buried bones that have been subjected to frequent perfusion by groundwater and accompanying compounds thought to be at least partially responsible for the PCR inhibition present in some aDNA extracts. Mitochondrial D-loop DNA sequences and restriction fragment length polymorphisms (RFLPs) were assayed in DNAs extracted from a subset of specimens. The study was conducted in a facility in which no DNA-based analyses of modern human populations had ever been performed so that the sole source of contamination was physical handling of the specimens by researchers.

2.1.1. Materials and Methods. DNA extraction and purification. Various subsets of approximately 20 specimens were used to test different combinations of extraction and purification methods. Extraction methods involved partial to complete decalcification of the skeletal material and purification methods included proteinase K digestion, organic extraction of proteins, ethanol precipitation of nucleic acids, Centricon filtration, and resin-based purification. The results of these initial optimization experiments can be summarized as follows: 1) Complete decalcification (immersion in 0.5 M ethylenediaminetetraacetate [EDTA], 4°C, for 1–2 weeks) of the skeletal material to release the maximal quantity of DNA resulted in stronger aDNA PCR signals and weaker contaminating DNA signals. During the initial days of bone decalcification, the highest quantity of inhibitory material was released while a greater quantity of endogenous DNA was released after this period. However, complete decalcification of skeletal material may be necessary only for older specimens as Fisher *et al.* (1993) reported that no decalcification was necessary for bones up to 125 years old. 2) Minimal purification of the crude DNA extracts using Centricon filters (Amicon, Inc., Beverly, MA) resulted in stronger aDNA PCR signals and weaker contaminating DNA signals. Additional purification was found to be unnecessary and, in some cases, disadvantageous. The most dramatic effect was seen with resin-based purification (Magic MegaPrep DNA Purification System, Promega, Madison, WI) which resulted in exclusive recovery of contaminating DNA. This result may reflect reduced binding to the resin of aDNA that is cross-linked to proteinaceous material thus allowing preferential purification of the contaminating DNA. The inability of DNA-protein complexes, such as those found in aDNA extracts, to interact properly with DNA-binding resins provides a cautionary note for purification protocols based on these resins, e.g. silica-based purifications (Höss and Pääbo 1993).

The optimal DNA extraction and purification protocol was determined to be complete decalcification of the skeletal fragment in a single 2 week step in 0.5 M EDTA at 4°C followed by concentration of the soluble material through Centricon filters. DNA extracts were prepared from 13 specimens utilizing this protocol. All DNA extracts were initially assayed for inhibition of the amplification reaction. Many aDNA extracts contain substances that are inhibitory to the DNA polymerase and should be tested for such effects in order to calculate an appropriate dilution of DNA extract for amplification reactions. After inhibition testing, the 13 DNA extracts were used in the following PCR optimization experiments.

DNA amplification. A standard PCR protocol for aDNA, including use of elevated quantities of DNA polymerase and increased number of amplification cycles, was utilized. Amplifications contained GeneAmp buffer (PE Applied Biosystems, Foster City, CA), 1.75 mM $MgCl_2$, 200 μM of each dNTP, 20 μg/ml bovine serum albumin, 1 μM of each primer (Integrated DNA Technologies, Inc., Coralville, IA) and 1.25 units AmplitaqGold (PE Applied Biosystems) for each 10 μl reaction. Reactions were subjected to 50 cycles of PCR; 30 s at 94 °C, 30 s at annealing temperatures ranging from 49–55 °C, and 45 s at 72 °C. A detailed description of the PCR protocol can be found in Kolman and Tuross (in press). Fewer than 40 PCR cycles resulted in undetectable levels of amplification. Primers varying in length from 20 base pairs (bps) to 36 bps were tested. Preliminary experiments comparing primer pairs of 20 bps/20 bps and 36 bps/25 bps demonstrated that longer primers yielded increased intensity and specificity of the PCR signal although increased generation of primer dimer products represented a potential problem.

Based on the optimization experiments, five modifications to the standard PCR protocol were determined to be important for robust, specific amplification of the aDNA extracts. 1) UV irradiation of all reaction components, including tubes, ddH_2O, buffer, $MgCl_2$, BSA, dNTPs, and DNA primers, at 254 nm for 20 min immediately prior to addition of DNA and enzyme (Sarkar and Sommer 1990) was necessary for consistently negative blank extraction and PCR controls. Commercially obtained DNA primers were frequently found to be contaminated and were routinely "decontaminated" by irradiation at 254 nm for 20 min. 2) A "heat-soak" step (Ruano *et al.* 1992) at 95 °C for 10–20 min prior to temperature cycling was essential for robust DNA amplification. This was achieved either by adding the primers, dNTPs, and enzyme after the "heat-soak" step or by using AmplitaqGold DNA polymerase that requires a 12 min 95 °C incubation for full activity. 3) Low levels of spermidine facilitated amplification of some DNA samples (Wan and Wilkins 1993). The spermidine concentration was critical because excessive spermidine was found to inhibit the PCR. Optimal final concentrations of 400–800 μM spermidine were determined empirically. These spermidine concentrations inhibited amplification of DNA samples that amplified in the absence of spermidine. Therefore, DNA samples were first tested for amplification in the absence of spermidine and those DNA samples with no amplification were tested again with spermidine. 4) Reamplification of a weak PCR product extracted from an agarose gel was unsatisfactory and routinely generated contamination. The contamination most likely derived from previously amplified DNA present in the electrophoresis equipment. Supporting this interpretation, successful amplification of human mitochondrial D-loop DNA was recovered from a piece of agarose in which no DNA had been electrophoresed. 5) Positive PCR controls were never performed in order to reduce the risk of contamination by modern DNA. If a series of PCRs was completely negative, it was simply repeated in order to determine if a reagent had been omitted or if the lack of amplification was accurate.

Analysis of PCR products. Restriction digests of PCR products with 10–20 units of the appropriate restriction enzyme (New England Biolabs, Beverly, MA) were performed using the recommended buffer and incubated overnight at 37 °C. Digestion products were analyzed by electrophoresis on 2% agarose gels and amplification products containing the 9 bp region were analyzed on 4% Metaphor (FMC BioProducts, Rockland, ME). DNA sequence was determined from cloned PCR products for all specimens and, in some cases, directly from the amplification reactions also. PCR products were cloned using a PCR-Script Cloning kit (Stratagene, La Jolla, CA). Sequencing was performed using a DNA Sequencing kit, FS version (PE Applied Biosystems) and sequencing reactions were analyzed on a 373 DNA Sequencer (PE Applied Biosystems). For a detailed cloning and sequencing protocol, see Kolman and Tuross (in press). DNA sequence data were collected between bp 16,211–16,400 of the mitochondrial D-loop control region I (numbering according to Anderson *et al.* [1981]). Both light and heavy strands were sequenced for all clones.

2.1.2. Results. Four haplogroups, labeled A, B, C, and D, were defined by Torroni *et al.* (1992) based on mitochondrial restriction/deletion polymorphisms and were proposed to represent founding New World haplogroups. Mitochondrial D-loop sequence data have been used independently to confirm these haplotype groupings (Horai *et al.* 1993). Diagnostic polymorphisms based on restriction/deletion data and D-loop sequence data are listed in Table 1. D-loop polymorphisms were assayed in all aDNA extracts and are presented below (Table 2). Restriction/deletion polymorphisms were assayed in a subset of samples and confirmed all conclusions based on sequence data (data not shown).

None of the 13 DNA extracts tested in the current study were found to be inhibitory at concentrations of undiluted extract representing 10% of the final reaction volume. Lack of inhibition most likely reflected the pristine condition and absence of discoloration in the studied bones.

In general, it was possible to obtain appropriately sized PCR products from all DNA extracts. However, it became clear early in the study that contamination of the aDNA samples with modern DNA of presumed European origin was a persistent problem. Blank extraction and PCR controls were negative for all reported results so the cause was not simple contamination of reagents or laboratory ware. Only select speci-

Table 1. Defining polymorphisms for New World mitochondrial founding haplogroups

Site[a]	Haplogroup A	B	C	D
Restriction/deletion sites				
*Hae*III:663	1	0	0	0
9 bp region	2	1	2	2
*Alu*I:13262	0	0	1	0
*Alu*I:5176	1	1	1	0
D-loop polymorphisms				
16217	T	C	T	T
16223	T	C	T	T
16290	T	C	C	C
16298	T	T	C	T
16319	A	G	G	G
16325	T	T	C/t	C
16327	C	C	T	C
16362	C	T	T	C

[a]Restriction/deletion defining sites are provided by Torroni *et al.* (1992) and D loop polymorphisms are provided by Horai *et al.* (1993). Presence and absence of a restriction site are indicated by 1 and 0, respectively. 1 and 2 indicates one and two copies of the 9 bp repeat, respectively. For D-loop sequence data, the defining sites are underlined. X/y indicates that X is the predominant nucleotide at that position although y does occur at a low frequency.

mens produced DNA samples with evidence of modern contamination and, in these samples, the contamination was not consistent in all amplification reactions. The contamination was observed initially as two sets of bands on an agarose gel when a single New World-specific polymorphism was assayed. One specimen exhibited an *Alu*I site at bp 13,262 that defines haplogroup C, one of the four founding haplogroups proposed for the New World (Torroni *et al.* 1992). However, gel electrophoresis revealed two sets of bands corresponding to presence and absence of the *Alu*I recognition site (Fig. 1A). The two sets of bands were unlikely to reflect partial digestion by the *Alu*I restriction enzyme since all other reactions digested completely. Instead the two sets of bands most likely corresponded to two sources of DNA, one an ancient group C haplotype and the other a modern variant lacking the *Alu*I site. Two other samples exhibited the 9 bp deletion that is diagnostic for haplogroup B. However, they both showed a band corresponding to the undeleted form of this marker in addition to the deletion (Fig. 1B). In this case, there was no possibility of a partial digest accounting for the two bands and the cause was clearly two sources of DNA. Multiple extractions from the same individual resulted in similar levels of contamination suggesting that the skeletal specimens themselves had been contaminated.

The contaminated specimens most likely contained at least two populations of amplifiable DNA molecules, one ancient in origin and the other of modern origin. Presumably, both types of DNA were present at extremely low concentrations and were unable to efficiently amplify alone thus explaining the lack of amplification in blank extraction and PCR controls. In this case, both DNAs would be needed to increase the DNA concentration to a critical level at which point they would compete for reagents. The result was agarose gel bands of varying intensity that corresponded to the two types of DNA.

Table 2. Mitochondrial D-loop haplotypes for 13 skeletal samples

Time period	Sample designation	Haplotypes[a]	N[b]	Summary
1–400 AD	Sample A	Reference (1)[c]	2	2 non-NW sequences
		256:T, 270:T, 399:G (2)	2	
	Sample B	217:C, 223:T (3)[c]	1	1 non-NW sequence
	Sample C	Haplogroup C	3	Haplogroup C + 1 non-NW sequence
		224:C, 311:C (4)	4	
	Sample D	217:C, 223:T (3)	3	3 non-NW sequences
		223:T, 225:T, 227:G, 278:T, 357:C (5)	8	
		223:T, 227:G, 243:C, 278:T, 337:T (6)	1	
800–950 AD	Sample E[d]	Haplogroup C[c]	4	Haplogroup C
	Sample F[d]	Reference (1)[c]	3	2 non-NW sequences
		291:T (7)	2	
	Sample G	Reference (1)	3	1 non-NW sequence
	Sample H	256:T, 270:T, 399:G (2)	2	1 non-NW sequence
	Sample I	Haplogroup C	1	Haplogroup C + 2 non-NW sequences
		Reference (1)	2	
		294:T (8)[c]	2	
	Sample J	Reference (1)[c]	4	1 non-NW sequence
1200–1600 AD	Sample K[d]	Haplogroup C[c]	10	Haplogroup C + 2 non-NW sequences
		217:C, 223:T (3)	1	
		218:G (9)	1	
	Sample L[d]	Haplogroup B[c]	4	Haplogroup B + 6 non-NW sequences
		Reference (1)	2	
		217:C, 223:T (3)	3	
		311:C (10)	2	
		218:T, 328:A, 362:C (11)[c]	3	
		286:T, 393:T (12)[c]	2	
		261:T (13)	3	
	Sample M[d]	Haplogroup B	2	Haplogroup B + 1 non-NW sequence
		294:T, 296:T, 304:C (14)[c]	4	

[a]Haplotypes are presented as polymorphisms relative to the human reference sequence (Anderson *et al.* 1981). Nucleotide positions are listed without the "16" prefix. The only New World haplotypes detected were haplogroups B (217:C) and C (223:T 298:C 325:C 327:T) (and are presented in parentheses).
Unique non-New World haplotypes are numbered sequentially.

[b]Number of independent clones that exhibited each haplotype.

[c]DNA sequences determined by direct sequencing of PCR products.

[d]Samples E, F, K, L, and M are presented in Kolman and Tuross (in press) as Samples 1, 5, 2, 3, and 4, respectively.

DNA sequence of the mitochondrial D-loop was determined in order to precisely define the nature and extent of contamination present in the DNA extracts. DNA sequence was determined from cloned PCR products for all specimens and, in some cases, directly from the amplification reactions also. This analysis provided confirmation of contamination based on the identification of more than one sequence in the majority of samples. Presumably, at least some of the original skeletal specimens were contaminated

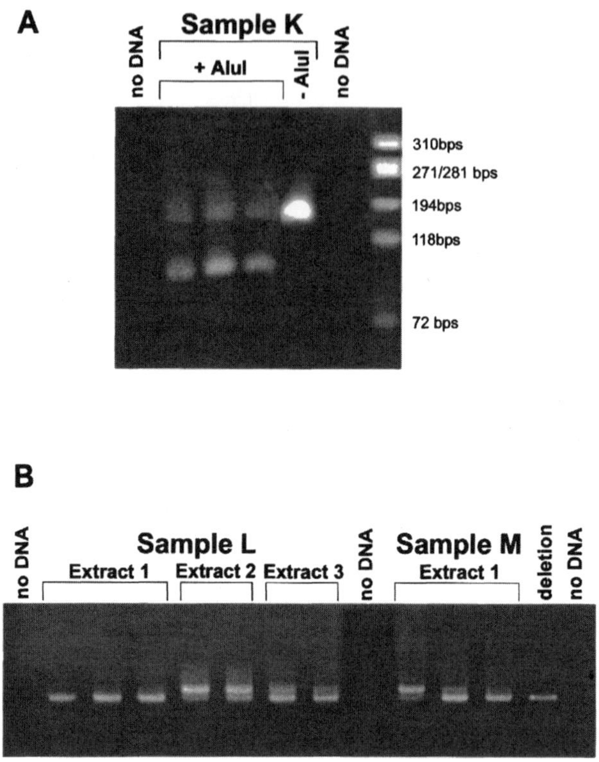

Figure 1. Two populations of DNA molecules present in select specimens. (A) Ethidium bromide-stained agarose gel showing the results of PCR analysis of Sample K with primers to assay the *Alu*I site at bp13262. Lanes one and six depict blank PCR controls, lanes two through four depict three amplification reactions using varying amounts of Sample K DNA extract followed by *Alu*I digestion, and lane five is an undigested PCR of Sample K. DNA size markers were run in lane seven. (B) Ethidium bromide-stained agarose gel showing the results of PCR analysis of Samples L and M with primers to assay the 9 bp deletion. Independent PCRs using varying amounts of DNA from three Sample L extracts and one Sample M extract are presented. Controls include three blank PCRs and a PCR of an individual with the 9 bp deletion (positive control PCR was performed at the Smithsonian Tropical Research Institute in Panamá and stored frozen for several years).

because multiple extracts from the same specimen gave identical results of contamination. Table 2 shows the 16 different haplotypes that were determined for 13 individuals. Two of the four founding haplogroups that have been proposed for the New World (Torroni *et al.* 1992) were detected, haplogroups B and C. Fourteen additional sequences were identified that have never been detected in New World populations. Since the majority of the 14 sequences have been detected in European populations (one of which was identical to the author, #14), they most likely reflect contamination by modern DNA of European origin. An in-depth analysis of five of the specimens, samples E, F, K, L, and M, is presented in Kolman and Tuross (in press).

2.1.3. Discussion. In general, no correlation could be made between the age of a specimen, the likelihood of detecting a New World haplotype, and the extent of contamination as measured by presence of non-New World haplotypes. Since contamination appears to be nearly ubiquitous in human aDNA studies, it would be extremely useful to have a criterion by which to classify the origin of a sequence as endogenous or contamination that is independent of the particular genetic markers assayed. However, this study makes clear that no simple criterion exists. Neither the most common or most uncommon sequence could be assumed to be endogenous as both extremes of frequencies were detected for New World and non-New World sequences in different specimens. It also has been suggested that direct sequencing of the PCR products would be more likely to reveal the sequence of the original, undamaged template as opposed to DNA sequences from individual clones that could contain mutations due to template damage. However, direct PCR sequence analysis identified the New World haplotype in only three out of seven cases where New World haplotypes were detected. It has also been proposed that shorter PCR products are more likely to represent aDNA because the damaged aDNA should contain a larger percentage of short DNA fragments relative to modern DNA (Handt *et al.* 1994; Richards *et al.* 1995). However, all 238 bp PCR products from Sample K yielded ancient haplogroup C whereas 132 bp products revealed a mixture of haplogroup C and contaminating sequences (data not shown). Thus, no simple, objective method exists to identify and eliminate contamination sequences and the only means by which contamination was detected in the current study was by assaying polymorphisms that differed between the two sources of DNA, i.e. New World indigenous groups and European populations.

In order to understand the extent and origin of modern DNA contamination, it is instructive to estimate the number of distinct sources of contamination or number of contamination events represented in the reported dataset. Some of the contaminating sequences detected in the current study differed from each other by only a single polymorphism, raising the possibility that polymerase errors were responsible for the differences and that multiple sequences represented only a single DNA source. To reduce this possibility, sequences that differed by a single nucleotide within the same specimen are presented in Table 2 only if they were detected in multiple clones. If we go one step further and eliminate all sequences differing by only a single polymorphism from any sequence in the data set (#7, 8, 9, 10, and 13), nine contaminating sequences remain. This number far overshadows the number of ancient sequences that were detected and provides a cautionary note when reviewing studies in which no contamination was reported and all detected sequences were treated as endogenous.

Considerable effort was made in the current study to detect DNA contamination through 1) analysis of multiple diagnostic markers capable of distinguishing between ancient DNA and probable contaminants and 2) DNA sequence analysis of cloned PCR products. The reported data set may represent an upper limit in the ability to detect DNA contaminants. Moreover, the fact that the study was conducted in a laboratory in which no analyses of modern human DNA had ever been performed means that the opportunity for contamination was significantly reduced relative to studies in which ancient and modern populations were analyzed in the same laboratory or in different rooms of the same laboratory. Therefore, although the ability to detect contamination was improved, the actual number of contaminants present in the current study may be reduced relative to other studies using less rigorous laboratory protocols.

The ability to detect contamination is directly determined by the genetic markers chosen for analysis. In ancient New World populations, four RFLP/deletions are commonly assayed because they were originally used to define founding haplogroups (Table 1) and have now been studied in thousands of New World individuals (Ward *et al.* 1991, 1993; Torroni *et al.* 1992, 1993; Kolman *et al.* 1995, 1997; Merriwether *et al.* 1995). However, these RFLP/deletions are not ideal markers for aDNA studies and may actually conceal contamination because they represent only a subset of all mitochondrial polymorphisms currently assayed in contemporary human populations. Since modern populations may carry less genetic variation than their ancestors or may be more distantly related to ancient populations than is currently recognized, it is a dangerous strategy to assay prehistoric populations for a restricted set of polymorphisms that have been culled from contemporary populations. To assay ancient specimens for only a few of the four diagnostic markers with the justification of damaged aDNA and commensurate increase in time required for aDNA analyses is to invite incorrect haplogroup assignments. In sum, unreported evidence of contamination or the use of research strategies that are unlikely to detect contamination should not be interpreted as absence of contamination or as proof of authenticity of the data.

2.2. Impact of Widespread Contamination With Modern Human DNA

Contamination of archaeological material and the complex interplay between endogenous ancient DNA and modern contamination reported here are unlikely to represent a unique situation. Numerous cases exist in the published literature that indicate similar problems have been previously encountered, only some of which were identified as contamination. As mentioned above, Krings *et al.* (1997) detected presumably modern human contamination in their determination of mitochondrial D-loop DNA sequence from a Neandertal specimen. A second example is provided by Handt *et al.* (1996) who identified deleted and undeleted forms of the 9 bp region in two out of three 600-year-old Paleoindian skeletal samples, similar to the result reported for Samples L and M. Kaestle (1997) identified cut and uncut PCR products at the *Hinc*II: bp13,259 site (also used to define haplogroup C) in a collection of western Nevada skeletons and provides a third example of contamination comparable to that reported for Sample K.

Despite a decade of aDNA research, contamination by modern DNA remains a problem because the many sources and modes of contamination are still not known or understood and, therefore, can not be controlled or eliminated. One early cause of contamination was aspiration of DNA through the barrel of automatic pipeting aids from one amplification reaction to another. A simple solution was the use of plugged pipet tips. However, remaining causes of contamination are unlikely to be so easily identified or simply solved. Negative blank extraction and PCR controls are not a reliable indicator of contamination because of the complex interdependence of endogenous and contaminating DNA necessary for successful amplification. Locating an aDNA laboratory outside of the main laboratory that is still utilized by the same researchers is unlikely to eliminate contamination. Previous work performed by my group on natural history specimens of cichlid fish provides an example of the pervasiveness of contamination; one year after moving all positive control goldfish DNA to another floor and wing of the building, goldfish contamination was still detected in ancient cichlid DNA PCRs (data not shown). Only because of the high level of variation present at the locus being studied, could the exact source and mode of the fish contamination be identified.

Clearly, aDNA investigation of human populations requires a reanalysis of research strategies to reflect the ubiquitous presence of contamination. Experiments must be designed with a goal of identifying all DNA contaminants in order to distinguish convincingly between endogenous and contaminating DNA. Each aDNA research project must be custom designed to assay polymorphisms that differentiate between the ancient population and potential sources of contamination. The reduced mitochondrial diversity of New World populations (Ward *et al.* 1991; Torroni *et al.* 1992; Merriwether *et al.* 1995; Kolman *et al.* 1996) and the large comparative database of mitochondrial sequences worldwide permits the straightforward identification of contaminating haplotypes in New World populations. Ancient DNA studies wherein the ancient populations and likely source(s) of contamination are genetically similar will be much more difficult to conduct. The increased levels of mitochondrial variation detected in contemporary African populations (Vigilant *et al.* 1991; Watson *et al.* 1997) relative to modern European populations (Richards *et al.* 1996) and reduced odds of a shared sequence between an ancient African specimen and a modern contaminant may mean that aDNA studies of African populations are more feasible. However, the inconsistent occurrence of contamination in the samples presented here illustrates the insidious nature of very low level contamination and the need for constant vigilance in the detection of such contamination.

Furthermore, the incredible amount of time and resources necessary for aDNA studies mandate that only questions that cannot be addressed with modern samples be undertaken. Stone and Stoneking's (1993, 1998) analysis of 108 prehistoric New World skeletons, assayed for mitochondrial RFLPs and D-loop sequence, represents the most comprehensive human aDNA population study published to date. Despite this wealth of data, however, no conclusions could be drawn that had not already been reported from studies of contemporary New World populations. While it is beneficial to obtain confirmation of previous results, aDNA studies must offer a more substantial return on the considerable investment of time and resources. To accomplish this, it is essential that researchers cast a critical eye when judging potential aDNA research projects. With respect to populations from the New World and Europe, the genetic markers presently available are not able to differentiate between populations from neighboring geographic regions. Unfortunately, those types of studies represent a majority of the aDNA investigations that are currently in progress. In addition to redesigning the research strategies employed for human aDNA studies, the theories being tested must be refocused to accommodate the reality of limited geographic genetic variability and to avoid the undesirable outcome of inconclusive or uninteresting results at the end of a costly project.

3. ANCIENT DNA ANALYSIS OF HUMAN PATHOGENS

The only means of controlling contamination is to prevent the introduction of contaminants. This solution is clearly not an option with analyses of human populations because of the presence of human researchers. However, study of human pathogens is an excellent option that resolves the contamination issue while maintaining anthropological interest. Analysis of ancient specimens in a laboratory where the pathogen has never been studied previously and investigation of a pathogen that does not commonly occur in human populations are obvious requirements for such a study. The need to avoid positive PCR controls is less clear and virtually unheard of in the aDNA field where the vast majority of amplification reactions may be negative. However, use of modern, undamaged DNA as a positive control represents a conscious introduction of a potential DNA contaminant. Addition of control DNA to PCRs or storage of control DNA in a separate laboratory is unlikely to eliminate contamination since it is typically the same researcher handling both modern and ancient DNAs. As explained above, lack of knowledge about the sources and modes of contamination means that they cannot be controlled or eliminated. Therefore, any experiments on modern pathogens, including positive controls, should not be performed in the facility in which the aDNA research is being conducted. Experiments to test primers and optimize reaction conditions should be done by a collaborative laboratory that is not housed in the same building and, ideally, is geographically distant from the lab conducting the aDNA research.

3.1. Molecular Identification of Venereal Syphilis in a Skeletal Specimen

My own studies demonstrate the value of a collaborative project wherein all work on ancient pathogens is conducted in a laboratory separate from the laboratory pursuing research on modern pathogens. N. Tuross and I have recently completed a study of a 200-year-old skeleton from Easter Island that displayed osseous lesions consistent with a diagnosis of treponemal infection (Kolman *et al.*, in press). All experiments on the Easter Island skeleton were performed at the Smithsonian Center for Materials Research and Education in Suitland, MD. Positive PCR controls were never used and no research on modern treponemes was ever conducted at the Smithsonian. Primer testing and reaction optimization were performed at the University of Washington in Seattle by A. Centurion-Lara and S. Lukehart.

3.1.1. Materials and Methods. Immunological analysis. Immunoglobulins of the IgG class were purified from powdered, decalcified bone and tested in an enzyme-linked immunosorbant assay against *Treponema pallidum* antigen (Clark Laboratories). Each sample was tested in two separate experiments in duplicate or triplicate as sample volume allowed. A second Easter Island specimen with no evidence of treponemal infection was processed as a negative control.

Nucleic acid analysis. DNA was isolated from bone fragments that had been decalcified in 0.5 M EDTA as described above. EDTA solution was removed after 1.5 weeks and replaced with fresh solution that was removed after an additional 1.5 weeks. The two EDTA extracts were processed separately. Extracts were dialyzed against distilled water and concentrated by partial freeze-drying.

DNA was amplified for the 5′ untranslated region of the treponemal 15-kilodalton lipoprotein (*tpp15*). This region has been determined to carry a single base polymorphism that differentiates *Treponema pallidum* subsp. *pallidum*, the causative agent of venereal syphilis, from four other human and non-human treponemes (Centurion-Lara *et al.* 1998). Amplification reaction components and thermal cycler conditions were as described above except that the PCR profile was 1 min at 94 °C, 2 min at 60 °C, and 1 min at 72 °C. Restriction digests were performed directly on amplification products using 10 units of *Eco*47III (New England Biolabs, Beverly, MA) and overnight incubation at 37 °C. PCR products were cloned and sequenced as described above. DNA sequence data were collected between –223 and –145 bps of the 5′ flanking region of *tpp15*. Both light and heavy strands were sequenced for all clones.

A complete description of the immunological and nucleic acid analysis is presented in Kolman *et al.* (in press).

3.1.2. Results. Purified IgG fraction demonstrated significant antigen binding that was indicative of treponemal activity (data not shown). The amount of binding observed with the negative control was not in the positive range established by the treponemal IgG controls.

PCR amplification was recovered only from the second EDTA DNA extract collected during days 12–24 of the bone decalcification procedure. Successful amplification produced a fragment of the expected size of 120 bps. *Eco*47III digestion produced a fragment consistent with the expected size of 92 bps (Fig. 2A). It was not possible to resolve

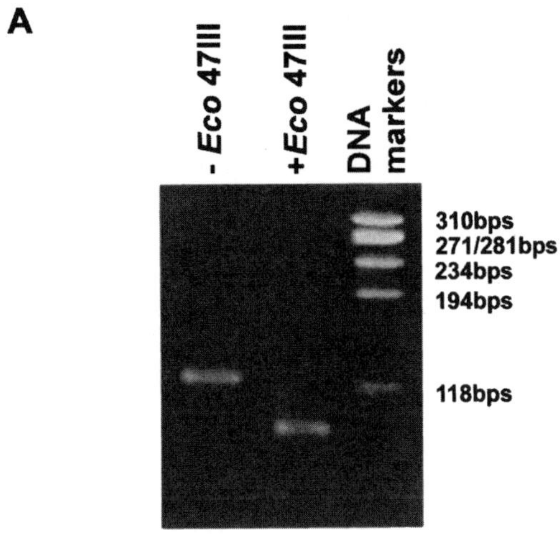

B

	*Eco*47III
Easter Island specimen	**CTC<u>AGCGCT</u>TTA**
T.p.pallidum	**CTC<u>AGCGCT</u>TTA**
T.p.pertenue	**CTCAGTGCTTTA**
T.p.endemicum	**CTCAGTGCTTTA**
Simian *Treponema* sp.	**CTCAGTGCTTTA**
T. paraluiscuniculi	**CTCAGTGCTTTA**

Figure 2. Analysis of DNA extracted from Easter Island specimen, RHO235. (A) DNA was amplified with primers specific for the 5′ untranslated region of *tpp15* and then digested with *Eco*47III. Undigested and digested PCR products were electrophoresed in the first and second lanes while DNA size markers were run in the third lane. (B) Comparison of DNA sequence from the Easter Island specimen and published sequences from three human pathogenic treponemes (*T.p.* subsp. *pallidum*, *T.p.* subsp. *pertenue*, *T.p.* subsp. *endemicum*), one simian treponeme, and one rabbit treponeme (Centurion-Lara *et al.* 1998). Sequence was determined for the Easter Island specimen directly from amplification products and from cloned PCR amplicons. The recognition site for *Eco*47III is underlined.

the smaller digestion fragment of 28 bps from primer dimer products. DNA sequence was determined directly from amplification fragments and from cloned amplicons. The cytosine at position −10 bps, which is diagnostic for *T. p.* subsp. *pallidum*, was present in all DNA sequences (Fig. 2B; Centurion-Lara *et al.* 1998).

3.1.3. Discussion. Immunological studies confirmed the presence of treponemally reactive IgG, but could not provide the sub-species classification necessary for a diagnosis of venereal syphilis. DNA sequence analysis enabled specific identification of *T.p.* subsp. *pallidum*, the causative agent of venereal syphilis. Due to the complete absence of treponemal DNA in our laboratory other than the ancient specimen, there is virtually no possibility that our results reflect contamination with modern treponemal DNA. A collaborative arrangement such as that described herein is critical for aDNA studies.

Identification of *T.p.* subsp. *pallidum* DNA in the skeletal specimen represents the first nucleic acid-based confirmation of a morphologically identified treponemal infection. Analysis of additional specimens with treponemal-type lesions is now warranted to aid in the reconstruction of the evolution of treponematoses. Debate over the origin of venereal syphilis has focused on whether or not the disease was endemic to the New World prior to European contact. Identification of *T. pallidum* subspecies DNA in archaeological specimens from critical geographic and temporal periods could provide the answer to this long-standing question and also furnish essential information on the evolutionary relationship of the other treponemal subspecies.

3.2. Previous Research

Table 3 lists the aDNA pathogen studies that have been published to date and attests to the fascinating anthropological and epidemiological questions that can be addressed by the study of ancient pathogenic DNA. Three studies attempted to detect *Mycobacterium tuberculosis* in ancient skeletal and lung tissue samples (Spigelman and Lemma 1993; Salo *et al.* 1994; Baron *et al.* 1996) and Salo *et al.* (1994) reported success in establishing a pre-European contact New World presence of tuberculosis. Other studies focused on identification of the bacterium that caused ancient bubonic plagues (Drancourt *et al.* 1998) and determination of the influenza strain responsible for the 1918 pandemic (Taubenberger *et al.* 1997). Jackson *et al.* (1998) presented confirmation of reports that the 1979 anthrax outbreak in Sverdlovsk, Russia was due to an accidental release of spores from a military microbiology facility. All of these studies represent innovative applications of aDNA technology. Protocols were developed to extract DNA from a variety of sources including skeletal and soft tissue samples, teeth, and formalin-fixed tissue specimens. All studies reported successful amplification of target DNA with success rates ranging from 7% to 100%. However, with the exception of Taubenberger *et al.* (1997), the possibility that DNA contaminants were analyzed and assumed to represent endogenous DNA can not be ruled out. This is because positive PCR controls were performed in all studies. With the exception of Taubenberger *et al.* (1997), identical DNA sequences were determined for the ancient extracts and the positive PCR controls. This result precludes convincing proof that ancient material was extracted and analyzed. Although Taubenberger *et al.*'s (1997) study utilized positive controls, DNA sequences derived from the formalin-fixed tissue sample were different from the controls for the three loci published providing support that positive control DNA had not contaminated the ancient specimen.

The presence of negative blank extraction and PCR controls is typically cited as evidence that no contamination was encountered in a study. However, lack of amplification in these reactions is not a reliable indicator of contamination because of the interdependence between endogenous and contaminating DNA as explained above. In pathogen studies, it is unknown if there is endogenous pathogen DNA in the extracts, but there is certainly endogenous human DNA that could act as carrier DNA to enhance amplification of a contaminant. In the negative controls, there is no DNA to serve as carrier and enhance amplification of the contaminant. Thus, negative controls may not show amplification even though contaminating DNA is present because the DNA concentration is too low. Furthermore, amplification of negative controls is never pursued as vigorously as amplification of ancient samples so the fact that negative controls generally yield negative amplification may simply reflect a difference in effort expended.

The study of ancient pathogens in human populations has great potential to elucidate the evolutionary history of human pathogens and to clarify the migration patterns and evolutionary relationship of human populations. However, aDNA technology by itself has only limited utility for diseases that leave no lesions. Large numbers of samples

Table 3. Summary of published human pathogen aDNA studies

Disease	Target pathogen	Result	Reference	Comments
Tuberculosis	*Mycobacterium tuberculosis*	Successful PCR in 4/11 samples.	Spigelman and Lemma 1993	Lack of analysis of PCR products and insufficient detail in Materials and Methods means possibility of contamination can not be eliminated.
		Successful PCR in 1/1 sample. Sequence analysis confirmed correct DNA sequence.	Salo *et al.* 1994	DNA sequences of ancient extract and positive control samples were identical meaning that possibility of contamination with control DNA can not be eliminated.
		Successful PCR in 3/3 samples.	Baron *et al.* 1996	Extensive exposure to modern *T. tuberculosis* DNA through bacterial cultures, experimentation with modern DNA, and positive controls in ancient PCR series means possibility of contamination with modern DNA can not be eliminated.
Leprosy	*Mycobacterium leprae*	Insufficient PCR product for analysis	Rafi *et al.* 1994	Impossible to draw any conclusions based on inadequate description of methods and lack of data.
Chagas' disease	*Trypanosoma cruzi*	Successful PCR in 27/27 samples.	Guhl *et al.* 1997	Lack of analysis of PCR products means possibility of contamination can not be eliminated.
"Spanish" influenza	Human influenza strain	Successful PCR of 9 fragments in 1/14 samples. Sequence analysis identified DNA sequences distinct from those in control strains.	Taubenberger *et al.* 1997	Results and conclusions appear solid. It would be useful to present the raw DNA sequence data in order to verify that contamination was not present since phylogenetic analysis shows high similarity between ancient and human positive control sequences at several of the tested regions.
Septicemic plague	*Yersinia pestis*	Successful PCR in 8/19 samples.	Drancourt *et al.* 1998	DNA sequences of ancient extracts and positive control samples were identical meaning that possibility of contamination with control DNA can not be eliminated.
Human anthrax	*Bacillus anthracis*	Successful PCR of various loci in 13/13 samples, including one locus which showed multiple alleles in different samples. Sequence analysis identified expected DNA sequences.	Jackson *et al.* 1998	Successful amplification of 400–800 bp amplicons is unusual for aDNA studies. Use of multiple positive controls means possibility of contamination can not be eliminated.

are inappropriate for screening using aDNA methodology because of the expected high rate of PCR failure and the increased risk of cross-contamination. An improved strategy would be to pre-screen samples using a non-DNA protocol and subsequently assay DNA from a smaller set of promising candidates. Immunological screening with antibodies generated against a particular pathogen could provide the initial screen as in our *T. p.* subsp. *pallidum* study. Protocols have been developed that provide highly purified extracts of immunological material (Tuross 1991). After immunological screening, only appropriately reactive specimens would be analyzed by PCR thus providing a significant reduction in the number of DNA samples to analyze. Each aDNA sample could then be treated with the utmost care as required for aDNA analysis.

4. SUMMARY

Contamination of ancient specimens and aDNA extracts with modern DNA remains a significant problem in aDNA analysis. Studies of human aDNA are uniquely sensitive to contamination due to the presence of human researchers and require novel solutions for identification of contamination. The most critical step in an aDNA study is the careful selection of polymorphic markers capable of discriminating between DNA from the ancient specimens and probable DNA contaminants. A common goal in aDNA studies is the identification of novel haplotypes that have become extinct in modern populations. However, identification of a novel haplotype can make it difficult to fulfill the recommendation to assay markers diagnostic for the ancient populations. In fact, the novelty of a sequence is often used in support of its antiquity. The current study makes clear that the novelty of a haplotype is insufficient proof of its authenticity and that every effort must be made to identify contaminating DNA in order to distinguish convincingly between ancient DNA and contamination.

Analysis of ancient human pathogens represents a promising application of aDNA technology because of the reduced risk of contamination. Subspecies- and strain-specific DNA polymorphisms are currently being detected for many pathogens with increasing speed due to the advent of PCR-based methodologies and whole genome sequencing (Centurion-Lara *et al.* 1998, Fraser *et al.* 1998). For relatively young diseases like treponematoses, analysis of ancient samples from distinct geographic regions may represent an alternative method to determine the origin and evolution of disease families that have not diverged significantly. The incredible amount of time and resources necessary for aDNA studies mandate that only questions which can not be addressed with modern samples be undertaken. Study of ancient pathogenic DNA represents an excellent use of archaeological material. The impressive database of polymorphic DNA data available for a wide variety of organisms assures the merit of a project assaying these markers in ancient populations, but only if extensive precautions have been taken to ensure the accuracy and reliability of the results.

5. ACKNOWLEDGMENTS

My sincere appreciation goes to N. Tuross for the opportunity to pursue aDNA research in an intellectually stimulating, constructively critical, and enormously enjoyable atmosphere.

6. REFERENCES

Anderson S, Bankier AT, Barrell BG, de Bruijn MHL, Coulson AR, Drouin J, Eperon IC, Nierlich DP, Roe BA, Sanger F, Schreier PH, Smith AJH, Staden R, Young IG (1981): Sequence and organization of the human mitochondrial genome. Nature 290:457–465.

Austin JJ, Ross AJ, Smith AB, Fortey RA, Thomas RH (1997): Problems of reproducibility–does geologically ancient DNA survive in amber-preserved insects? Proc Royal Soc Lond B Biol Sci 264:467–474.

Baron H, Hummel S, Herrmann B (1996): *Mycobacterium tuberculosis* complex DNA in ancient human bones. J Archeol Sci 23:667–671.

Cano RJ, Poinar HN, Pieniazek NJ, Acra A, Poinar GO (1993): Amplification and sequencing of DNA from a 120–135-million-year-old weevil. Nature 363:536–538.

Centurion-Lara A, Castro C, Castillo R, Shaffer JM, Van Voorhis WC, Lukehart SA (1998): The flanking region sequences of the 15-kDa lipoprotein gene differentiate pathogenic treponemes. J Infect Dis 177:1036–1040.

DeSalle R, Barcia M, Wray C (1993): PCR jumping in clones of 30 million-year-old DNA fragments from amber preserved termites (*Mastotermes electrodominicus*). Experientia 49:906–909.

DeSalle R, Gatesy J, Wheeler W, Grimaldi D (1992): DNA sequences from a fossil termite in Oligo-Miocene amber and their phylogenetic implications. Science 257:1933–1936.

Drancourt M, Aboudharam G, Signoli M, Dutour O, Raoult D (1998): Detection of 400-year-old *Yersinia pestis* DNA in human dental pulp: An approach to the diagnosis of ancient septicemia. Proc Natl Acad Sci USA 95:12637–12640.

Fisher DL, Holland MM, Mitchell L, Sledzik PS, Webb Wilcox A, Wadhams M, Weedn VW (1993): Extraction, evaluation, and amplification of DNA from decalcified and undecalcified United States Civil War bone. J Forensic Sci 38:60–68.

Fraser CM, Norris SJ, Weinstock GM, White O, Sutton GG *et al.* (1998): Complete genome sequence of *Treponema pallidum*, the syphilis spirochete. Science 281:375–388.

Golenberg EM, Giannassi DE, Clegg MT, Smiley CJ, Durbin M, Henderson D, Zurawski G (1990): Chloroplast DNA sequence from a Miocene *Magnolia* species. Nature 344:656–658.

Guhl F, Jaramilla C, Yockteng R, Vallejo GA, Cárdenas-Arroyo F (1997): *Trypanosoma cruzi* DNA in human mummies. Lancet 349:1370.

Handt O, Krings M, Ward RH, Pääbo S (1996): The retrieval of ancient human DNA sequences. Am J Hum Genet 59:368–376.

Handt O, Höss M, Krings M, Pääbo S (1994): Ancient DNA: Methodological challenges. Experientia 50:524–529.

Hedges SB, Schweitzer MH (1995): Detecting dinosaur DNA. Science 268:1191–1192.

Horai S, Kondo R, Nakagawa-Hattori Y, Hayashi S, Sonoda S, Tajima K (1993): Peopling of the Americas, founded by four major lineages of mitochondrial DNA. Mol Biol Evol 10:23–47.

Höss M, Pääbo S (1993): DNA extraction from Pleistocene bones by a silica-based purification method. Nucleic Acids Res 21:3913–3914.

Jackson PJ, Hugh-Jones ME, Adair DM, Green G, Hill KK, Kuske CR, Grinberg LM, Abramova FA, Keim P (1998): PCR analysis of tissue samples from the 1979 Sverdlovsk anthrax victims: The presence of multiple *Bacillus anthracis* strains in different victims. Proc Natl Acad Sci USA 95:1224–1229.

Kaestle F (1997): Molecular analysis of ancient Native Americans from western Nevada. In Molecular Analysis of Ancient DNA. Nevada Historical Society Quarterly 40:85–96.

Kolman CJ, Tuross N (1999): Ancient DNA analysis of human populations. Am J Phys Anthrop (in press).

Kolman CJ, Centurion-Lara A, Lukehart SA, Owsley DW, Tuross N: Identification of *Treponema pallidum* sub-sp. *pallidum* in a 200-year old skeletal specimen. J Infect Dis (in press).

Kolman CJ, Bermingham E (1997): Mitochondrial and nuclear DNA diversity in the Chocó and Chibcha Amerinds of Panamá. Genetics 147:1289–1302.

Kolman CJ, Sambuughin N, Bermingham E (1996): Mitochondrial DNA analysis of Mongolian populations and implications for the origin of New World founders. Genetics 142:1321–1334.

Kolman CJ, Bermingham E, Cooke R, Ward RH, Arias T, Guionneau-Sinclair F (1995): Reduced mtDNA diversity in the Ngöbé of Panamá. Genetics 140:275–283.

Krings M, Stone A, Schmitz RW, Krainitzki H, Stoneking M, Pääbo S (1997): Neandertal DNA sequences and the origin of modern humans. Cell 90:19–30.

Merriwether DA, Rothhammer F, Ferrell RF (1995): Distribution of the four founding lineage haplotypes in native Americans suggests a single wave of migration for the New World. Am J Phys Anthropol 98:411–430.

Neel JV, Biggar RJ, Sukernik RI (1994): Virologic and genetic studies relate Amerind origins to the indigenous people of the Mongolia/Manchuria/southeastern Siberia region. Proc Natl Acad Sci USA 91:10737–10741.

Pääbo S (1989): Ancient DNA: Extraction, characterization, molecular cloning, and enzymatic amplification. Proc Natl Acad Sci USA 86:1939–1943.

Poinar HN, Hss M, Bada JL, Pääbo S (1996): Amino acid racemization and the preservation of ancient DNA. Science 272:864–866.

Rafi A, Spigelman M, Stanford J, Lemma E, Donoghue H, Zias J (1994): *Mycobacterium leprae* DNA from ancient bone detected by PCR. Lancet 343:1360–1361.

Richards M, Côrte-Real H, Forster P, Macaulay V, Wilkinson-Herbots H, Demaine A, Papiha S, Hedges R, Bandelt H-J, Sykes B (1996): Paleolithic and neolithic lineages in the European mitochondrial gene pool. Am J Hum Genet 59:185–203.

Richards MB, Sykes BC, Hedges REM (1995): Authenticating DNA extracted from ancient skeletal remains. J Archeol Sci 22:291–299.

Ruano G, Pagliaro EM, Schwartz TR, Lamy K, Messina D, Gaensslen RE, Lee HC (1992): Heat-soaked PCR: An efficient method for DNA amplification with applications to forensic analysis. BioTechniques. 13:266–274.

Salo WL, Aufderheide AC, Buikstra J, Holcomb TA (1994): Identification of *Mycobacterium tuberculosis* DNA in a pre-Columbian Peruvian mummy. Proc Natl Acad Sci USA 91:2091–2094.

Sarkar G, Sommer SS (1990): Shedding light on PCR contamination. Nature 343:27.

Sidow A, Wilson AS, Pääbo S (1991): Bacterial DNA in Clarkia fossils. Phil Trans R Soc Lond B Biol Sci 333:429–433.

Spigelman M, Lemma E (1993): The use of the polymerase chain reaction (PCR) to detect *Mycobacterium tuberculosis* in ancient skeletons. Inter J Osteoarchael 3:137–143.

Stone AC, Stoneking M (1998): mtDNA analysis of a prehistoric Oneota population: Implications for the peopling of the New World. Am J Hum Genet 62:1153–1170.

Stone AC, Stoneking M (1993): Ancient DNA from a pre-Columbian Amerindian population. Am J Phys Anthropol 92:463–471.

Taubenberger JK, Reid AH, Krafft AE, Bijwaard KE, Fanning TG (1997): Initial genetic characterization of the 1918 "Spanish" influenza virus. Science 275:1793–1796.

Torroni A, Schurr TG, Cabell MF, Brown MD, Neel JV, Larsen M, Smith DG, Vullo CM, Wallace DC (1993): Asian affinities and continental radiation of the four founding Native American mtDNAs. Am J. Hum Gene 53:563–590.

Torroni A, Schurr TG, Yang C-C, Szathmary EJE, Williams RC, Schanfield MS, Troup GA, Knowler WC, Lawrence DN, Weiss KM, Wallace DC (1992): Native American mitochondrial DNA analysis indicates that the Amerind and Nadene populations were founded by two independent migrations. Genetics 130:153–162.

Tuross N (1991): Recovery of bone and serum proteins from human skeletal tissue: IgG, osteonectin and albumin. In (Ortner DJ and Aufderheide AD, eds) *Human Paleopathology: Current Synthesis and Future Options*. pp. 51–54.

Vigilant L, Stoneking M, Harpending H, Hawkes K, Wilson AC (1991): African populations and the evolution of human mitochondrial DNA. Science 253:1503–1507.

Wan C-Y, Wilkins TA (1993): Spermidine facilitates PCR amplification of target DNA. PCR Methods Applic. 3:208–210.

Ward RH, Redd A, Valencia D, Frazier B, Pääbo S (1993): Genetic and linguistic differentiation in the Americas. Proc Natl Acad Sci USA 90:10663–10667.

Ward RH, Frazier BL, Dew-Jager K, Pääbo S (1991): Extensive mitochondrial diversity within a single Amerindian tribe. Proc Natl Acad Sci USA 88:8720–8724.

Watson E, Forster P, Richards M, Bandelt H-J (1997): Mitochondrial footprints of human expansions in Africa. Am J Hum Genet 61:691–704.

Woodward SR, Weyand NJ, Bunnell M (1994): DNA sequence from cretaceous period bone fragments. Science 266:1229–1232.

14

INTERSPERSED REPEAT INSERTION POLYMORPHISMS FOR STUDIES OF HUMAN MOLECULAR ANTHROPOLOGY

Prescott L. Deininger[1,2], Stephen T. Sherry[3], Gregory Risch[3], Chadwick Donaldson[3], Myles B. Robichaux[3], Himla Soodyall[4], Trefor Jenkins[4], Fang-miin Sheen[5], Gary Swergold[5], Mark Stoneking[6], and Mark A. Batzer[3,7]

[1]Tulane Cancer Center SL-66 and
the Dept. of Environmental Health Sciences
Tulane University Medical Center
1430 Tulane Ave, New Orleans, LA 70112
[2]Laboratory of Molecular Genetics
Alton Ochsner Medical Foundation
1516 Jefferson Highway
New Orleans, LA 70121
[3]Depts. of Pathology
Biometry and Genetics
Biochemistry and Molecular Biology
Louisiana State University Medical Center
1901 Perdido St., New Orleans, LA 70112
[4]Department of Human Genetics
The South African Instate for Medical Research
University of Witwatersrand
Johannesburg 2000, South Africa
[5]Division of Molecular Medicine
Department of Medicine
Columbia University
622 West 168th Street
New York, NY 10024
[6]Dept. of Anthropology
Pennsylvania State University
University Park, PA 16802
[7]Stanley S. Scott Cancer Center
Neuroscience Center of Excellence
Louisiana State University Medical Center
1901 Perdido St., New Orleans, LA 70112

Genomic Diversity: Applications in Human Population Genetics
edited by Papiha, Deka, and Chakraborty, Kluwer Academic / Plenum Publishers, New York, 1999.

ABSTRACT

The recent insertion of mobile elements, of the Alu and L1 families, in the human genome provides a distinct class of polymorphism in the human genome. Because the insertion of these elements in the genome is so rare and once they are inserted they are stable; they represent a unique group of markers that are identical by descent. This type of marker is among the most informative in ascertaining relationships between individuals and populations. The assays for these markers are extremely robust, easy to perform, and readily adaptable to mass analysis and automation. In addition, as the insertion alleles are all newly arisen, the ancestral allele is always the allele missing the insertion. This information allows estimations of the roots of trees, which are not possible with all types of markers. The greatest potential for these markers is with upcoming developments that will allow the identification of new insertions in many different genomes simultaneously. These procedures will allow investigators to isolate markers that are particularly informative for specific populations and allow development of panels of markers tailored for particular populations.

1. INTRODUCTION

There are a number of genetic markers that have proven useful for molecular studies of human populations and their relationships. The most widely used involve the mitochondrial genome, whose rapid evolution and ease of characterization make it particularly advantageous for these studies (Soodyall *et al.* 1996; Rogers and Jorde 1995; Stoneking 1993; Vigilant *et al.* 1991). However, it is always best to test evolutionary questions with multiple types of markers. In addition, mitochondria are limited to maternal inheritance, which allows more rapid fixation of traits and may be biased in certain situations. Nuclear markers will play an increasingly important role in studying human populations.

The advent of PCR technologies has allowed the development of a number of different types of nuclear DNA markers that can be assayed rapidly and economically. The typical problem with studies of nuclear diversity is that most human nuclear DNA shows little or no variation and is therefore uninformative. Thus, the key to studies involving the nuclear genome is to identify regions either with high rates of variability, or regions in which specific mutational events have occurred at an informative time in human evolution. There are some regions of the human genome which undergo relatively rapid accumulation of mutations. Although these regions can be excellent markers, one has to factor in any selective pressures possibly occurring in those regions which may make the evolution non-random. In addition, parallel mutations or back mutations at specific sites can weaken the reliability of any single marker (see Fig. 1). A good example of highly variable markers is the use of VNTRs and microsatellites (Goldstein *et al.* 1995; Bowcock *et al.* 1994; McComb *et al.* 1995; Deka *et al.* 1991). Poly CA and other microsatellite markers show distinct length variations in human populations and are easy to assay. In addition, any given locus may have many different alleles, providing a high level of individual discrimination from a single locus. However, because of the rapid expansion and contraction of these sequences, identical lengths of poly CA in two individuals does not necessarily indicate identity by descent, but instead may simply represent convergence to this length independently in the lineages leading to those two individuals (Fig. 1). Thus, any single marker may falsely point towards a close relationship between two individu-

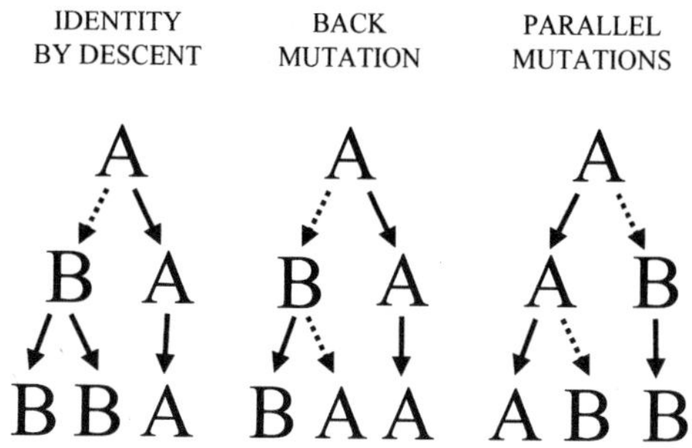

Figure 1. Identity by descent mutations. This figure represents three basic patterns of mutation that can occur at a locus in the human population. The "A" allele stands for the ancestral allele for the locus. The "B" allele stands for a mutation relative to that ancestral allele. The solid arrow shows the inheritance of an allele without mutation, while the dotted lines represent a mutation occurring. In the Identity by Descent panel, the relationships between individuals in the third generation (bottom row) would be appropriately predicted with this locus. In the Back Mutation and Parallel Mutations panels, a second mutation occurs which either erases the original mutation in some individuals or creates an identical allele in otherwise unrelated individuals. In either case, the third generation would not have the appropriate relationships predicted with this locus.

als, and it is only through analysis of many such markers that reliable inferences about population histories can be made.

Markers that are identical by descent have benefits for molecular anthropology studies over markers that may be subject to parallel or back mutations (Fig. 1). Each marker provides more definite information about a lineage of descent. The types of markers which are likely to be identical by descent are distinctive mutations which occur at such a rare frequency that there is almost no likelihood of the same event happening independently in different individuals. The major drawback to such markers is that, because they are rare, it is difficult to identify a large number of them in the human genome. We have developed specific techniques that have allowed us to identify a number of Alu and L1 element insertions in the human genome that serve as markers that are identical by descent and have used them to study human population genetics.

1.1. Identification of Recently Integrated Alu Elements

Alu elements are a family of closely related, approximately 300 bp, elements that have been inserted throughout the human genome via an RNA-mediated transposition mechanism to a copy number in excess of 500,000 (Fig. 2, reviewed in Deininger and Batzer 1993; Deininger and Batzer 1995). Most Alu insertions occurred early in primate evolution, and there are only about 2,000 Alu insertions that are specific for the human genome (Batzer *et al.* 1991; Batzer *et al.* 1995). Analysis of all 2,000 recent Alu insertions found in a given genome would demonstrate that about 75% of them are present throughout the human population. This indicates that they inserted early in human evolution and were completely fixed in the population prior to the radiation of humans from Africa. About another 10% show only very limited diversity with the Alu insertion being present in that locus in almost all humans. Only about 25% of recent Alu insertions show a really useful level of diversity, generally present in 10–90% of all chromosomes tested (Batzer *et al.* 1996b). This means that any given individual may have 400 highly informative Alu insertion sites. Because these variable sites will be somewhat different in

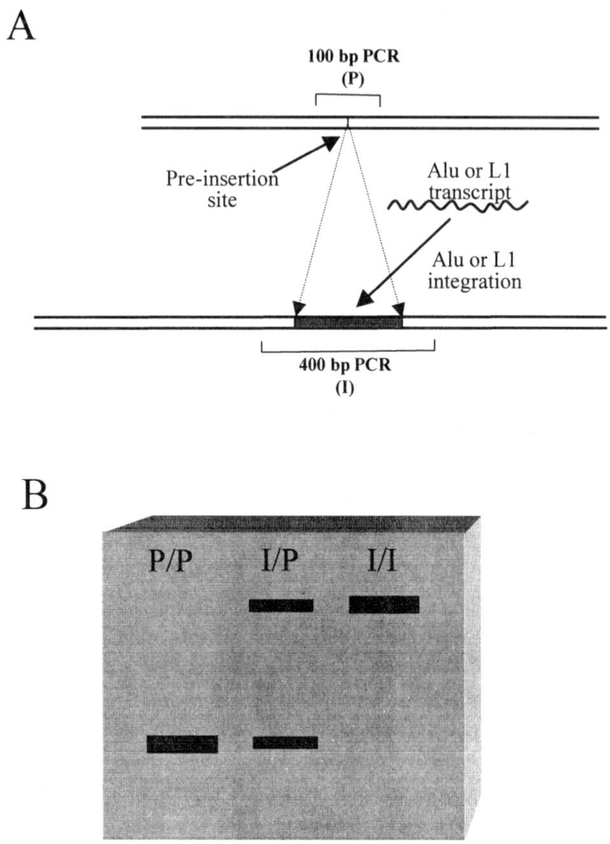

Figure 2. Schematic diagram for PCR-based analysis of polymorphic Alu insertions. Panel A) A pre-insertion site in the human genome is marked. PCR amplification with primers flanking that site will result in a small fragment of about 100 bp (P). Alu or L1 elements use an RNA intermediate to create a new copy that is integrated into a site in the genome. Once, inserted the Alu element increases the fragment size formed by the flanking PCR primers to about 400 bp (I). L1 insertions are larger and more variable in size and require modifications to the assay as shown in Fig. 3. Panel B) A schematic diagram is shown of agarose gel electrophoresis of a hypothetical insertion locus in three individuals and the ability to define the genotype from the assay. In one, only the P allele is found (P/P), the next is heterozygous for the P and I alleles (P/I) and the last is homozygous for the I allele (I/I).

different individuals, there are hundreds of such sites that could be identified in the human population.

The key to identifying this relatively small number of useful sites from the midst of the 500,000 elements is that most of the recent inserts belong to two specific subfamilies of Alu elements. These subfamilies, Ya5 and Yb8, are defined by either 5 or 8 specific base changes that distinguish their family members from the bulk of Alu repeats (Shen *et al.* 1991; Batzer *et al.* 1996a). We have shown that these specific base changes can readily be used to isolate clones containing members of these subfamilies (Batzer and Deininger 1991; Batzer *et al.* 1990; Batzer *et al.* 1991). Following sequence analysis, it is then possible to define PCR primers that allow the rapid assessment of the state of Alu insertion at any given locus.

The PCR assay for Alu insertion polymorphisms is very robust and easy to perform.

The PCR primers are designed to unique sequences flanking the specific Alu insertion (Fig. 2). Thus, if the Alu insertion is present, the band identified by agarose gel electrophoresis may be about 400 bp in length, while the pre-insertion locus will show a band of about 100 bp. Heterozygotes will show bands of both sizes. Thus, homozygotes of both types, as well as heterozygotes, are readily distinguished.

Extensive studies of the evolution of Alu elements at orthologous primate loci have demonstrated that Alu elements are essentially never neatly deleted from the genome (reviewed in Deininger and Batzer 1993 and 1995). There are rare cases in which an Alu element is part of a deletion, and sporadic cases of recombination between Alu elements that create deletions. These cases would almost always be distinguishable from the original pre-insertion allele by either a different fragment size or by loss of amplification and departure from Hardy-Weinberg equilibrium expectations due to deletion of one of the primer sites. Although we cannot say that it never occurs, in extensive studies of many Alu insertion loci (Batzer *et al.* 1994, 1996b; Novick *et al.* 1998; Stoneking *et al.* 1997), it is clear that it must be extremely rare for an Alu element to be either inserted at identical places independently in different individuals or to be neatly removed, creating a back mutation.

Besides identity by descent, there is an additional benefit of the stability of the Alu insertions. Because they are not removed after insertion, the pre-insertion allele is always known to be the ancestral allele. Thus, it is possible to define the ancestral state of each locus when analyzing data., facilitating rooting of population networks using minimal assumptions (Batzer *et al.* 1994, 1996b; Stoneking *et al.* 1997).

1.2. Isolation of Markers

We have utilized several independent approaches to isolation of Alu insertion loci. Our primary approach was to isolate nearly 700 independent lambda clones carrying Ya5 Alu loci from the HeLa cell line (Batzer *et al.* 1991). These loci were then sequenced and PCR primers picked from flanking sequences for use in the assay (Fig. 2). Loci that showed 10–90% presence in human chromosomes from a preliminary pop-tube type analysis of 20 individuals from African-American, Hispanic-American and European American populations were then analyzed in detail.

An alternative approach was used to characterize all of the Ya5 and Yb8 loci found on chromosome 19 (Arcot *et al.* 1998). High-density grids of the cosmid array of chromosome 19 were screened to identify Ya5 and Yb8 Alu inserts. Because the cosmids on the array have been assembled into a contig, each Alu can be assigned a physical location on the chromosome. These Ya5 and Yb8 Alu inserts were then sequenced and flanking PCR primers were picked for polymorphism assays as described above. Three out of 26 Alus were found to show useful levels of polymorphism for identity by descent markers.

A number of markers have been found by other investigators, either using similar approaches (Matera *et al.* 1990), or by chance detection of Alu inserts while studying specific loci. One limitation of the methods discussed above is that they almost all focus on the DNA of a single individual. Thus, only a subset of potential markers can be located in this manner. In addition, the markers will be underrepresented in terms of markers that are more prevalent in other populations than those from which the original DNA originated. Thus, the ideal would be to isolate markers from a diverse group of human populations.

L1 elements represent a longer insertion element that is also present in high copy number in the human genome. There are also a significant number of L1 insertion loci that can be utilized as similar identity by descent markers. Dr. G. Swergold has recently developed a PCR-based assay for isolating polymorphic L1 elements from the human genome (unpublished). We have developed a modification of our Alu insertion assay to effectively assay/genotype these L1 elements (Fig. 3). The key difference between these two assays is that the L1 inserts are too long to routinely PCR amplify across the entire insert. Thus, in the top panel of the figure, we would expect only amplification of the smaller product between the two dark arrows. It is only if the L1 element is not present that we get amplification between the two flanking markers. The combination of the results from the two reactions allows us to distinguish heterozygotes, containing fragments of both sizes, from homozygotes, containing only one of the predicted fragments. Alu and L1 markers can effectively distinguish populations when as few as nine markers are assayed. The range of marker diversity values and similarities between Alu and L1-associated genetic variation will be demonstrated in the results and discussion.

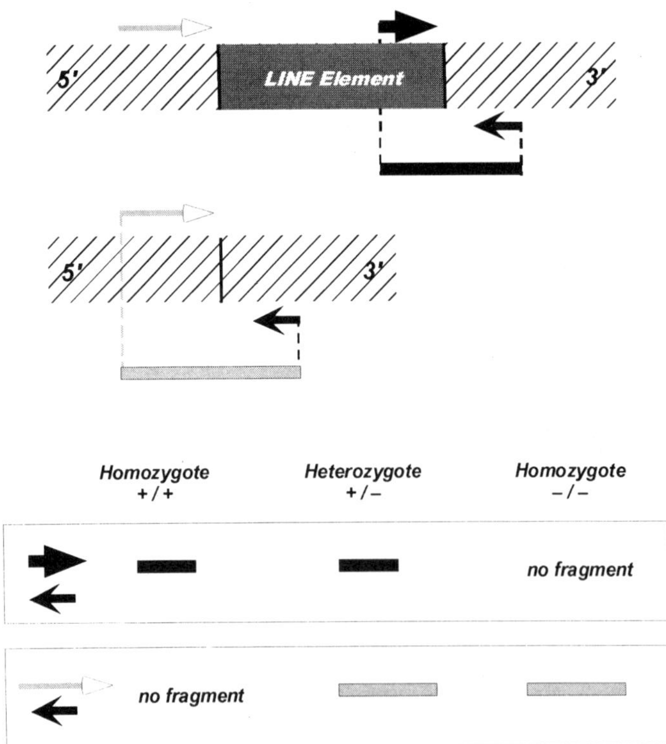

Figure 3. Schematic diagram of the PCR based analysis of polymorphic LINE insertions. Two PCR reactions are required to genotype samples for each L1 element as shown above. The first amplification utilizes both 5′ and 3′ flanking unique sequence primers for the PCR (lower portion of figure). A PCR product is produced only if the LINE element is absent. The second PCR contains the Ta subfamily specific primer and 3′ flanking unique sequence primer, and will only amplify a fragment if the L1 element is present (upper portion of the figure).

2. METHODS

2.1. Population Samples

The following previously-described populations were studied: Bantu Speakers / Sotho/Tswana (Soodyall *et al.* 1996), Caucasian-American (Batzer *et al.* 1996b), African-American (Batzer *et al.* 1994), Hispanic-American (Batzer *et al.* 1996b), Swiss (Batzer *et al.* 1996b), Breton (Batzer *et al.* 1996b), French (Batzer *et al.* 1996b), Turkish Cypriot (Batzer *et al.* 1994), Alaska Native (Batzer *et al.* 1994), Greenland Native (Batzer *et al.* 1994).

2.2. Typing of Interspersed Repeat Insertion Polymorphisms

PCR amplification conditions, genotyping by agarose gel electrophoresis, and primer sequences for all eight Alu loci (TPA25, PV92, FXIIIB, APO, ACE, D1, A25 and B65) were described previously (Arcot *et al.* 1995a; Arcot *et al.* 1995b; Batzer *et al.* 1996b). The TPA25, FXIIIB, ACE and APO loci are also known as PLAT, F13B, APOAI and DCPI respectively. All eight Alu loci are present in humans and absent from orthologous positions within the genomes of (at least) fifteen non-human primate genomes (Arcot *et al.* 1995a; Arcot *et al.* 1995b; Batzer *et al.* 1996b). The LID1 L1 element insertion polymorphism was genotyped using primers and conditions described previously (Sheen *et al.* submitted). This L1 element was also absent from orthologous positions within the genomes of fifteen non-human primate DNA samples. The observed numbers of each genotype for each locus and population are available upon request from either P.D. or M.A.B.

2.3. Data Analysis

Unbiased estimates of heterozygosity and Fst values, corrected for sample size, were calculated using equations in Nei (1987). A principal coordinates (PC) analysis (Cavalli-Sforza *et al.* 1994) of the allele frequencies at the nine interspersed repeat loci was performed using a program provided by H. Harpending.

3. RESULTS

3.1. Population Statistics

An average of 740 individuals from 12 worldwide populations was typed for each of the nine interspersed repeat insertion polymorphisms. The allele frequencies, sample sizes, heterozygosity values and population statistics for each population are shown in Table 1. All loci were polymorphic in all populations, with the exception of A25 in Syrians and FXIIIB in Bantu Speakers which were both fixed for the absence of the interspersed repeat (Table 1).

The average heterozygosity for each locus was substantial, with several values (ACE, TPA25, D1, and B65) approaching the theoretical maximum heterozygosity of 0.5 for a bi-allelic locus (Table 1). The Fst values (a measure of the amount of between population differentiation) were also substantial (Table 1). The Fst values ranged from 0.025 for D1 to 0.392 for FXIIIB indicating that from 97.5% (D1) to 61% (FXIIIB) of the vari-

Table 1. Population genetic characteristics of nine Alu loci

Population	n	LID1 (+) freq	h	n	A25 (+) freq	h	n	TPA25 (+) freq	h
African American	72	0.08	0.142	67	0.07	0.139	68	0.43	0.493
Alaskan Natives	50	0.23	0.358	42	0.05	0.092	46	0.26	0.390
Bantu Speakers	50	0.06	0.114	48	0.08	0.154	47	0.30	0.423
Bretons	69	0.24	0.367	72	0.17	0.280	67	0.32	0.439
Caucasian American	72	0.24	0.371	71	0.14	0.244	70	0.54	0.501
French	72	0.22	0.348	70	0.09	0.158	70	0.60	0.483
German	72	0.22	0.348	70	0.14	0.247	70	0.51	0.503
Greenland Natives	50	0.27	0.398	50	0.18	0.298	50	0.46	0.502
Hispanic American	72	0.17	0.289	71	0.15	0.254	72	0.56	0.497
Swiss	65	0.38	0.473	71	0.15	0.254	69	0.50	0.504
Syrians	70	0.17	0.286	66	0.00	0.000	70	0.51	0.503
Turkish Cypriot	59	0.21	0.337	59	0.08	0.156	59	0.40	0.483
Average frequency and h		*0.21*	*0.32*		*0.11*	*0.19*		*0.45*	*0.48*
Total frequency and h	*773*	*0.21*	*0.330*	*757*	*0.11*	*0.197*	*758*	*0.46*	*0.497*
Fst		0.033			0.040			0.041	

Population	n	ACE (+) freq	h	n	APO (+) freq	h	n	FXIIIB (+) freq	h
African American	68	0.44	0.497	69	0.59	0.488	68	0.07	0.137
Alaskan Natives	46	0.45	0.500	40	0.98	0.049	41	0.92	0.155
Bantu Speakers	48	0.33	0.449	44	0.73	0.401	47	0.00	0.000
Bretons	66	0.28	0.407	68	0.90	0.186	62	0.16	0.273
Caucasian American	71	0.50	0.504	63	0.93	0.134	62	0.06	0.122
French	71	0.41	0.487	68	0.95	0.098	68	0.29	0.412
German	69	0.46	0.501	64	0.87	0.232	61	0.05	0.094
Greenland Natives	48	0.59	0.488	50	0.99	0.020	41	0.79	0.341
Hispanic American	72	0.53	0.501	71	0.97	0.055	59	0.50	0.504
Swiss	70	0.39	0.480	51	0.94	0.112	65	0.16	0.273
Syrians	70	0.40	0.483	68	0.93	0.137	69	0.28	0.402
Turkish Cypriot	59	0.36	0.467	56	0.96	0.086	41	0.01	0.024
Average frequency and h		*0.43*	*0.48*		*0.89*	*0.17*		*0.27*	*0.23*
Total frequency and h	*758*	*0.43*	*0.490*	*712*	*0.89*	*0.195*	*684*	*0.25*	*0.375*
Fst		0.021			0.146			0.392	

Population	n	PV92 (+) freq	h	n	D1 (+) freq	h	n	B65 (+) freq	h
African American	69	0.20	0.326	68	0.47	0.502	69	0.36	0.461
Alaskan Natives	41	0.29	0.419	38	0.34	0.456	43	0.49	0.506
Bantu Speakers	48	0.28	0.409	48	0.21	0.333	48	0.46	0.502
Bretons	71	0.27	0.395	59	0.47	0.502	67	0.51	0.504
Caucasian American	71	0.23	0.352	70	0.21	0.339	71	0.58	0.489
French	69	0.16	0.270	66	0.39	0.478	68	0.49	0.504
German	65	0.10	0.181	63	0.31	0.431	69	0.35	0.457
Greenland Natives	50	0.71	0.416	50	0.48	0.504	50	0.31	0.432
Hispanic American	68	0.51	0.503	65	0.35	0.456	64	0.02	0.046
Swiss	66	0.16	0.270	71	0.34	0.451	71	0.57	0.494
Syrians	68	0.18	0.293	69	0.29	0.415	70	0.31	0.434
Turkish Cypriot	59	0.15	0.261	55	0.31	0.431	58	0.39	0.479
Average frequency and h		*0.27*	*0.34*		*0.35*	*0.44*		*0.40*	*0.44*
Total frequency and h	*745*	*0.26*	*0.385*	*722*	*0.35*	*0.453*	*748*	*0.40*	*0.482*
Fst		0.115			0.025			0.083	

ation in these loci is found within populations. These values compare favorably to those previously reported for eight of these loci and a more geographically diverse set of populations (Stoneking *et al.* 1997).

3.2. Population Relationships

In order to access the population relationships a principal coordinates (PC) analysis (Cavalli-Sforza *et al.* 1994) of the allele frequencies at the nine polymorphic interspersed repeat loci was performed (Fig. 4). The first two coordinates, which provide the most information for a two-dimensional depiction of population relationships, account for 80% of the variance in the data. In the PC analysis, populations with similar genetic makeup cluster together. In this analysis, three distinct clusters comprised of Caucasians, Africans and Amerinds are formed. It is interesting to note that the Hispanic-American population resides between the source Amerind and Caucasian populations. The clusters of populations are concordant with geographic affiliations for each of the populations. The root included in the analysis consisted of a hypothetical ancestral population in which the allele frequencies of all nine interspersed repeat loci were set to zero. In this analysis the root is closest to the African population clusters in PC1 suggesting an African origin for these loci. These data are concordant with previous studies involving the analysis of mtDNA, and a variety of nuclear DNA polymorphisms that indicate an African origin for the species (Cann *et al.* 1987; Goldstein *et al.* 1995; Jorde *et al.* 1995; Nei *et al.* 1996; Relethford and Harpending 1994 and 1995; Tishkoff *et al.* 1996; Vigilant *et al.* 1991). The Amerind populations appear to be closest to the root in the second PC. However, this association is presumably artificial since the analysis did not include populations from the Pacific that have previously been associated with the root in the second PC based on the analysis of Alu insertion polymorphisms (Stoneking *et al.* 1997).

4. DISCUSSION

Alu and L1 insertion markers make exceptional markers for studies of human populations for several reasons. First, because they represent identity by descent markers, the data they provide concerning population relationships is extremely reliable. They also show very high levels of heterozygosity for biallelic markers presumably as a result of the ascertainment method and initial pop-tube based screening for polymorphic loci. These two properties combine to make all of the insertion markers we have isolated extremely informative as evidenced by the Fst values (Table 1) measured for each of these markers. Therefore, the analysis of fewer markers provides more useful information than with other types of markers.

We have developed an assay that allows us to add the use of L1 insertion markers, greatly increasing the pool of available insertion loci that can be used in these studies. Because L1 and Alu elements have some tendency to insert in different regions of the human genome (Arcot *et al.* 1998), it may prove advantageous to combine both types of markers in order to sample different chromosomal regions.

We have chosen to analyze the data presented in this paper using a PC plot (Fig. 4), but the data can readily be presented with any of the other methods for analyzing such data, including a number of different types of trees (Perna *et al.* 1992; Novick *et al.* 1998; Batzer *et al.* 1994,1996b; Stoneking *et al.* 1997). These data show the ability of these markers to discriminate between African, Caucasian and Native American popu-

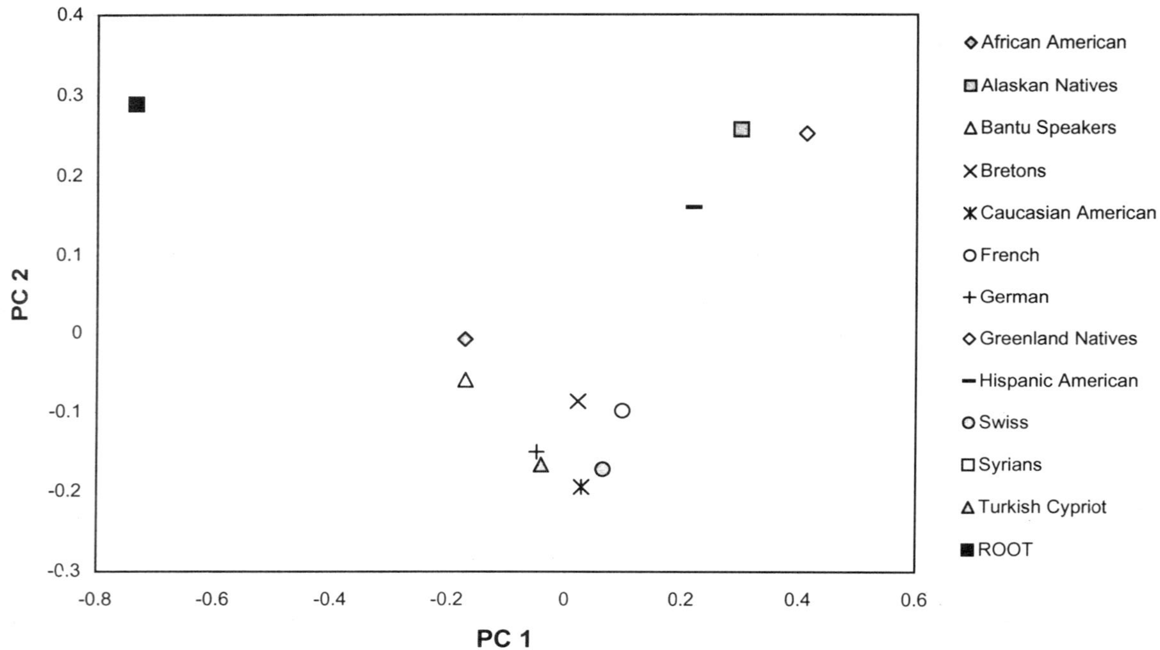

Figure 4. Plot of the first two principal coordinates of the allele frequencies of nine interspersed repeat loci. The same relations were observed when the hypothetical ancestral population (root) was removed from the analysis, indicating that the ancestral population is not distorting the relationships.

lations. Their relationship to the root is also consistent with a previous report utilizing Alu insertion polymorphisms (Stoneking *et al.* 1997). In addition, we have found that these markers are consistent in their ability to detect admixtures between different populations, as seen with the Hispanic American population.

One of the greatest potentials for these types of markers is that experimental techniques can be devised that should allow such markers to be isolated from specific population groups. This can certainly be accomplished using the basic procedures that have been used to isolate these markers in the past (Batzer *et al.* 1991). However, there are several different PCR-based procedures that are being developed for display of L1 elements (Sheen *et al.* unpublished) and Alu elements (Kass *et al.* unpublished) that allow the simultaneous display of Alu insertion polymorphisms in multiple individuals. This not only allows these markers to be identified more easily and efficiently, but should facilitate the identification of markers that are informative for specific population groups.

5. ACKNOWLEDGMENTS

This research was supported by grants from NIH (GM45668) to P.D., NSF (SBR-9610147) to M.A.B., NIH NRSA (GM19110) to S.T.S., and NSF (SBR-9423118) to M.S.

6. REFERENCES

Arcot, SS, JJ Fontius, PL Deininger, MA Batzer. 1995a. Identification and analysis of a "young" polymorphic Alu element. Biochem. Biophys. Acta 1263:99–102.

Arcot, SS, Z Wang, JL Weber, PL Deininger, MA Batzer. 1995b. Alu repeats: a source for the genesis of primate microsatellites. Genomics 29:136–141.

Arcot, SS, AW Adamson, G Risch, J LaFleur, JE Lamerdin, AV Carrano, MA Batzer. 1998. High-resolution cartography of recently integrated chromosome 19-specific Alu fossils. J. Mol. Biol. 281:843–855.

Batzer, MA, PL Deininger. 1991. A human-specific subfamily of Alu sequences. Genomics 9:481–487.

Batzer, MA, GE Kilroy, PE Richard, TH Shaikh, TD Desselle, CL Hoppens, PL Deininger. 1990. Structure and variability of recently inserted Alu family members. Nucleic Acids Res. 18:6793–6798.

Batzer, MA, VA Gudi, JC Mena, DW Foltz, RJ Herrera, PL Deininger. 1991. Amplification dynamics of human-specific (HS) Alu family members. Nucleic Acids Res. 19:3619–3623.

Batzer, MA, M Stoneking, M Alegria-Hartman, H Bazan, DH Kass, TH Shaikh, GE Novick, PA Ioannou, WD Scheer, RJ Herrera, PL Deininger. 1994. African origin of human-specific polymorphic Alu insertions. Proc. Natl. Acad. Sci. USA 91:12288–12292.

Batzer, MA, CM Rubin, U Hellmann-Blumberg, M Alegria-Hartman, EP Leeflang, JD Stern, HA Bazan, TH Shaikh, PL Deininger, CW Schmid. 1995. Dispersion and insertion polymorphism in two small subfamilies of recently amplified human Alu repeats. J Mol. Biol. 247:418–427.

Batzer, MA, PL Deininger, U Hellmann-Blumberg, J Jurka, D Labuda, CM Rubin, CW Schmid, E Zietkiewicz, E Zukerkandl. 1996a. Standardized nomenclature for Alu repeats. J Mol. Evol. 42:3–6.

Batzer, MA, SS Arcot, JW Phinney, M Alegria-Hartman, DH Kass, SM Milligan, C Kimpton, P Gill, M Hochmeister, PA Ioannou, RJ Herrera, DA Boudreau, WD Scheer, BJB Keats, PL Deininger, M Stoneking. 1996b. Genetic variation of recent Alu insertions in human populations. J Mol. Evol. 42:22–29.

Bowcock, AM, A Ruiz-Linares, J Tomfohrde, E Minch, JR Kidd, LL Cavalli-Sforza. 1994. High resolution of human evolutionary trees with polymorphic microsatellites. Nature 368:455–457.

Cann, RL, M Stoneking, AC Wilson. 1987. Mitochondrial DNA and human evolution. Nature 325:31–36.

Cavalli-Sforza, LL, P Menozzi, A Piazza. 1994. The History and Geography of Human Genes. Princeton Univ. Press, Princeton, New Jersey.

Deininger, PL, MA Batzer. 1993. Evolution of retroposons. Evol. Biol. 27:157–196.

Deininger, PL, MA Batzer. 1995. SINE master genes and population biology. In The Impact of Short Interspersed Elements (SINEs) on the Host Genome (ed. RJ Maraia), pp. 43–60. RG Landes, Georgetown, Texas.

Deka, R, R Chrakroborty, RE Ferrell. 1991. A population genetic study of six VNTR loci in three ethnically defined populations. Genomics 11:83–92.

Goldstein, DB, AR Linares, LL Cavalli-Sforza, MW Feldman. 1995. Genetic absolute dating based on microsatellites and the origin of modern humans. Proc. Natl. Acad. Sci. USA 92:6723–6727.

Jorde, LB, MJ Bamshad, WS Watkins, R Zenger, AE Fraley, PA Krakowiak, KD Carpenter, H Soodyall, T Jenkins, AR Rogers. 1995. Origins and affinities of modern humans: a comparison of mitochondrial and nuclear genetic data. Am. J Hum. Genet. 57:523–538.

Matera, AG, U Hellmann, MF Hintz, CW Schmid. 1990. Recently transposed Alu repeats result from multiple source genes. Nucleic Acids Res. 18:6019–6023.

McComb, J, N Blagitko, AG Comuzzie, MS Schanfield, RI Sukernik, WR Leonard, MH Crawford. 1995. VNTR DNA variation in Siberian indigenous populations. Hum. Biol. 67:217–229.

Nei, M 1972. 1987. Molecular Evolutionary Genetics. Columbia Univ. Press, New York.

Nei, M, N Takezaki. 1996. The root of the phylogenetic tree of human populations. Mol. Biol. Evol. 13:170–177.

Novick, GE, CC Novick, J Yunis, E Yunis, P Antunez de Mayolo, PL Deininger, M Stoneking, MA Batzer, RJ Herrera. 1998. Polymorphic Alu insertions and the Asian origin of native American populations. Human Biology 70:23–39.

Perna, NT, MA Batzer, PL Deininger, M Stoneking. 1992. Alu insertion polymorphism: a new type of marker for human population studies. Hum. Biol. 64:641–648.

Relethford, JH, HC Harpending. 1994. Craniometric variation, genetic theory, and modern human origins. Am. J Phys. Anthropol. 95:249–270.

Relethford, JH, HC Harpending. 1995. Ancient differences in population size can mimic a recent African origin of modern humans. Cur. Anthro. 36:667–674.

Rogers, AR, LB Jorde. 1995. Genetic evidence on modern human origins. Hum. Biol. 67, 1–36.

Shen, MR, MA Batzer, PL Deininger. 1991. Evolution of the master Alu gene(s). J Mol. Evol. 33:311–320.

Soodyall, H, L Vigilant, AV Hill, M Stoneking, T Jenkins. 1996. mtDNA control-region sequence variation suggests multiple independent origins of an "Asian-specific" 9-bp deletion in sub-Saharan Africans. Am. J Hum. Genet. 58:595–608.

Stoneking, M 1993. DNA and recent human evolution. Evol. Anth. 2:60–73.

Stoneking, M, JJ Fontius, S Clifford, H Soodyall, SS Arcot, N Saha, T Jenkins, MA Tahir, PL Deininger, MA Batzer. 1997. Alu insertion polymorphisms and human evolution: evidence for a larger population size in Africa. Genome Research 7:1061–1071.

Tishkoff, SA, E Dietzsch, W Speed, AJ Pakstis, JR Kidd, K Cheung, B Bonne-Tamir, AS Santachiara-Benerecetti, P Moral, M Krings, S Paabo, E Watson, N Risch, T Jenkins, KK Kidd. 1996. Global patterns of linkage disequilibrium at the CD4 locus and modern human origins. Science 271:1380–1397.

Vigilant L, M Stoneking, H Harpending, K Hawkes, AC Wilson. 1991. African populations and the evolution of human mitochondrial DNA. Science 253:1503–1507.

15

WORLDWIDE DISTRIBUTION OF A POLYMORPHIC *Alu* INSERTION IN THE PROGESTERONE RECEPTOR GENE

Adriana Antúnez-de-Mayolo[1], Gabriela Antúnez-de-Mayolo[1], Emmanuel Thomas[1], Erika P. Reategui[1], Michael D. Brown[2], and Rene J. Herrera[1]

[1]Department of Biological Sciences
Florida International University
University Park, Miami, FL 33199
[2]Department of Genetics and Molecular Medicine
Emory University, Atlanta, GA 30322

ABSTRACT

Alu repetitive elements represent the largest family of short interspersed repetitive elements (SINEs) in humans as there are an excess of 500,000 copies per haploid genome. Of this, only a few thousand are thought to be polymorphic exhibiting both the insertion and lack of insertion alleles. The probability of independent *Alu* insertion events at the same site is remote. The ancestral state of an *Alu* repeat represents the absence of the insertion while the presence of an insertion, at a particular locus, is the derived state. In the present study, 1,533 unrelated individuals from 38 worldwide populations have been screened for a polymorphic *Alu* element using PCR technology with a novel pair of primers. This *Alu* insertion is located within intron G of the human progesterone receptor gene. Our results indicate that this *Alu* polymorphic insertion is population specific since it is absent in most of the sub-Sahara African and Central/East Asian groups examined. Furthermore, this element is very rare in sub-Sahara Africa; hence, it may constitute the first *Alu* polymorphism known to have resulted from a retroposition outside Africa subsequent to modern man migration out of Africa approximately 200,000 years ago. The low frequency of this insertion in Northern Asia and its absence in Central and South Asia may suggest that native American's ancestors might have had contributions in their gene pool from Caucasian groups.

Address for correspondence and reprints: Dr. Rene J. Herrera, Department of Biological Sciences, Florida International University, University Park, Miami, FL 33199, Phone number: (305) 348-1258, Fax number: (305) 348-1259

Genomic Diversity: Applications in Human Population Genetics
edited by Papiha, Deka, and Chakraborty, Kluwer Academic / Plenum Publishers, New York, 1999.

1. INTRODUCTION

The *Alu* family of short interspersed repetitive elements (SINEs) first appeared in primate genomes about 65 million years ago (Deininger and Daniels 1986; Deininger and Slagel 1988; Deininger *et al.* 1992) and is distributed in an excess of 500,000 copies per haploid genome (Deininger and Batzer 1993; Novick *et al.* 1996). A subfamily of the *Alu* repeats have been termed human-specific (HS) (Batzer *et al.* 1990; Batzer and Deininger 1991) as the members of this subdivision have recently been inserted into the human genome, within the last 6 million years (Batzer *et al.* 1991; Batzer and Deininger 1991).

Several features make *Alu* repeats excellent genetic markers: the stability of the insertion, the lack of a known mechanism for their complete and exact removal from their specific chromosomal site of insertion, and the fact that some HS *Alu* insertions have not yet reached fixation. This last feature indicates that not all chromosomes will carry an *Alu* element at a certain locus. Additionally, the ancestral state of an *Alu* repeat is known to be the absence of the element at a particular site while the presence of the repeat at that locus is considered the forward mutational change or derived state. These elements are selectively neutral since there is usually no selection pressure for the presence or absence of such sequences. For recent reviews on polymorphic *Alu* insertions see Novick *et al.* 1996 and Szmulewicz *et al.* 1998. A polymerase chain reaction (PCR) based *Alu* insertion polymorphism assay (Batzer *et al.* 1991; Novick *et al.* 1994) as well as multiplex PCR systems (Thomas and Herrera 1998) have been developed to analyze population phylogeny. Other primate lineages posses their own set of polymorphic *Alu* insertions (Szmulewicz *et al.* 1999) which are excellent genetic markers for population genetics and identification studies.

The issue of origin, timing, point of entry, and number and routes of migration of Native American ancestors into the New World has remained controversial. Different polymorphic systems including mitochondrial DNA, restriction fragment length polymorphism (RFLPs), variable number of tandem repeats (VNTRs), point mutations in *Alu* sequences, and polymorphic *Alu* insertions (O'Rourke *et al.* 1992; Schanfield 1992; Kidd, Pakstis *et al.* 1993; Torroni *et al.* 1994; Kidd and Kidd 1996; Knight *et al.* 1996; Novick *et al.* 1998) have been used to study the aforementioned questions.

Several hypotheses have been proposed to explain the routes and number of migrations into the New World. The three-migration theory based on linguistic, dental, and genetic information, proposes three separate waves of migration which gave rise to three main clusters: Eskaleut, Na-Dene and Amerind (Greenberg *et al.* 1986; Wallace and Torroni 1992; Cavalli-Sforza *et al.* 1994). The first migration, occurring 35,000 to 15,000 years ago, came from Siberia and spread over most of the New World. The next settlement occurred on the northwestern coast of North America, between 15,000 and 10,000 years ago, and is attributed to Na-Dene-speaking people. The third wave gave rise to the Eskaleut cluster about 10,000 years ago. The original locations of the Na-Dene and Eskaleut are not completely clear; however, it is possible that they had common origins in Asia. A second hypothesis, known as the single-migration model (Rogers *et al.* 1991), states that most of North America was covered with ice during the last glaciation; however, some areas were still suitable for habitation. Migratory groups populated these separate areas, using the ice bridge between the Old and New Worlds. The isolation might have encouraged differentiation and genetic drift.

Additional studies have focused on the origins of the Native American's ancestors.

Anthropological analysis supports the theory that the Northern Asian ancestors reached the New World by crossing over Beringia during the last glaciation 20,000 years ago (Fladmark 1983). Studies of mtDNA and RFLP variation have shown that Mongolians have a larger proportion of the four haplogroups (defined by Wallace and Torroni 1992) found in the New World than the geographically more proximate Siberian populations. This suggests that Mongolians (or a common ancestor to both contemporary Mongolians and American aboriginals) are the most likely founders of the New World (Merriwether *et al.* 1996). Dental analyses of living and ancient populations have provided data that support two migrations from Southeast Asia: one moved to Polynesia and a second one to China (Greenberg *et al.* 1986; Turner 1989). After spreading into China and Mongolia the ancestors migrated to Northeast Asia into the Siberian Arctic, eventually moving across the Bering land bridge into Alaska and the rest of the New World. According to Turner (1989) the American ancestors were from the area between the western border of Lake Baykal and the northern border of China. Recently, morphological (Morell 1998a) and mtDNA (Brown *et al.* 1998; Morell 1998b) evidence seems to suggest that some of the Native American founders may have been Caucasian in origin.

In the present study, we analyze a group of 38 populations worldwide using an *Alu* insertion (PR) located within intron G in the hormone-binding domain of the human progesterone receptor gene in chromosome 11 (McKenna *et al.* 1995). This HS *Alu* insertion is polymorphic. Yet, in most sub-Sahara Africa and Southern/Central Asian populations the insertion allele is absent (the lack of insertion allele is fixed) making this genetic marker population specific. Since the complete removal of the *Alu* sequence is highly unlikely and has not been shown to occur, this *Alu* element may represent the first known to have retrotransposed after man left Africa. In addition, although our data is based on a single locus, the distribution of this HS *Alu* insertion among worldwide populations suggests that Native American founders had a Northern Asian component and possibly a Caucasian contribution.

2. MATERIALS AND METHODS

2.1. Populations Sampled

A total of 38 groups were analyzed in the present study; their geographic location is shown in Fig. 1. Additionally, their names and sample sizes are provided in Table 1. The Vizcaya DNA samples were donated by Marian Martinez de Pancorvo, University of the Basque Country, Spain. The Swedish and Finnish populations were provided by Dr. Nils-Goran Larsson and Dr. Marja-Liisa Savontaus, respectively. Samples from Galicia, Spain were kindly donated by Dr. Jose Luis Caeiro. The Madras group was obtained from Dr. C. Damodaran. The Bahrain, Jordanian and African-American from Indianapolis samples were kindly provided by Dr. Gabriel E. Novick, Indianapolis. The Maya Campeche, Quechua, Surui and Karitiana groups were donated by Judy R. Kidd, Yale University, New Haven, Connecticut. The Navajo group of 37 individuals was provided by Dr. Ralf Bertrand from Colorado College. The other populations examined in this study are part of the laboratory repositories of one of the authors. This study was approved by the Human Subject Review Board of Florida International University and appropriate informed consent was obtained from human subjects. Allelic frequencies for this *Alu* insertion in the German and Irish populations were previously published (McKenna *et al.* 1995).

2.3. PCR Amplification and Gel Electrophoresis

Amplifications of the target locus were carried out in 15 ul reactions as previously described for other *Alu* insertions (Novick *et al.* 1995). The *Alu* repeat, found within intron G of the human PR gene, was amplified using a novel flanking PCR primer pair: 5′-CAGACAGCCAAGATTATGAGCTA-3′ (5′-end primer) and 5′-AGTATTTTCTTGCTAAATGTCTGTTT-3′ (3′-end primer). Every sample was subjected to 35 cycles of the following amplification conditions: 1 minute at 94 °C (denaturing), 1.5 minutes at 58 °C (annealing) and 4 minutes at 65 °C (extension).

The PCR products (15 ul) were electrophoresed in 3% agarose Tris-acetate/EDTA [50X: 242 g Tris base, 100 ml 0.5 M EDTA (pH 8.0), 57.1 ml glacial acetic acid per liter] gels. DNA fragments were visualized by UV fluorescence after staining with ethidium bromide. 0.5 ug of *Hae* III digested ΦX174 DNA was used as the molecular weight marker. Band patterns were scored in the following manner: homozygous individuals for the insertion exhibited one band approximately 550 bp long, homozygous individuals that lack the insertion in the human PR gene had a band approximately 200 bp long, heterozygous individuals had bands of both fragment sizes (Fig. 2).

2.4. Data Analyses and Statistical Methods

Allelic frequencies were determined using the gene counting method (Ausebel *et al.* 1987). These were subsequently tested for Hardy-Weinberg equilibrium using the chi-square test for goodness of fit.

3. RESULTS

In this study, a total of 1,533 unrelated individuals were analyzed from 38 populations from around the world. The locations of populations are described in Table 1. *Alu*

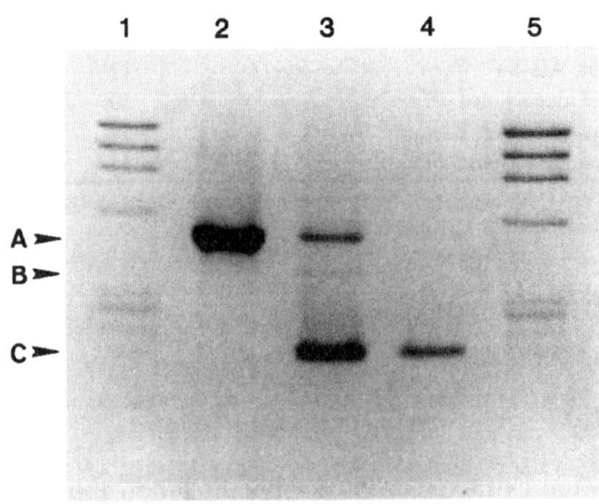

Figure 2. Gel electrophoresis of *Alu* PCR products for the human PR locus. Lanes 1 and 5 represent the molecular size marker *Hae* III digested ΦX174 DNA. Lane 2 contains the homozygous insertion allele band. The individual on lane 3 is heterozygous and lane 4 homozygous lack of insertion band. The band for the insertion could be seen at the level of arrow A (approx. 550 bp) and the lack of insertion at the level of arrow C (approx. 200 bp). Arrow B shows a heterodimer band.

insertion frequencies as well as the geographic location for these populations are shown in Fig. 1.

The presence of the *Alu* insertion ranged from a frequency of 0.295 in the Wayuu group from Northern Colombia to zero in some South and Central Asian (Chinese, Korean and Southeast Asia) and sub-Sahara African populations (Cameroon, !Kung and Kwhe). Overall, the Europeans exhibited the highest frequency of insertion ranging from 0.105 (Sweden) to 0.282 (Guscopia, Basque Country, Spain) with the exception of the Germans (0.069). In Asia, the populations from China, Korea and South East Asia were fixed for the lack of insertion allele. However, to the northern and western areas of Asia, the insertion allelic frequencies increased gradually with the exception of the Kungurtug. This Mongolian group exhibited a low frequency (0.021) when compared to neighboring populations like the Tadzhik and the Ulan-Ude (0.083 and 0.070, respectively).

The Native American populations with the highest insertion frequencies were detected in the central part of the continent. Populations to the north or south exhibited progressively lower insertion frequencies. In meso-America, the Maya Campeche population located in the Campeche Province, in the Yucatan Peninsula of Mexico exhibited a relatively low frequency for the *Alu* insertion (0.063) in an area where the insertion frequencies of the neighboring groups were much higher. There are two other populations that exhibited lower frequencies than expected in relation to their neighboring populations. These were the Doco group (Northern Colombia) and the African Americans from Indianapolis. Most likely, the low frequencies are a result of the fact that both populations have a large sub-Sahara African component. The lower insertion frequency in the African American population from Indianapolis (0.012) relative to the Doco's insertion frequency (0.050) is expected since African American populations have an approximately 10% contribution from other races, mainly Caucasian, and the Docos are only 40–50% Negroid (E. Yunis, personal communication).

All populations analyzed here conformed to Hardy-Weinberg expectations ($p > 0.050$) except for the Basque group from Vizcaya ($p = 0.035$), Maya Buctzotz ($p = 0.040$), Wayuu ($p = 0.047$) and the North Africans ($p < 0.001$).

The *Gst* value was determined for this *Alu* insertion (0.080) to measure interpopulation variability. This indicates that there are significant differences among the different groups analyzed with respect to the frequencies for this polymorphism.

4. DISCUSSION

In this study, we have shown that this polymorphic *Alu* insertion is population specific with the lack of insertion allele absent in most of the South/Central Asian and sub-Sahara African populations examined. This HS *Alu* polymorphism was not found in three of the four sub-Sahara African (Cameroon and two groups from South Africa), and the South and Central Asian (Korea, China and South East Asia) populations analyzed. These groups were fixed for the lack of insertion allele. On the other hand, the PR *Alu* sequence is not completely absent in the sub-Saharan groups analyzed. The sample obtained from Rwanda (0.020) shows the presence of this element. However, its frequency is small (two out of fifty individuals were heterozygous at this loci) which could be attributed to admixture with other ethnic groups (eg, Caucasian). Since the presence of this element is very rare in sub-Sahara Africa, this *Alu* element may constitute the first repeat known to have occurred after modern man left Africa. All other known polymorphic *Alu* insertions are much more common in sub-Sahara Africa (Batzer *et al.* 1996). Therefore, these results may suggest migration out of Africa and settlement in Southeast and Central Asia prior to the insertional event. On the other hand, it is possible that drift may account for the loss of the insertion allele in sub-Sahara Africa. Another possibility is that the insertion occurred in a population contributing to the migration out of Africa, whose descendants however, were not examined in this study. Genetic flow from Europe subsequent to the original founder migration out of Africa toward Asia may account for the presence of this *Alu* in the Middle East and West Asia.

Figure 1 illustrates two main high frequency foci for the presence of the PR *Alu* element. One of them is in Europe and the other in the central region of the New World. From our data, it is unclear where the PR *Alu* insertion allele originated. It is possible that it may have originally retrotransposed in Europe. A decreasing cline is seen in Fig. 1 as we move from Europe to Western Asia, then Northern Asia, and into India and South/Central Asia where some populations are fixed for the lack of insertion allele. The settlement of modern man in Southern and Eastern Asia, and Europe prior to the insertional event could explain the complete absence of this *Alu* insertion in the populations examined from China, Korea and Southeast Asia and its presence in high frequencies in Europe if the insertional event occurred in Europe. It is interesting that the Basque, considered a relatively ancient European population, exhibits the highest insertion frequency in Europe. This may reflect just genetic drift resulting from geographical and reproductive isolation. Furthermore, the Basque may be direct descendants of the group where the PR insertion took place.

Since the Mongoloid groups from South and Central Asia seem to lack the *Alu* insertion, the theory of an exclusive genetic flow from South Asia (Turner 1989) is unlikely. Considering the overall high insertion frequencies in Native American populations, a component of these migrants are probably from Northern Asia or further west.

Previously reported mtDNA data on the origin of Native American populations established that the American ancestors were derived from North Asian groups (Schurr *et al.* 1990; Ballinger *et al.* 1992; Torroni *et al.* 1992; Wallace and Torroni 1992). However, migration wave(s) from Mongolian and/or Siberian groups may not have been the only ones contributing to the high frequencies of this progesterone receptor insertion seen in America. The considerable higher frequencies in New World populations could be the result of inbreeding and/or genetic drift. On the other hand, the migration into the New World might have had contributions from other populations with a much higher frequency for this *Alu* insertion (e.g., Caucasian groups).

It is possible that genetic drift, inbreeding, admixing between native groups and gene flow from recent European colonization could be responsible for the overall higher insertion frequencies observed in New World populations as compared to groups from North Asia and Mongolia. Higher *Alu* insertion frequencies are seen in geographically distant populations like the Tlingit in Alaska, Sioux of North-Central United States, the Maya in Buctzotz (Mexico), Surui of Brazil and Wayuu from Colombia. Native American populations like the Tlingit from Alaska, Surui and Karitiana from Brazil have been known to be reproductively isolated from neighboring groups; therefore, their high insertional frequencies can not be accounted by admixture. Linguistically different groups like the Na-Dene represented by the Tlingit from Alaska and the Navajo of South West United States and the Amerind represented by several North, Central and South American groups also exhibit relatively high insertion frequencies. It is possible that different linguistic groups could have picked up the insertion allele during migrations through Siberia. Therefore, several, independent genetic drift events leading to increments in the insertion frequency are required to explain the distribution of the examined populations with high insertion frequency.

The idea of migration wave(s) moving into the New World from Northern Asia having a Caucasian component may explain the high frequencies of the insertion allele in America. There is precedence for potential Caucasian genetic contribution to American native populations. Mitochondrial and morphological data may be indicative of Caucasian genetic input from founder populations. Mitochondrial data suggests the possibility that some Native American ancestors were of Caucasian origin (Brown *et al.* 1998). The presence of a minor haplogroup X in Europe and Asia Minor and not in Asia suggests a link between Eurasians and Native Americans (Brown *et al.* 1998; Morell 1998b). These mtDNA studies identified a consensus haplogroup X motif that characterizes a minor lineage in some European and Native American populations. Depending on the number of founders, time estimates for the arrival of the X haplogroup in America are 12,000–36,000 years ago, indicating a penetration contemporary to the original Native American founders. In addition, recent skeletal findings dating back 10,000 years ago also point to European rather than Asian features (Morell 1998a). The presence of European DNA in the ancestor group(s) of existing Native American populations could account for the presence of the insertion allele in Mongolia and North Asia and its higher frequencies in American populations as compared to its absence in South and Central Asian populations. Caucasian groups could have migrated across Mongolia and Northern Asia into the New World leaving behind in their path insertion frequencies higher than Oriental populations to the south.

5. ACKNOWLEDGMENTS

Ms. Lourdes Andino for technical support; Timothy Rolynds for critical advice and Douglas Wallace for providing DNA from the two South African populations. The work was supported by the National Institutes of Health (RR08205) to R.J.H.

6. REFERENCES

Ausebel FM, Brent R, Kingston UE, Moore DD, Seldman JG, Smith JA, Struhl K (1987) Current Protocols in Molecular Biology. John Wiley and Sons, Inc. New York.

Ballinger SW, Schurr TG, Torroni A, Gan YY, Hodge JA, Hassan K, Chen K-H, Wallace DC (1992) Southeast Asian mitochondrial DNA analysis reveals genetic continuity of ancient Mongoloid migrations. Genetics 130:139–152.

Batzer MA, Arcot SS, Phiney JW, Alegria-Harman M, Kass DH, Milligan SM, Kimpton C, Giu P, Hochmeister M, Ioannou PA, Herrera RJ, Boudreau DA, Shceer WD, Keats BJB, Deininger PL, Stoneking M (1996) Genetic variation of recent *Alu* insertions in human populations. J Mol Evol 42:22–29.

Batzer MA, Deininger PL (1991) A human-specific subfamily of *Alu* sequences. Genomics 9:481–487.

Batzer MA, Gudi VA, Mena JC, Foltz DW, Herrera RJ, Deininger PL (1991) Amplification dynamics of Human-Specific (HS) *Alu* family members. Nucleic Acids Res 19:3619–3623.

Batzer MA, Kilroy GE, Richard PE (1990) Structure and variability of recently inserted *Alu* family members. Nucleic Acids Res 18:6793–6798.

Brown MD, Hosseini SH, Torroni A, Bandelt HJ, Allen JC, Schurr TG, Scozzari R, Cruciani F, Wallace DC (1998) mtDNA haplogroup X: An ancient link between Europe/Western Asia and North America? Am J Hum Genet 63:1852–1861.

Cavalli-Sforza LL, Menozzi P, Piazza A (1994) The History and Geography of Human Genes. Princeton University Press, Princeton.

Deininger PL, Batzer MA (1993) Evolution of retroposons. Evolutionary Biology, MK Hect, RJ MacIntyre, MT Clegg, eds. New York: Plenum Press 157–196.

Deininger PL, Batzer MA, Hutchinson III CA, Edgell MH (1992) Master genes in mammalian repetitive DNA amplification. Trends Genet 8:307–312.

Deininger PL, Daniels GR (1986) The recent evolution of mammalian repetitive elements. Trends Genet 2:76–80.

Deininger PL, Slagel VK (1988) Recently amplified *Alu* family memebers share a common parental *Alu* sequence. Molec Cell Biol 8:4566–4569.

Fladmark KR (1993) Early Man in the New World. Beverly Hills, CA: Sage Publications.

Greenberg, JH, Turner II CG, Zegura SL (1986) The settlement of the Americas: Comparison of the linguistic, dental and genetic evidence. Curr Anthropol 4:477–497.

Kidd KK, Kidd JR (1996) Nuclear prospective on human evolution. In Molecular Bioloby and Human Diversity, AJ Boyce, CGS Mascie-Taylor, eds. London, England: Cambridge University Press 242–264.

Kidd JR, Pakstis AJ, Kidd KK (1993) Global levels of DNA variation. In Proceedings of the 4th International Symposium on Human Identification. Madison, WI: Promega Corp 21–30.

Knight A, Batzer MA, Stoneking M, Tiwari HK, Scheer WD, Herrera RJ, Deininger PL (1996) DNA sequence of *Alu* elements indicate a recent single origin for modern humans. Proc Natl Acad Sci USA 4360–4364.

McKenna NJ, Kieback DG, Carney DN, Fanning M, McLinden J, Headon DR (1995) A germline *Taq*I restriction fragment length polymorphism in the progesterone receptor gene in ovarian carcinoma. Br J Cancer 71:451–455.

Merriwether DA, Hall WW, Vahlne A, Ferrell RE (1996) mtDNA Variation Indicates Mongolia May have been the Source for the Founding Population for the New World. Am J Hum Genet 59:204–212.

Morell V (1998a) Kennewick Man's Trials Continue. Science 280:190–192.

Morell V (1998b) Genes May Link Ancient Eurasians, Native Americans. Science 280:520.

Novick GE, Batzer MA, Deininger PL, Herrera RJ (1996) The Mobile Genetic Element *Alu* in the Human Genome. Bioscience 46:32–41.

Novick GE, Novick CC, Yunis J, Yunis E, Martinez K, Duncan GG, Troup GM, Deininger PL, Stoneking M, Batzer MA, Herrera RJ (1995) Polymorphic human-specific *Alu* insertions as markers for human identification. Electrophoresis 16:1596–1601.

Novick GE, Novick CC, Yunis J, Yunis E, Antunez de Mayolo P, Scheer WD, Deininger PL, Stoneking M, Batzer MA, Herrera RJ (1995) On the Oriental origin of Native American populations. Hum Biol 70:23–39.

Novick GE, Menendez CM, Novick CC, Duncan GG, Yunis J, Yunis E, Deininger PL, Batzer MA, Herrera RJ (1994) The use of polymorphic *Alu* insertions as a methodological alternative in human paternity testing and child identification. International Pediatrics 9:60–68.

O'Rourke DH, Mobarry A, Suarez BK (1992) Patterns of genetic variation in native Americans. Hum Biol 64:417–434.

Rogers RA, Rogers LA, Hoffmann RS, *et al.* (1991) Native American biological diversity and the biogeographic influence of Ice Age refugia. J Biogeogr 18:623–630.

Schanfield MS (1992) Allotypes (GM and KM) indicate multiple founding populations of native Americans: Evidence of at least four migrations to the new world. Hum Biol 64:381–402.

Schurr TG, Ballinger SW, Gan YY, Hodge JA, Merriwether DA, Lawrence DN, Knowler WC, Weiss KM, Wallace DC (1990) Ameridian mitochondrial DNAs have rare Asian mutations at high frequencies suggesting they derived from four primary maternal lineages. Am J Hum Genet 46:613–623.

Szmulewicz MN, Novick GE, Herrera RJ (1998) Effects of *Alu* insertions on gene function. Electrophoresis 19:1260–1264.

Szmulewicz MN, Andino LM, Reategui EP, Wooley-Baker T, Jolly CJ, Disotell TR, Herrera RJ (1999). An *Alu* insertion in a baboon hybrid zone. In Press in the American Journal Physical Anthropology.

Thomas E, Herrera RJ (1998). Multiplex Polymerase Chain Reaction of *Alu* polymorphic insertions. Electrophoresis 19:2373–2379.

Torroni A, Chen Y-S, Semino O, Santachiara-Beneceretti AS, Scott CR, Lott MT, Winter M, Wallace DC (1994) mtDNA and Y-chromosome polymorphisms in four native American populations from southern Mexico. Am J Hum Genet 54:303–318.

Torroni A, Schurr TG, Yang CC, Szathmary EJE, Williams RC, Schanfield MS, Troup GA, Knowler WC, Lawrence DN, Weiss KM (1992) Native American mitochondrial DNA analysis indicates that the Amerind and the Nadene populations were founded by two independent migrations. Genetics 130:153–162.

Turner II CG (1989) Teeth and Prehistory in Asia. Scientific American 88–96.

Wallace DC, Torroni A (1992) American Indian prehistory as written in the mitochondrial DNA: A review. Hum Biol 64:403–441.

16

STATISTICAL ISSUES REGARDING THE USE OF MICROSATELLITE LOCI FOR MOLECULAR ANTHROPOLOGICAL STUDIES

Ranajit Chakraborty

Human Genetics Center
The University of Texas
School of Public Health
Houston, Texas

ABSTRACT

The class of polymorphic loci that are most abundant in the genome, yet have large number of segregating alleles, are called repeat loci. Of that class, the microsatellites have been the most popular in their use for gene mapping, studying evolutionary relatedness of populations, identifying the source of biological specimens and parentage testing. In this presentation the biological features of repeat polymorphism are briefly reviewed to illustrate that interpretation of data on repeat polymorphism in general, and microsatellites in particular, requires some statistical considerations that are different from the ones used in the context of blood group and protein polymorphism studies. In addition, issues with regard to sampling of individuals for using microsatellite loci for biological anthropological studies are discussed along with the choice of measures of genetic variation that derive the maximum information provided by this class of loci.

1. INTRODUCTION

Ever since the discovery of the first genetic polymorphism in the human genome, biological anthropological studies relied heavily on genetic data to determine affinities between anthropologically defined populations (see Roychoudhury and Nei 1988; Cavalli-Sforza *et al.* 1994). An often-quoted interpretation of such data is that at a per locus basis, the proportion of genetic variation that can be attributed to genetic-differentiation of human populations is small (Lewontin 1972; Nei and Roychoudhury 1974; Barbujani *et al.* 1997). Nevertheless, genetic variation between individuals have been shown to be useful for the purposes of allocating individuals in different populations (Smouse *et al.* 1982), for

Genomic Diversity: Applications in Human Population Genetics
edited by Papiha, Deka, and Chakraborty, Kluwer Academic / Plenum Publishers, New York, 1999.

resolving parentage dispute cases (Walker 1983), for forensic identification of individuals (NRC 1996), for reconstructing the history of human migration (Cavalli-Sforza *et al.* 1994) and for studying the past demographic history of human populations (Harpending *et al.* 1993; Polanski *et al.* 1998; Kimmel *et al.* 1998). Although most of these applications were initiated at a time when genetic variations were being studied by using blood groups and enzyme protein markers, the invent of recombinant DNA technology prompted revisiting these issues for several reasons. First, on an absolute scale, the number of polymorphic DNA markers is several orders of magnitude larger than the traditional serological and biochemical markers, as a consequence of which the detection of inter-individual variation through the use of DNA polymorphism became more efficient. Second, at the DNA level, at the polymorphic regions of the genome, the spectrum of genetic variation is also larger (e.g., the number of segregating alleles/haplotypes is larger), so that genetic distinction between individuals can now be made more precisely than that at protein and blood group loci. Third, DNA studies enjoy an operational advantage as well, because polymorphism at the DNA level can be detected even when the gene products are not necessarily retained in intact conditions. Thus, DNA studies are possible from a wide variety of biological specimens, including that from remains of populations of the past (e.g., hair roots, semen, saliva stains, bones and tissue remains, etc.) from some of which blood grouping and protein-enzyme typing are not generally possible.

However, the use of DNA polymorphism for molecular anthropological studies introduced several analytical issues, some of which were not relevant for classical markers. In addition, for many applications new analytical tools are necessary to deal with the detailed nature of the molecular data. A comprehensive discussion of all analytical issues of molecular data pertinent to molecular anthropological studies is beyond the scope of this presentation. In stead, here I present examples of some of them, in particular of the ones which relate to the use of the most popular set of DNA markers that are currently being employed in the area of molecular anthropology. These markers, generally called repeat polymorphisms, being abundant in the genome, are widely used in gene mapping, and the ease of genotyping these loci prompted researchers in utilizing them for all of the applications mentioned above.

In the following sections, first I present a brief synopsis of the biological criteria of genetic variations of repeat polymorphisms, from which analytical issues that are unique to such markers are recognized. Based on these criteria, some statistical comments are made, particularly in relation to size of databases, allele frequency computations and their use for detecting population substructure and for measuring genetic distances between populations. For simplicity, mathematical details are avoided in this presentation, but the sources are cited for readers interested in such backgrounds. References to specific problems are made for the main purpose of guiding readers to appropriate interpretation of results. Finally, some comments are made in reference to other types of molecular data (e.g., Single Nucleotide Polymorphism (SNP) and nucleotide sequences) for some of which the statistical methods developed for studying repeat polymorphism may be applicable.

2. REPEAT POLYMORPHIMS—A BRIEF HISTORICAL SYNOPSIS AND UNIQUE FEATURES

RFLP Analysis of VNTR Loci: The discovery of Wyman and White (1980), which detected the first hypervariable locus through DNA typing, is a seminal contribution in human molecular population genetics, because it truly opened investigations on a new class of polymorphism, not recognized before. This class of polymorphic loci differs from the others discovered earlier in a number of ways. First, the polymorphism at these loci are due to copy number repeats of short stretches of nucleotide sequences that are tandemly located in the genome, and thus, allelic variation at these loci is not merely due to nucleotide substitutions and/or simple insertion/deletion events. Second, alleles designated by the size of the repeat region allowed an ordered distinction of alleles which have been later shown to offer more incisive analyses of data for the purposes of determining relatedness between individuals and evolutionary dynamics of populations (Chakraborty *et al.* 1991; Chakraborty and Jin 1993a,b). Third, this class of loci, called Variable Number of Tandem Repeats (VNTR) by Nakamura *et al.* (1987), were soon discovered to be abundant in the genome and hence, they gained popularity for the purpose of gene mapping because of the ease of detecting recombination events in pedigree studies. Almost parallel to the development of probes of individual VNTR loci, simultaneously detected length variations of DNA segments at many homologous loci, led to the concept of DNA Fingerprinting (Jeffreys *et al.* 1985), which in a sense introduced a new area of investigation, called DNA Forensics (NRC 1992; 1996). The VNTR loci, characterized in these studies, are repeat loci at which the core motif is generally large (tens to hundreds of nucleotides long), with copy numbers on any chromosomal segment also being large (on occasions numbering hundreds). Thus, variations at these regions were being scored by the Southern blot restriction fragment length polymorphism (RFLP) analysis, which gave a quasi-continuous data on allelic variation at each locus.

Characteristic Features of VNTR Loci scored by RFLP Analysis: The unique biological features of data on VNTR loci scored by RFLP analysis are: (i) even in inbred populations, large number of alleles segregate at most loci, with some having over 200 possible alleles (Jeffreys *et al.* 1993); and (ii) alleles are distinguishable by their repeat sizes. However, since for most of them the position of the flanking restriction sites are not generally recorded, allele lengths are to be mathematically represented by $a + br$, where a is the total distance of the restriction sites (summing the 3′ and 5′ ends of the repeat sequence) flanking the repeat region, b is the length of each repeat motif, and r is the number of copies of the repeat motif unit. In this representation, length variation (i.e., the variation of $a + br$) among chromosomes is caused by variation in r, the number of repeat units, with a and b being constants (of course, specific to the locus and restriction enzyme used for digestion). Since the length of the fragments are estimated through comparisons of fragments of known molecular weights tested on the same autoradiograph, allele-length data are necessarily quasi-continuous, inherently consisting of measurement errors (i.e., subject to lane-to-lane, or gel-to-gel variation, combined with calibration variation during digitized automated estimation of fragment lengths); (iii) protocols of separation of fragment lengths (e.g., voltage used during electrophoresis, density of the polyacrylamide used in the gel preparation, time of electrophoretic runs, etc.) necessarily result in two other features that are extremely important for interpretation of single-band profiles at any locus. While ordinarily such profiles should be regarded as homozygotes, they are not necessarily so, because heterozygotes with one or a few copy number differences of repeat units may not be separated enough on a gel to be visualized as a two banded-pattern. More importantly, the optimization of band separations may result in nondetectability of fragments of either too small or too large sizes. In other words, in addition to true homozygotes, single-banded patterns may also be observed for heterozygotes with two alleles of nearby repeat numbers (a phenomenon described as allelic coalescence, Devlin *et al.* 1990), or heterozygote with one allele being outside the measurable range of allele sizes (also called non-detectability of alleles, Chakraborty *et al.* 1994; Steinberger *et al.* 1993); (iv) As a consequence, population genetic analysis of VNTR fragment sizes scored by Southern blot RFLP analysis requires some form of binning the fragment sizes in categories, consequently, leading to an operational definition of alleles, either through a sliding window of categories (called floating bin; see Weir 1992), or through a pre-prescribed fixed set of categories (called fixed bin, Budowle *et al.* 1991).

Some critics labeled these features of VNTR fragment length polymorphism as limitations (Lander 1989; Mueller 1991; Geisser and Johnson 1992). However, it should be noted that such concepts (i.e., allelic coalescence and non-detectability of alleles) were implicit in many blood group and enzyme-protein loci (see e.g., Martin 1983). Thus, the operational definition of alleles by binning VNTR fragment lengths is not necessarily a new concept, although the biological explanations of their occurrence are different for VNTR loci. In addition to the above features, the probes that detected fragment length variation simultaneously at more than one locus (called the multilocus probes, such as the 33.15 and 33.6 minisatellite probes of Jeffreys *et al.* 1985; the Bkm probe of Aggarwal *et al.* 1994; or the F10 probe of Pena *et al.* 1990) result in characteristics that make data interpretation even more complex. First, the genetic profile revealed by any of these multilocus probes is a multi-banded pattern (often compared with bar-code inscriptions on commercial products), from which the allelic or locus identity of bands cannot be inferred. Second, in addition to allelic coalescence of band locations, co-migration of alleles from different loci cannot be ruled out for multilocus DNA fingerprint data. Third, without characterization of individual loci underlying each multilocus probe, allelic independence across loci cannot be generally assumed (since some of the loci may indeed be linked). Undoubtedly, these limitations somewhat restrict the use of DNA fingerprinting data for some applications. Nevertheless, the high discriminatory power of genetic profiles observed through any of the above-stated multilocus probes have been shown to be extremely useful for the purposes of forensic identification, parentage testing, establishing stated relationships, as seen through empirical as well as theoretical studies (Jeffreys *et al.* 1985; Chakraborty and Jin 1993a, b; Pena and Chakraborty 1994). At this point, however, it should be mentioned that although the technique of DNA fingerprinting by multilocus minisatellite probes gained an initial popularity in forensic identification, parentage testing as well as evolutionary studies (see Burke *et al.* 1991; Pena *et al.* 1993 for various such applications), for some natural limitations and further developments, they were soon replaced first by single locus probes (studied by RFLP analysis) and subsequently by PCR-based studies of shorter repeat sequences, called the microsatellites or short tandem repeats.

The major limitations of multilocus probes, recognized in this context are: (i) RFLP analysis by multilocus probes require large quantities of DNA, (ii) high molecular intact DNA is needed to have all relevant fragments recognized in the RFLP analysis; (iii) recognition of bands for specific fragments of lower molecular weights require high stringency hybridization, as a consequence of which, often a fraction of the autoradiograph is generally difficult to interpret; and (iv) RFLP analysis, by itself, is time-consuming and difficult to automate for large scale studies.

Microsatellite Loci scored by PCR analysis: Interestingly, before the formal invent of DNA fingerprinting, the presence of simple repeat sequence polymorphisms in the human genome was recognized as "polypyrimidinic stretches" (Birnboim and Straus 1975). However, the generality of these general classes of sequences in the eukaryotic genomes was not known at that time. Independently, tetranucleotide (e.g., GATA-sequences) and dinucleotide (GT or CA) repeats were also described (see e.g., Epplen *et al.* 1982; Hamada *et al.* 1982) before the formal invention of DNA fingerprinting by Jeffreys (1985). Litt and Luty (1989) were the first to coin the term "microsatellite", to distinguish this class of tandem repeat loci (for which the core sequences are of length 1 to 6 base pairs long) from the minisatellites. Edwards *et al.* (1991) used the term "short tandem repeat" (STR) to describe the same class of loci (see Tautz 1993 for a more detailed historical perspective of this subject). Soon, however, several distinctive features of the microsatellites emerged, which made these more popular than the minisatellites for applications of DNA technologies for diversity studies. First, since these loci can be scored by examining length polymorphism of relatively smaller PCR-products (generally smaller than 200 to 300 base pairs), allele size determina-tion results into a discrete allele system, as opposed to quasi-continuous data of fragment sizes (characteristic of minisatellite data). Second, the smaller PCR-products also allowed feasibility of direct sequencing each allele to detect microvariation of alleles (i.e., sequence heterogeneity and/or presence of insertion/deletion within the repeat elements). Third, development of multiplexed PCR-conditions allowed simultaneous detection of genotypes at a number of loci together, allowing automation and time-saving features of genotyping that preserved characterizing the alleles for each locus individually (in direct contrast of non-recognition of allelic relationships of bands in an autoradiograph generated by a multilocus minisatellite probe in RFLP analysis). Finally, the PCR-amplification process requires far lesser quantities of DNA for typing and assessment of polymorphism based on DNA fragments in shorter product sizes meant that degraded low molecular weight DNA could also be subjected to typing for microsatellite loci.

Characteristics of Microsatellite Loci: These advantages quickly led to identification of literally thousands of polymorphic microsatellites on the human genome, and validation of the genotyping procedures in samples of an wider sources of biological samples (e.g., saliva, blood stain, semen, vaginal swabs, tissue or bone samples from body remains, etc.). As a consequence, the major characteristic features of microsatellite loci that emerged from these studies are: (i) it is estimated that there could be upto 100,000 polymorphic microsatellite loci in the human genome, a number far greater than the minisatellites; (ii) these are well dispersed in the genome, and not generally clustered near the telomeric regions of chromosomes as is the case of minisatellites. Indeed, the current 2 cM map of the human genome is based primarily on the microsatellites; (iii) in terms of number of polymorphic loci, although the reason is still unclear, the abundance is inversely related to the size of the core motif; i.e., there are more dinucleotide loci and the tri-, or tetra-nucleotides; the larger polymorphic repeats being comparatively far rarer; (iv) both empirical as well as experimental data now support that, by and large, mutations at microsatellite loci are driven by the replication slippage type of mechanism, in contrast with a more complex "recombinatorial" mechanism of production of new alleles at minisatellites (Valdes *et al.* 1993; Shriver *et al.* 1993; Jeffreys *et al.* 1993; Di Rienzo *et al.* 1994; Kimmel and Chakraborty 1996). Of course, the presence of microvariants at many microsatellites do not rule out the possibility of more than one mutation mechanisms (e.g., slippage and insertion/deletion) simultaneously working on these loci; (v) in terms of variability within populations, minisatellites are, however, more variable than microsatellites (i.e., number of segregating allele as well as per locus heterozygosity higher for minisatellites), making the microsatellites somewhat less discriminatory in applications for parentage testing and forensic identification.

3. STATISTICS OF REPEAT LOCI: ISSUES AND SOLUTIONS

Sampling, Sample Size Requirement and Allele Frequency Determination: Genetic variation within and across populations is generally studied from samples of unrelated individuals selected without any prior knowledge of genotypes. Such samples are called random samples, although the practice of any precise random sampling scheme is rare in human genetic studies. For this reason, human population genetic samples are sometimes referred to as convenient samples (NRC 1996). The question then arises as to what extent a sample collected through such convenient sampling procedure represents the population from which it is gathered.

Before answering this question, it may be worthwhile to recall the intent for which human genetic samples are collected. As mentioned above, samples from populations are collected to estimate the extent of genetic variation in populations. Therefore, the randomness of sampling should relate to the genetic polymorphism at underlying loci being investigated. Therefore, any sampling procedure in which the subjects selected for sampling does not depend on the genotypes at such loci can be called "random". In this sense, sampling in genetic context is somewhat different from that of political or social sciences. In particular, with regard to the human genome diversity studies, this difference is even more conspicuous, since a great majority of genetic loci studied at the level of DNA polymorphism do not have any functional significance. Of course, this does not mean that a sample of hypertensive individuals would reflect the allele frequencies at the angiotensine gene locus of the general population accurately, nor a group of patients being treated at a mental institute would reflect the true population frequencies at the dopamine receptor 2 locus. In other words, for genetic loci at which phenotype-genotype association studies have demonstrated evidence of clear nonrandom association, selection of subjects should be done to avoid inherent obvious ascertainment biases. For other loci, as far as possible, prior knowledge of genotypes has to be avoided to guard against any systematic bias.

For human populations, strict random sampling may not be feasible, even when needed. This is so, because the concept of population in the human context is somewhat fluid. Events like birth, death and migration continually change the composition of human populations even within a short period of time. Hence, even if the sample was drawn by following a strict random sampling scheme at a certain point of time, because of the changes of the population composition, at any other point of time that sample would no longer remain to be random. Further, there are several empirical studies (e.g., National Center for Health Statistics 1980; Chakraborty *et al.* 1995) that show that convenient samples recover information on genetic polymorphism at traditional as well as DNA loci rather accurately. Use of strict random sampling, sometimes, may create problems in data interpretation as well. For example, consider an isolated population that can merely be a large kindred with close biological relationships between many of its living members. If every individual is to be given an equal chance of being included in the sample (an important premise of random sampling), then the sample would constitute a non-trivial number of related individuals. An immediate consequence of this would be some systematic departure of genotype frequencies from some rules that may be postulated from the biological features of the polymorphic loci. Presence of relatives in a sampled database does not systematically bias allele frequencies, as long as the loci studied are not subject to forces of natural selection (i.e., viability and/or fertility differences). But, sampling of relatives may make the genotype distribution considerably different from the ones expected under the Hardy-Weinberg assumption. As a consequence, human geneticists generally select unrelated individuals without any a priori knowledge of genotypes, taking the precaution that the sources of sample collection do not inherently bias genotypes to be scored for the study.

Another point of contention in sampling is the assignment of population of origin of individuals. Preferably, detailed pedigree information and birth-place of individuals may be used for defining populations. In population-based genetic epidemiological studies, when pedigree data are collected, generally the population description of individuals meets this ideal requirement. However, often blood or sample collection is made from volunteers at blood bank, hospital registries, or work places, in which either the sample collector assigns population of origin from last names and physical appearance of individuals, or the subjects themselves are asked questions as to which population they would belong. In absence of detailed pedigree information, this is generally not so precise, yet, data collected in this fashion generally meets the purpose of broad grouping of populations. In the Human Genome Diversity work, the general practice is to collect birth-place data of four grandparents, and assign individual's population of origin based on this information. Whatever the practice followed, because of population mixtures in recent history in human populations, no definition of population is truly free of error, since genetic admixture is often undetected without detailed multi-generation pedigree information with birthplaces known for all members of the pedigree for several generations. In contrast, there are several genetic markers at which genotypes can reveal past history of population mixtures. In other words, even when population-of-origin is determined from pedigree histories, signature of past genetic admixture can be detected through the presence of population-specific alleles or haplotypes in individuals (Chakraborty *et al.* 1992; Shriver *et al.* 1997).

For each population selected for study, the next question of interest is how many individuals are to be surveyed for conducting the study. Obviously, the answer depends on the purpose of study. For biological anthropology, for the purpose of characterizing the genetic composition of population and for inquiring inter-population genetic difference, the sample size requirement issue had been addressed for the traditional serological and biochemical markers in the work of Nei (1978). He showed that for estimating average heterozygosity within populations and to measure genetic distance, a sample of 20 to 25 individuals per population should be sufficient when the study is composed of a large number of loci. Further, he showed that the sampling uncertainties of such parameters of genetic variation are controlled more efficiently by increasing the number of loci randomly selected from the genome, rather than by increasing the number of individuals in the study. Molecular variation at DNA level, however, showed that the sample size requirement may be somewhat different. For example, for microsatellites, where the number of segregating alleles is comparatively larger, small sample sizes more often result in missing a larger fraction of alleles in the sample, even when they are segregating in the population. Chakraborty (1992) used the mutation-drift equilibrium theory of infinite allele model of mutations to conclude that when average heterozygosity at a locus is 70% or higher, probably 100 to 150 individuals per population would be required to have an adequately precise information of intra-population variation. Under this formulation, he further computed the minimum threshold allele frequency below which allele frequencies would be imprecise for any given sample size. Other investigators approached the sample size question in a somewhat different fashion. For example, for databases to be used for forensic identification purposes, Evett and Gill (1991) showed that samples of size 100 to 120 individuals per locus provide robust estimation of DNA profile match probabilities. Harding (1998) used a technique of sequential sampling to examine the stability of allele frequency estimates in a sample of any given size.

The differences of sample size requirement for repeat polymorphism and the traditional markers are caused not only due to their differences in terms of the number of segregating alleles. Other features of polymorphism at repeat loci are also to be accounted in this context. For example, since mutations are more frequent for repeat loci (than that at traditional loci), and since the alleles can be labeled by their repeat numbers, several alternative measures of within- and between-population variability may also be considered for analysis. Their precise estimation, consequently, may depend upon knowing allele frequency distributions more accurately. For example, allele size variance has been shown to be a better measure of within-population variation at microsatellite loci than the simple gene diversity (Kimmel *et al.* 1996; Kimmel and Chakraborty 1996). Consequently, allele size variance estimation requires a larger sample size, since rare alleles affect this statistic more sensitively than gene diversity.

Choice of Measures of Variation and their Estimation: The biological features of polymorphism of repeat loci, discussed earlier, necessitated introduction of alternative measures of genetic variation, at the level of within as well as between populations. First, it should be mentioned that as the alleles at microsatellite loci are codominant, the allele frequency estimates can be obtained in a much simpler manner, by the gene count method (Li 1976). However, since occasionally some alleles may be preferentially amplified, some loci may exhibit non-detectable alleles. In the presence of nondetectable alleles, it has been shown that the gene count estimates of allele frequencies are overestimates of the actual allele frequencies, the amount of bias depending upon the frequency of nondetectable alleles in the population (Chakraborty *et al.* 1992). Therefore, for some applications, where a conservative (i.e., biased upwards) estimate of genotype frequencies are needed (as is the case of estimating DNA match probabilities), the presence of nondetectable alleles does not necessarily invalidate the gene count estimates of allele frequencies at microsatellite loci. Of course, the presence of nondetectable alleles introduces a departure of genotype frequencies that may be otherwise ascribed to population substructure (Chakraborty *et al.* 1994).

Another simple example is that of describing the genetic variation within a population. Ideally, when every mutation is assumed to be new and not seen before in the population, an appropriate measure of genetic variation is the probability with which two alleles drawn from the population would be different from each other. In terms of allele frequencies, this is mathematically equivalent to the complement of sum of squares of all segregating allele frequencies in the population, called gene diversity (Nei 1973). In a random mating population, the gene diversity is also equivalent to the proportion of heterozygote individuals in the population. For repeat polymorphism, in contrary, the mutational process is such that two alleles of the same repeat length may not be evolutionarily identical (i.e., the result of forward-backward mutations, deriving from ancestral alleles of different repeat size). Furthermore, alleles of dissimilar sizes have extra information regarding their evolutionary distance (i.e., alleles that differ more in their repeat lengths are generally evolutionarily more distant than the ones more similar in sizes). These considerations lead to an alternative measure of within population genetic variability, the allele size variance. Empirical as well as theoretical data now suggest that the allele size variance is a better measure of within population genetic variability for repeat loci than the gene diversity statistic, since the former can be more easily related to the factors of maintenance of genetic variation, such as mutation rate and genetic drift (Kimmel and Chakraborty 1996; Kimmel *et al.* 1996).

At this point it is worth noting that there had been some concerns that since several microsatellites have been claimed to be under the influence of expansion bias of mutations, not all measures of genetic variation may appropriately reflect their pattern of variation (Rubinsztein *et al.* 1995). This concern, however, does not apply to gene diversity or allele size variance, since these measures are symmetric functions of differences of allele sizes (non-identity for gene diversity and squared allele size difference for allele size variance) that are used for measuring genetic variation (see Kimmel *et al.* 1996 and Kimmel and Chakraborty 1996 for a mathematical proof of this property).

The opportunity of constructing alternative measures of genetic variation has also resulted in investigations through which more incisive questions may be answered with data from repeat loci. For example, contrasts of gene diversity and allele size variance from data on the same set of loci, collected from the same populations, have been used to draw inference about the past population size fluctuations (Kimmel *et al.* 1998) and mutation rate differences among different groups of repeat loci, grouped by their repeat motifs (Chakraborty *et al.* 1997), as well as by their genomic locations (Deka *et al.* 1999).

Further, the information that alleles at repeat loci can be ordered and distinguished by their repeat lengths has resulted into studying population substructure effects not only by gene diversity (Nei 1973), but also through a formal analysis of variance of repeat lengths of alleles (Slatkin 1995; Kimmel *et al.* 1996). More recent investigations on this issue suggest that the ANOVA-based population substructure analysis is more powerful for detection of substructuring effect than that of the gene diversity analysis, because in the latter analysis the full information of allelic dissimilarities at microsatellite loci are not used (Chakraborty *et al.* unpublished data).

Such investigations also reveal that although a number of genetic distance measures have been proposed and used for microsatellite loci (see e.g., Kimmel *et al.* 1996; Goldstein *et al.* 1995; Takezaki and Nei 1996), not all of which are equally effective for different purposes of population genetic analysis. For example, Goldstein *et al.*'s (1995) genetic distance, under the assumptions of constancy of population sizes, is expected to increase more linearly with the time of divergence of populations than the gene diversity-based measures of distance. However, the former is subject to larger sampling errors, and hence less reliable, when estimated from a small number of loci, or by pooling data on microsatellite loci with differential effects of allele size constraints (Deka *et al.* 1999).

It is obvious that for most of the measures of genetic variation, the substitution of the sample statistics in the parametric definition of the measure do not yield an unbiased estimates of the measures intended to be evaluated. For example, Goldstein's genetic distance (squared difference of mean allele sizes in two populations) is not unbiasedly estimated from the squared difference of the observed mean allele sizes in samples drawn from the populations. Bias correction is critical, and can be easily accomplished. Such bias correction is critically important for microsatellite loci, since they exhibit a high degree of within population variation.

The hypervariability of microsatellite loci also introduced some features in testing specific genetic hypotheses that were not so relevant for the traditional markers. For example, one indicator of the presence of population substructure is that the genotype frequencies at a locus would deviate from the Hardy-Weinberg expectations based on allele frequencies. For traditional loci, where the number of segregating alleles is not so large, all genotypes are generally seen in a sample in sufficient frequencies so that a direct comparison of the observed and expected genotype frequencies can be made by using large sample distributions of the test statistics. For microsatellite loci, this is often not the case. As a consequence, exact tests based on permutation-based principles were developed to solve the problems caused by sparse data that is inherent to the features of microsatellite variation (Chakraborty *et al.* 1991; Guo and Thompson 1992). Sparse data, in some occasions, made the problems of hypotheses testing simpler for microsatellite loci. For example, for testing mutual independence of several loci, to be judged from multilocus genotype data, it can be shown that the likelihood ratio test statistic is simply dictated by the summed number of heterozygous loci in the sample (Zaykin *et al.* 1995). As opposed to the distribution of a likelihood ratio test statistic, that of the summed heterozygous loci is simpler and it can be readily computed iteratively without computer simulations (Chakraborty 1993).

In summary, the choice of measures of genetic variation, their estimation as well as means of hypotheses testing are dependent on the biological features of microsatellite loci. Thus, the methods of data analyses that were most appropriate for traditional loci are not necessarily so for all microsatellites. Nonetheless, the detailed nature of genetic variation, revealed by the microsatellites, makes it possible to answer several questions that could not be attempted with the classical markers. These aspects of microsatellite variation make this class of loci even more useful for biological anthropological studies.

4. DISCUSSION AND CONCLUSIONS

Repeat polymorphism is fundamentally a new class of polymorphism which differ from the traditional loci not only in terms of their mutation mechanism, but also in terms of the pattern of genetic variation that they exhibit. As a consequence, methods of interpretation of data on repeat polymorphism are somewhat unique and they differ from those of classical loci. As discussed above, these differences do not necessarily make the repeat loci more problematic for applications to solve biological and anthropological questions. In contrast, the unique features of repeat loci act as advantages of their applications for biological anthropological studies.

As more and more data on microsatellite variation in global populations are emerging, it is becoming even more clear that none of these loci are exactly identical in their biological properties. They differ by their rates of mutations, causes of which are not yet clear. Available data suggest that the mutation rate difference is dependent on repeat motif as well as genomic locations. Mutation patterns are also different from one locus to the other, for some mutations are mostly expansion or contraction of one or a few repeats, while for others larger jump of allele sizes as well as bias of contraction/expansion is predominant (Deka *et al.* 1999; see also Deka and Chakraborty, this volume). Mutation rate and pattern differences result into some analyses of microsatellites that make them particularly useful for some studies that cannot possibly done for any other polymorphic loci. For example, extensive diversity of multilocus genotypes at microsatellite loci has been used successfully for individual identification to the extent that is impossible with any other types of loci (Chakraborty *et al.* 1999). Such analyses also showed that while newer methods of detecting genetic variation may reveal even a larger class of polymorphic loci (e.g., single nucleotide polymorphisms, the SNPs), they alone can not replace the microsatellites for all applications.

In conclusion, therefore, it may be stated that with over 10,000 microsatellites now mapped on the human genome, the human genome diversity can now be examined in a manner much detailed than what has been accomplished to this date. The SNPs and sequence polymorphisms will add to such details, not necessarily as a replacement of the microsatellite loci, but probably as additional tools. The technology advancement that is likely to arise through the development of SNPs and sequence-based polymorphism studies will indeed help in understanding how the microsatellites evolve from which the pattern of microsatellite variation that we see in the genome will become probably clearer. Thus, future studies of molecular anthropology are likely to be composed of depicting the broad picture of the human genome diversity through the use of sequence-based SNPs, on which the microsatellite polymorphisms can be superimposed to provide the finer details. As a consequence, we should have a much deeper understanding of how the different human populations differ and how the inter-individual variation provides the uniqueness of each genome.

5. REFERENCES

Aggarwal RK, Majumder KC, Lang JW, Singh L (1994) Generic affinities among crocodilians as revealed by DNA fingerprinting with a Bkm-derived probe. Proc. Natl. Acad. Sci. USA 91:10601–10605.

Barbujani G, Magagni A, Minch E, Cavalli-Sforza LL (1997) An apportionment of DNA diversity. Proc. Natl. Acad. Sci. USA 94:4516–4519.

Birnboim HC, Straus NA (1975) DNA from eukaryotic cells contain unusually long pyrimidine sequences. Can. J. Biochem. 53:640–643.

Budowle B, Giusti AM, Waye JS, Baechtel FS, Fourney RM, Adams DE, Presley LA, *et al.* (1991) Fixed-bin analysis for statistical evaluation of continuous distributions of allelic data from VNTR loci, for use in forensic comparisons. Amer. J. Hum. Genet. 48:841–855.

Burke T, Dolf G, Jeffreys AJ, Wolff R (1991) DNA Fingerprinting: Approaches and Applications, Birkhauser Verlag, Basel.

Cavalli-Sforza LL, Menozzi P, Piazza A (1994) The History and Geography of Human Genes. Princeton University Press, Princeton.

Chakraborty R (1992) Sample size requirements for addressing the population genetic issues of forensic use of DNA typing. Hum. Biol. 64:141–159.

Chakraborty R (1993) A class of population genetic questions formulated as the generalized occupancy problem. Genetics 134:953–958.

Chakraborty R, de Andrade M, Daiger SP, Budowle B (1992) Apparent heterozygote deficiencies observed in DNA typing data and their implications in forensic applications. Ann. Hum. Genet. 56:45–57.

Chakraborty R, Fornage M, Gueguen R, Boerwinkle E (1991) Population genetics of hypervariable loci: analysis of PCR-based VNTR polymorphism within a population. In: DNA Fingerprinting: Approaches and Applications (T Burke, G Dolf, AJ Jeffreys, R Wolff, eds.), Birkhauser Verlag, Basel, pp. 127–143.

Chakraborty R, Jin L (1993a) Determination of relatedness between individuals by DNA fingerprinting. Hum. Biol. 65:875–895.

Chakraborty R, Jin L (1993b) A unified approach to study hypervariable polymorphisms: Statistical considerations of determining relatedness and population distances. In: DNA Fingerprinting: State of the Science (SDJ Pena, R Chakraborty, JT Epplen, AJ Jeffreys, eds.), Birkhauser Verlag, Basel, pp. 153–175.

Chakraborty R, Jin L, Zhong Y, Deka R (1995) intra- and inter-population variation at VNTR, Short tandem repeat and polymarker loci and their implications in forensic and paternity testing. In: Proceedings of the Fifth International Symposium on Human Identification 1994, Promega Corporation, Madison, WI, pp. 29–41.

Chakraborty R, Kamboh MI, Nwankwo M, Ferrell RE (1992) Caucasian genes in American Blacks: new data. Amer. J. Hum. Genet. 50:145–155.

Chakraborty R, Kimmel M, Stivers DN, Deka R, Davison LJ (1997) Relative mutation rates at di-, tri-, and tetra-nucleotide microsatellite loci. Proc. Natl. Acad. Sci. USA 94:1041–1046.

Chakraborty R, Stivers DN, Su B, Zhong Y, Budowle B (1999) The utility of short tandem repeat loci beyond human identification: implications for development of new typing systems. Electrophoresis (In Press).

Chakraborty R, Zhong Y, Jin L, Budowle B (1994) Nondetectability of restriction fragments and independence of DNA fragment sizes within and between loci in RFLP typing of DNA. Amer. J. Hum. Genet. 55:391–401.

Deka R, Guangyun S, Smelser D, Zhong Y, Kimmel M, Chakraborty R (1999) Rate and directionality of mutations and effects of allele-size constraints at anonymous, gene-associated and disease-causing trinucleotide loci. Mol. Biol. Evol. (In Press).

Devlin B, Risch N, Roeder K (1990) No excess of homozygosity at loci used for DNA fingerprinting. Science 249:1416–1420.

Di Rienzo A, Peterson AC, Garza JC, Valdes AM, Slatkin M *et al.* (1994) Mutational process of simple-sequence repeat loci in human populations. Proc. Natl. Acad. Sci. USA 91:3166–3170.

Edwards A, Civitello A, Hammond HA, Caskey CT (1991) DNA typing and genetic mapping with trimeric and tetrameric tandem repeats. Amer. J. Hum. Genet. 49:746–756.

Epplen JT, McCarrey JR, Sutou S, Ohno S (1982) Base sequences of a cloned snake W-chromosome DNA fragment and identification of a male putative mRNA in the mouse. Proc. Natl. Acad. Sci. USA 79:3798–3802.

Evett IW, Gill P (1991) A discussion of the robustness of methods for assessing the evidential value of DNA single locus profiles in crime investigations. Electrophoresis 12:226–230.

Geisser S, Johnson W (1992) Testing Hardy-Weinberg equilibrium on allelic data from VNTR loci. Amer. J. Hum. Genet. 51:1084–1088.

Goldstein DB, Ruiz-Linares A, Cavalli-Sforza LL, Feldman MW (1995) Genetic absolute dating based on microsatellites and the origin of modern humans. Proc. Natl. Acad. Sci. USA 92:6723–6727.

Guo SW, Thompson EA (1992) Performing the exact test of Hardy-Weinberg proportion for multiple alleles. Biometrics 48:361–372.

Hamada H, Petrino MG, Kakunaga T (1982) A novel repeated element with Z-DNA forming potential is widely found in evolutionary diverse eukaryotic genomes. Proc. Natl. Acad. Sci. USA 79:6465–6469.

Harding HWJ (1998) DNA database size. Jour. Forens. Sci. 43:248–249.

Harpending HC, Sherry ST, Rogers AR, Stoneking M (1993) The genetic structure of ancient human populations. Curr. Anthrop. 34:483–496.

Jeffreys AJ, Wilson V, Thein SL (1985) Individual-specific "fingerprints" of human DNA. Nature 316:76–79.

Jeffreys AJ, Monckton DG, Tamaki K, Neil DL, Armour JAL, MacLeod A, Collick A, Allen M, Jobling M (1993) Minisatellite variant repeat mapping: Application to DNA typing and mutation analysis. In: DNA Fingerprinting: State of the Science (SDJ Pena, R Chakraborty, JT Epplen, AJ Jeffreys, eds.), Birkhauser Verlag, Basel, pp. 125–139.

Kimmel M, Chakraborty R (1996) Measures of variation at DNA repeat loci under a generalized stepwise mutation model. Theor. Pop. Biol. 50:345–367.

Kimmel M, Chakraborty R, Stivers DN, Deka R (1996) Dynamics of repeat polymorphisms under forward-backward mutation model: within- and between-population variability at microsatellite loci. Genetics 143:549–555.

Kimmel M, Chakraborty R, King JP, Bamshad M, Watkins WS, Jorde LB (1998) Signatures of population expansion in microsatellite repeat data. Genetics 148:1921–1930.

Lander E (1989) DNA fingerprinting on trial. Nature 339:501–505.

Lewontin RC (1972) The apportionment of human diversity. Evol. Biol. 6:381–398. Li CC (1976) First Course in Population Genetics. Boxwood Press, Pacific Grove, CA.

Litt M, Luty JA (1989) A hypervariable microsatellite revealed by in vitro amplification of a dinucleotide repeat within the cardiac muscle actin gene. Amer. J. Hum. Genet. 44:397–401.

Martin W (1983) Consideration of "silent genes" in statistical evaluation of blood group findings in paternity testing. In: Inclusion Probabilities in Parentage Testing (RH Walker, ed.) Amer. Assoc. Blood Banks, Arlington, VA, pp. 245–257.

Mueller LD (1991) Population genetics of hypervariable human DNA. In: Forensic DNA Technology (MA Farley, JJ Harrington, eds.). Lewis Publishers, Chelsea, MI, pp. 51–62.

Nakamura Y, Leppert M, O'Connell P, Wolff R, Holm T, Culver M, Martin C, Fujimoto E, Huff M, Kumlin E, White R (1987) Variable number of tandem repeat (VNTR) markers for human gene mapping. Science 237:1616–1622.

National Center for Health Statistics (1980) Selected Genetic Markers of Blood and Secretions for Youths, 12–17 years of age, United States. US Department of Health, Education and Welfare, Publication No. (PHS) 80–1664, Hyattsville, MD.

National Research Council (1992) DNA Technology in Forensic Science. US National Academy Press, Washington DC.

National Research Council (1996) The Evaluation of Forensic DNA Evidence. US National Academy Press, Washington DC.

Nei M (1973) Analysis of gene diversity in subdivided populations. Proc. Natl. Acad. Sci. USA 70:3321–3323.

Nei M (1978) Estimation of average heterozygosity and genetic distance from a small number of individuals. Genetics 89:583–590.

Nei M, Roychoudhury AK (1974) Genetic variation within and between the three major races of man, Caucasoids, Negroids, Mongoloids. Amer. J. Hum. Genet. 26:421–443.

Pena SDJ, Chakraborty R, Epplen JT, Jeffreys AJ (1993) DNA Fingerprinting: State of the Science. Birkhauser Verlag, Basel.

Pena SDJ, Chakraborty R (1994) Paternity testing in the DNA era. Trends Genet. 10:204–209.

Pena SDJ, Macedo AM, Braga VMM, Rumjanek FD, Simpson AJG (1990) F10, the gene for the glycin-rich major eggshell protein of Schistosoma mansoni recognizes a family of hypervariable minisatellites in the human genome. Nucleic Acid Res 18:7466.

Polanski A, Kimmel M, Chakraborty R (1998) Application of t time-dependent coalescence process for inferring the history of population size changes from DNA sequence data. Proc. Natl. Acad. Sci. USA 95:5456–5461.

Roychoudhury AK, Nei M (1988) Human Polymorphic Genes: World Distribution. Oxford University Press, New York.

Rubinsztein DC, Amos W, Leggo J, Goodburn S, Jain S, *et al.* (1995) Microsatellite evolution—evidence for directionality and variation in rate between species. Nature Genet. 10:337–343.

Shriver MD, Jin L, Chakraborty R, Boerwinkle E (1993) VNTR allele-frequency distributions under the stepwise mutation model—a computer simulation approach. Genetics 134:983–993.

Shriver MD, Smith M, Jin L, Marcini A, Akey JM, Deka R, Ferrell RE (1997) Forensic ethnic affiliation estimation by use of population-specific allele DNA markers. Amer. J. Hum. Genet. 60:957–964.

Slatkin M (1995) A measure of population subdivision based on microsatellite allele frequencies. Genetics 139:457–462.

Smouse PE, Spielman RS, Park MH (1982) Multiple-locus allocation of individuals to groups as a function of the genetic variation within and differences among human populations. Amer. Nat. 119:445–463.

Steinberger EM, Thompson LD, Hartmann JM (1993) On the use of excess homozygosity for subpopulation detection. Amer. J. Hum. Genet. 52:1275–1277.

Takezaki N, Nei M (1996) Genetic distances and reconstruction of phylogenetic trees from microsatellite DNA. Genetics 144:389–399.

Tautz D (1993) Notes on the definition and nomenclature of tandemly repetitive DNA sequences. In: DNA Fingerprinting: State of the Science (SDJ Pena, R Chakraborty, JT Epplen, AJ Jeffreys, eds.), Birkhauser Verlag, Basel, pp. 21–28.

Valdes AM, Slatkin M, Freimer NB (1993) Allele frequencies at microsatellite loci: the stepwise mutation model revisited. Genetics 133:737–749.

Walker RH (1983) Inclusion Probabilities in Parentage Testing. American Association of Blood Banks, Arlington, VA.

Weir BS (1992) Independence of VNTR alleles defined as floating bins. Amer. J. Hum. Genet. 51:992–997.

Wyman AR, White R (1980) A highly polymorphic locus in human DNA. Proc. Natl. Acad. Sci. USA 77:6754–6758.

Zaykin D, Zhivotovsky L, Weir BS (1995) Exact tests for association between alleles at arbitrary numbers of loci. Genetica 96:169–178.

EPILOGUE

This century began with the rediscovery of the Mendelian principles of inheritance, following which the science of genetics bloomed, found its applications in understanding the pattern of human genetic variation, and is culminating with the sight of a full sequence of the human genome. A genome sequence, as such, will not address the questions of the extent of genetic variability in humans, but will certainly be the point of reference against which variability will be measured. In addition, the technologies that have already been developed in the context of the human genome analysis are shown to be useful in the area of human population genetics, which were unconceivable at the beginning of this century. In congruence with the title of this volume, the sixteen chapters illustrate how information on diversity of the human genome is currently being used in addressing some of the issues in understanding the evolutionary histories and relationships of modern humans.

The first four chapters have described the art of using the newly discovered repeat polymorphisms (VNTRs and STRs) in understanding the evolutionary biology of human populations; and their implications on the stability of the human genome. Chapters 5–12 have introduced the utility of the uniparentally transmitted Y-chromosome and mtDNA polymorphisms in tracing the lineage-specific evolutionary histories of contemporary populations. The next three chapters describe the utility of ancient DNA technology and the Alu insertion polymorphism in molecular anthropology. The final chapter discusses statistical issues in the analysis and interpretation of the biological features of the newer classes of genetic markers.

Admittedly, the case studies discussed in this volume do not encompass all methodologies, nor do they answer all questions relevant to human genome diversity. Nevertheless, some of the contributions presented here relate that when the same question is addressed using different types of markers, the answers could be discordant. For example, Hutz *et al.* (chapter 2) found different patterns of intra- and inter-population variability in five American Indian tribes of Brazil at nDNA, mtDNA and protein markers. Likewise, although worldwide diversity studies of Y-chromosome polymorphism are still comparatively limited, data available in the literature shows discordant patterns of apportionment of gene diversity in autosomal, mtDNA and Y-chromosome genomes (Seielstad *et al.* 1998). With the full reference sequence of the human genome to be in hand within the coming year or two, we anticipate that more and more of such apparently discordant features of genetic variation will be reported. In the post-genome-sequence millenium, therefore, one question of relevance will be: What contributes to such disparities of the human genome diversity and how do we resolve such differences?

Genomic Diversity: Applications in Human Population Genetics
edited by Papiha, Deka, and Chakraborty, Kluwer Academic / Plenum Publishers, New York, 1999.

At this point of time, we cannot answer these questions in full, but there are already hints of possible answers in the data available now. For example, as indicated in chapters 4 and 16, different types of markers (e.g., the repeat loci and the protein blood group markers) involve genetic variations caused by mutations of different rates and patterns. Even within some classes of these loci, rates and patterns of mutational changes are not similar (Chakraborty *et al.* 1997). Further, biparentally transmitted autosomal markers are subject to recombination and a somewhat weaker genetic drift than the uniparentally transmitted mtDNA and Y-chromosome markers. As a consequence, differences in biological features of different classes of genetic markers will have to be taken into account to interpret data, perhaps with different methods of analyses, even when they are utilized to address the same question.

In addition, evolutionary history of modern humans is complicated by factors such as the extent of isolation of populations, population movement during colonization of new territories, and admixture. While all of these events leave signatures on the genetic composition of the extant populations, the ability to detect such signatures with a specific set of markers depend upon how fast the genetic loci evolve and the time depth of the past events. For example, several studies have shown that the microsatellites and nucleotide sequence polymorphism data are well suited to detect effects of recent expansions of populations (Harpending *et al.* 1993; Kimmel *et al.* 1998). At the same time, such studies also indicate that the effects of past demographic histories are erased more quickly for loci at which the rates of mutations are higher. Likewise, studies on genetic effects of admixture have shown that the history of admixture is better predicted through the use of genetic markers at which the ancestral populations contributing to admixture have very dissimilar allele frequencies (Reed 1973; Chakraborty 1986; Parra *et al.* 1998). Thus, the genetic markers which, on a worldwide level, have a rather homogeneous allele frequency distribution will have relatively lesser power in detecting effects of admixture than the ones that carry population-specific variants. In sharp contrast, loci at which allele frequency distributions are homogenous on a global level are claimed to be ideal for gene mapping purposes using population-based disease-gene association studies. Thus, at present as an answer to the above question, we may qualitatively mention that not all types of markers are equally powerful to answer the same question. In fact, a combined approach would be required, in which the locus-specific distinctive biological features are to be integrated in formulating the data interpretation methods.

With the full genome sequence available, we anticipate another development in understanding human genome diversity, which is currently not practically possible. This relates to the context-dependence of genetic diversity at a locus. There are already some indications that loci with similar biological features have a different pattern of inter-individual variation depending upon the genomic context of their locations. For example, Deka *et al.* (1999) have found that the disease-causing trinucleotide repeats have the highest level of within-population variability and the gene-associated trinucleotides the lowest, an observation that appears to be uniform in diverse populations of the world. The genomic context feature is also important for understanding the diversity at haplotype level. For example, haplotypes defined by several loci spanning apparently equal physical distance show different patterns of diversity when they are chosen from different regions of the genome. In part, this difference is caused by uneven rates of recombination within distances of similar magnitude, varying from region to region of the genome. Thus, mechanisms that causes discordance of genome diversity can be better understood when such studies would be done in reference to the full genome sequence data.

It is becoming evident that single nucleotide polymorphisms (SNPs) will soon become the most abundant polymorphic systems for human genome studies. With an estimated number of one SNP per kb of the human genome (Wang *et al.* 1998), it is anticipated that there may be several millions of SNPs that can be placed on the reference genome sequence. Micro-chip array based technologies will offer the opportunity to study these markers for genome diversity, which will make context- and function-specific analyses feasible. Under such a framework, the concept of locus or gene is likely to be less important and sequence-based definition of variants/haplotypes will be the units of diversity studies. In addition, the rapid progress in the fields of bioinformatics and computational biology is likely to yield newer methods of data analysis, providing a better understanding of genomic diversity of human populations than what is possible now.

REFERENCES

Chakraborty R (1986) Gene admixture in human populations: models and predictions. Yearbook Phys Anthrop 29:1–43.

Chakraborty R, Kimmel M, Stivers DN, Davison LJ, Deka R (1997) Relative mutation rates at di-, tri-, and tetranucleotide microsatellite loci. Proc Natl Acad Sci USA 94:1041–1046.

Deka R, Guangyun S, Smelser D, Zhong Y, Kimmel M, Chakraborty R (1999) Rate and directionality of mutations and effects of allele-size constraints at anonymous, gene-associated and disease-causing trinucleotide loci. Mol Biol Evol 16:1166–1177.

Harpending HC, Sherry ST, Rogers AR, Stoneking M (1993) The genetic structure of ancient human populations. Curr Anthrop 34:483–496.

Kimmel M, Chakraborty R, King JP, Bamshad M, Watkins WS, Jorde LB (1998) Signatures of population expansion in microsatellite repeat data. Genetics 148:1921–1930.

Parra EJ, Marcini A, Akey J, Martinson J, Batzer M, Cooper R, Forrester T *et al.* (1998) Estimating African American admixture proportions by use of population-specific alleles. Am J Hum Genet 63:1839–1851.

Reed TE (1973) Number of gene loci required for accurate estimation of ancestral population proportions in individual human hybrids. Nature 244:575–576.

Seielstad MT, Minch E, Cavalli-Sforza LL (1998) Genetic evidence for a higher female migration rate in humans. Nature Genet 20:278–280.

Wang DG, Fan J-B, Siao C-J, Berno A, Young P, Sapolsky R, Ghandour G *et al.* (1998) Large-scale identification, mapping, and genotyping of single-nucleotide polymorphisms in the human genome. Science 280:1077–1082.

Corresponding Authors

Surinder S. Papiha (Chapter 1)
Department of Human Genetics
University of Newcastle upon Tyne
19-20 Claremont Place
Newcastle upon Tyne NE2 4AA
United Kingdom

Francisco M. Salzano (Chapter 2)
Departamento de Genética
Instituto de Biociências, UFRGS
Caixa Postal 15053
91501-970 Porto Alegre, RS
Brazil

R.J. Mitchell (Chapter 3)
Department of Genetics and Human Variation
La Trobe University
Bundoora, Victoria
Australia 3083

Ranjan Deka (Chapter 4)
Department of Environmental Health
University of Cincinnati Medical Center
3223 Eden Avenue
Cincinnati, OH 45267-0056
USA

Chris Tyler-Smith (Chapters 5 and 8)
Department of Biochemistry
University of Oxford
South Parks Road
Oxford OX1 3QU
United Kingdom

Peter Forster (Chapter 6)
McDonald Institute
University of Cambridge
Downing Street
Cambridge CB2 3ER
United Kingdom

S. Qasim Mehdi (Chapter 7)
Biomedical and Genetic Engineering Laboratories
P.O. Box 2891
Islamabad 44000
Pakistan

Paolo Francalacci (Chapter 9)
Dip Zoologia E Anthropologia Biologica
Università di Sassari
Via Regina Margherita 15
07100 Sassari
Italy

Richard Villems (Chapters 10 and 11)
Estonian Biocentre
23RIIA Str.
51010 Tartu
Estonia

D. Andrew Merriwether (Chapter 12)
Department of Anthropology
University of Michigan
500 S. State Street, 1020 LSA Building
Ann Arbor, MI 48109-1382
USA

Connie J. Kolman (Chapter 13)
NIH/NIAAA/LNG
12420 Parklawn Dr.
Park Bldg, Rm 451 MSC 8110
Rockville, MD 20852
USA

Prescott L. Deininger (Chapter 14)
Tulane Cancer Center
1430 Tulane Avenue
New Orleans, LA 70112-2699
USA

Rene J. Herrera (Chapter 15)
Department of Biological Sciences
Florida International University
University Park
Miami, FL 33199
USA

Ranajit Chakraborty (Chapter 16)
Human Genetics Center
University of Texas Health Science Center
6901 Bertner Avenue
Houston, TX 77030
USA

INDEX